METALS AND WELDING

The Authors

Willard R. Anderson
Professor Emeritus
Department of Agricultural Engineering
Iowa State University – Ames, Iowa

Thomas A. Hoerner
Professor Emeritus
Department of Agricultural Engineering
and Agricultural Education
Iowa State University – Ames, Iowa

V. J. Morford
Professor Emeritus
Department of Agricultural Engineering
Iowa State University – Ames, Iowa

HOBAR PUBLICATIONS
A DIVISION OF FINNEY COMPANY

Notice to the Reader

Safety is an important part of this manual. The information and material in this manual are accurate and true to the best of our knowledge. All recommendations are made without guarantee on the part of the authors or Hobar Publications. The authors and publisher disclaim any liability in connection with the use of information contained in this book or the application of such information.

The publisher advises the reader of this book to consult and work with an experienced metalworker or instructor to reinforce the material described within the book. The reader is expressly warned to consider and adopt all safety precautions regarding metalworking and welding, and to avoid all potential hazards. By following the instructions contained herein, the reader willingly assumes all risks in connection with such instructions.

Hobar Publications
A Division of Finney Company
8075 215th Street West
Lakeville, Minnesota 55044
Phone: (800) 846-7027
www.finney-hobar.com

ISBN-10: 0-913163-19-8
ISBN-13: 978-0-913163-19-1

First Printing	1970
Second Printing	1971
First Revision	1973
Fourth Printing	1974
Second Revision	1976
Third Revision	1978
Seventh Printing	1981
Fourth Revision	1988
Ninth Printing	2002
Tenth Printing	2007

INTRODUCTION

This manual, **Metals and Welding**, is designed to give you broad understanding and varied experiences in the metal and welding area. Courses in metals and welding should not be judged on the basis of what is made in the laboratory but rather on the basic skills and understandings that are developed. Projects or jobs have been selected that require a minimum amount of materials yet develop basic shop techniques and judgments that may be applied to construction, repair and maintenance problems common to agriculture or industry.

It is essential that you have a complete understanding of the related information needed to perform each laboratory activity. This information will be given by the instructor during the demonstration, lecture and recitation periods and through the study of this manual and suggested readings.

While the laboratory program outlined in this manual will require a careful budgeting of your time, it is hoped that quality of workmanship will not be sacrificed for speed. Accuracy and thoroughness are of prime importance. It should be noted that this laboratory outline only provides for a basic understanding or a beginning in developing techniques related to metals and welding. Once the basic understandings and techniques are achieved, much repeated practice is required to master the skills related to the metals welding area.

Each unit is concluded with **Classroom Exercises** with various forms of questions to aid the student in understanding the subject matter presented. In additon, each unit includes a wide variety of **Laboratory Exercises** to assist the student in further developing those basic skills related to Metals and Welding.

In addition to this manual, the successful teaching of a Metals and Welding unit can be made easier and more meaningful through the use of supplemental teaching aids such as worksheets, transparencies, filmstrips and microcomputer programs available from Hobar Publications. Further, a complete line of tools and equipment related to metals and welding is also available from Hobar Publications.

TABLE OF CONTENTS

PERSONAL SAFETY

Personal Safety - The human body has physical limitations in dealing with an unnatural environment. Chemicals, heat, fires, noise and mechanical equipment require that defenses in the form of personal protective equipment be used. This equipment has been used very successfully in many industries. A large percentage of accidents involve body parts which could have been shielded by protective equipment. The use of the following protective equipment. (see examples in Figure 1) might eliminate many injuries:

Safety Helmets or Hard Hats give head protection from flying or falling objects. A bump cap is a light weight, low-cost edition. It does not provide heavy impact protection.

Safety Eyeware must provide eye and/or face protection from injury due to flying objects, chemicals, excessive light and dust or chaff. Especially important times are when applying anhydrous ammonia, welding and operating portable or stationary power tools such as drills, grinders and saws. Ventilated plastic cover goggles, chemical goggles with hooded vents and face shields should be standard protective equipment. Glasses clear or tinted worn while at work should be industrial quality with impact-resistant lens set in approved frames. Physicians or opticians will assist in providing adequate eye protection.

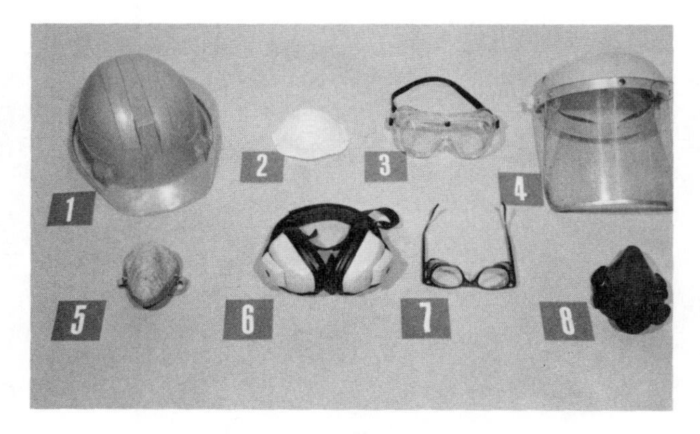

Figure 1. Safety Protection Equipment - Hard Hats, Safety Glasses and Goggles, Face Shields, Ear Muffs or Plugs, Dust Masks and Respirators can Provide Protection Against Shop and Laboratory Injuries.

Hearing Protection Devices eliminate noise-induced hearing loss. This loss of hearing usually due to long hours operating machinery and equipment in which the noise level is greater than 90 decibels. Industry has attempted to reduce the noise level of the machines more recently manufactured. Ear muffs or plugs give comfortable and effective protection from high noise level conditions.

Respirators and Dust Masks give protection against dust, paint, chaff, molds, chemicals and some gases. Respiration protection is available to handle most air hazards. The filter mask protects against non-toxic dusts and sprays as dust, chaff, spray paint and molds. The chemical cartridge cleans the air of chemicals and ammonia. This equipment does not provide protection in areas of high concentration of toxic gases or where there is little or no oxygen. Stay out of those places unless you use a special respirator provided with compressed air or oxygen.

Hand and Finger Injuries are very common. It seems that hands and fingers are always close to action and hazards. There are no hand protective devices that will prevent their being snapped up by a running machine but protection can be provided against cuts, abrasions, chemicals, burns and skin irritations with proper gloves. However gloves should not be worn where there is danger of the glove being caught and the hand drawn into the machine. Protective creams provide protection against grease, chemicals and paint.

Feet get stepped on, run over and have heavy objects dropped on them or are punctured by sharp objects. A survey indicates that 15% of the work injuries involved the feet and toes. Such injuries can be greatly reduced by the use of safety shoes with metal reinforced toe caps and special soles.

The Body needs special protection when welding or doing special repair jobs or handling rough, toxic or irritating chemicals. A variety of aprons, padding, rubber or plastic garments and knee pads can be found for most tasks. A cotton, ankle-length trouser without cuffs is preferred.

Personal Safety also includes the full-time use of all guards and machinery shields. If one is lost or damaged, it must be repaired or replaced before the machine is operated.

UNIT I
MEASURING INSTRUMENTS

Measuring instruments play a very important role in the development and learning of skills related to metal work. Almost all projects in the metal area require some form of measuring. Plans and blueprints are developed with various measurements such as: length, width, height, thickness, diameter and gage size. To accurately construct or complete an operation from a plan, one must be able to correctly and accurately use a number of measuring instruments. Dimensions on plans may be given in a variety of units such as feet, inches, fraction of inches, thousandths, millimeters or by gage or wire number. In using a metal lathe, you could be using the micrometer to check the diameter of a shaft in thousandths or you might be using a common steel rule to check the width of the cutting edge on a cold chisel forged in the gas furnace.

It is also necessary that a person be able to read drawings and blueprints as well as follow directions and read specifications. In the metal shop, you will be working with decimals and fractions and will need to know how to convert from one to the other. From this discussion, you should be able to understand that to work in the metal shop, it is a must that you become competent in using a wide variety of measuring instruments.

Measuring instruments commonly used in the metal shop are illustrated in Figure 1. The most common measuring instrument is the micrometer caliper, number 1. Instrument number 2 is referred to as an inside micrometer. The inside micrometer could be used for measuring the inside diameter of a pipe or the inside diameter of the arbor of a pulley. The telescoping gage, number 3, and small hole gage, number 4, are used with the micrometer caliper for the inside measurements. For measuring the depth of holes or grooves, item number 5, the depth micrometer, could be used. The vernier caliper, number 6, can be used for both inside and outside measurements. It can be read in both fractions and thousandths of an inch. Instrument number 7 is referred to as a sheet metal gage. It is commonly found in the metal shop and is used to determine gage numbers and thickness of sheet metal or steel plates. Instrument number 8, the steel rule, is very common for measuring in metal work such as the diameter of the eye on the eye bolt or the length of the tool gage. The pitch gage, number 9, is used for determining the pitch of a bolt or the number of threads per inch. Item number 10, the combination square, has many and varied uses in the metal shop, squaring metal and making 45° angles to name two common uses. The bevel protractor, number 11, is commonly used when laying out or checking angles.

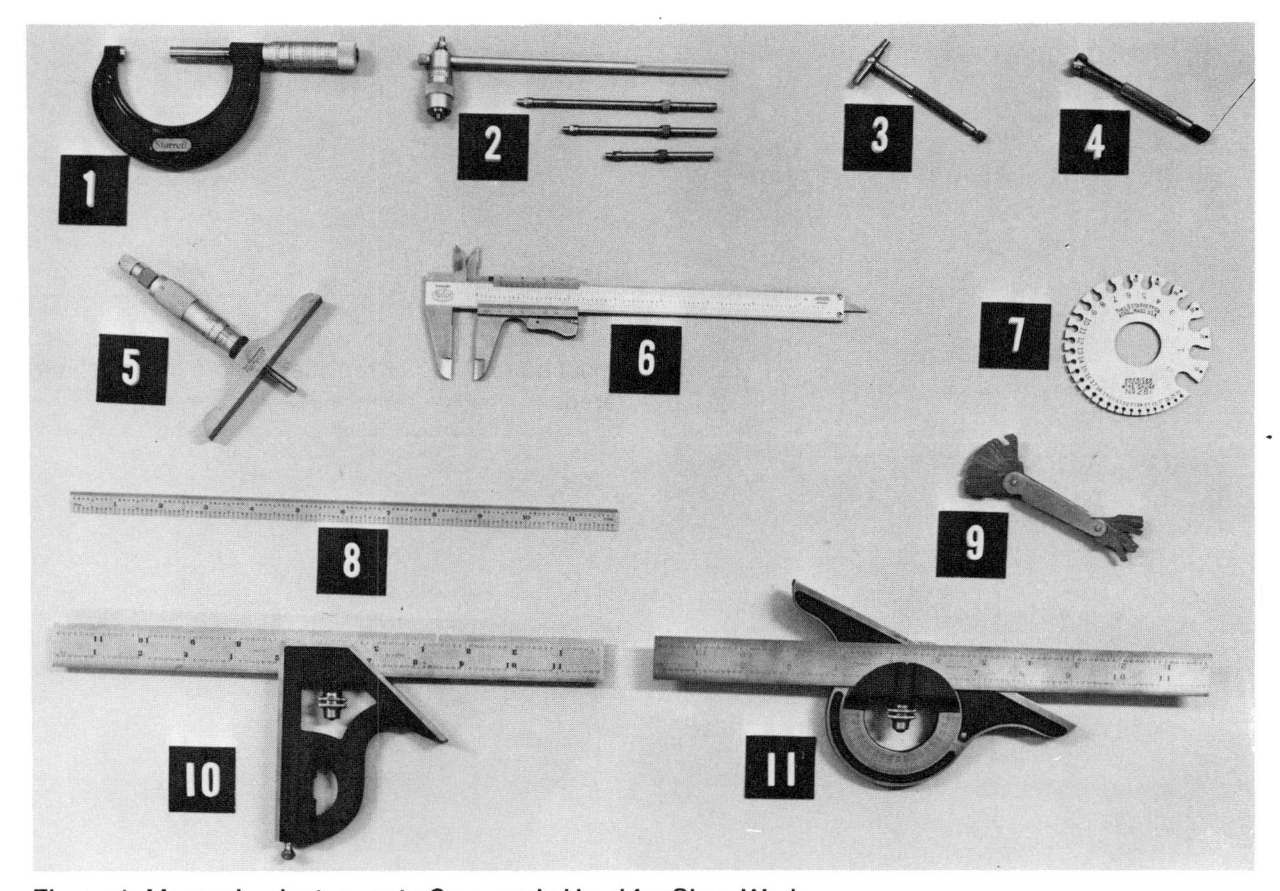

Figure 1. Measuring Instruments Commonly Used for Shop Work.

To become competent in the construction of metal projects, a person must becomes proficient in using a variety of measuring instruments. Most measurements are listed in thousandths of an inch; therefore, it is essential to have a basic understanding of a decimal and its fractional equivalent. Common fractions must be changed to their equivalent decimal fraction since instruments such as the micrometer are read in thousandths of an inch; for example 1/8 of an inch equals 0.125 of an inch. All decimals must be carried to three places such as 1/2 inch is equal to 0.500 of an inch. Fractions are changed to decimals by dividing the numerator by the denominator and computing to the three-place decimal. Fractions may be proper with the numerator less than the denominator (1/8) or improper where the numerator is larger than the denominator (9/8). The common fraction is one which has been reduced to its lowest term, for example, 1/8 is the common fraction for 2/16.

THE STEEL RULE

The steel rule as illustrated in Figure 2 is one of the more commonly used measuring instruments in the metal shop. It is used in laying out projects, checking diameter of rod, length of stock and the inside diameter of holes or pipe to name a few of its many uses.

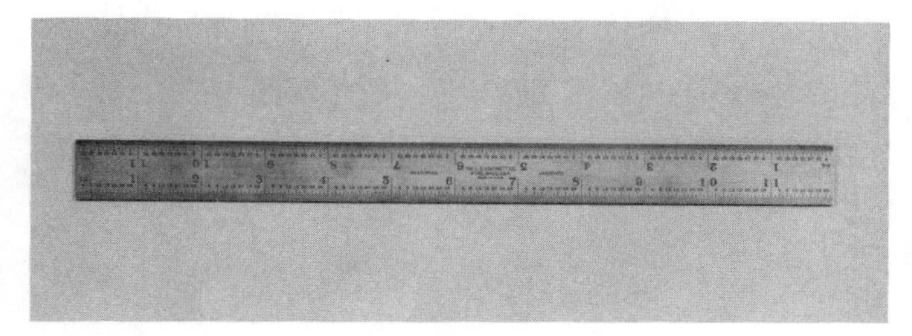

Figure 2. The Common Steel Rule.

The steel rule, also referred to as a machinist's rule, is sold in many widths, thicknesses and lengths. The most common lengths are 6 or 12 inches; however, it is manufactured in lengths of 1/4" to 4 feet. The edge of the steel rule is divided by fine lines into different fractions of an inch, such as 4ths, 8ths, 16ths, 32nds and 64ths. As shown in Figure 3, the smallest division is usually 1/64, the next larger 1/32, and the next 1/16 and so on up to the 1 inch division.

Some steel rules are divided into 100ths of an inch or decimals rather than fractions. In this case, 1/4 of an inch would be equal to 0.250 of an inch, 1/16" would be equal to .0625 or 625 ten thousandths of an inch. The divisions on the rule are called graduations. Some rules have divisions or graduations on the ends for measuring in small space. Another common rule used in the metals shop is made of brass rather than steel. The brass rule is commonly used for measuring hot metals as for checking the width of the cold chisel as it is being forged.

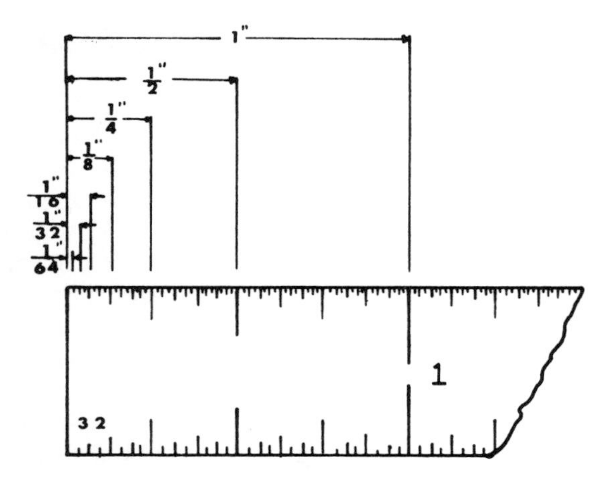

Figure 3. Divisions on the Steel Rule.

As shown on Figure 4, to measure with the steel rule, stand the rule up on its edge on the work so that the lines on the rule touch the thickness; always measure to the center of the lines.

It is a good habit to measure from the one inch line as the end of the rule could be worn. The steel rule should be handled with care to keep the edges from becoming nicked or worn round.

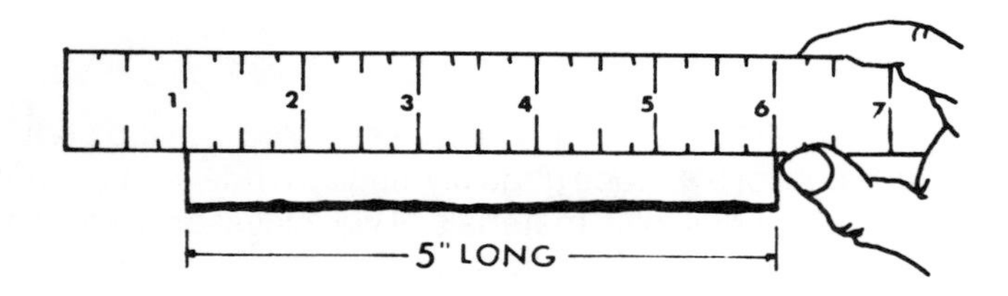

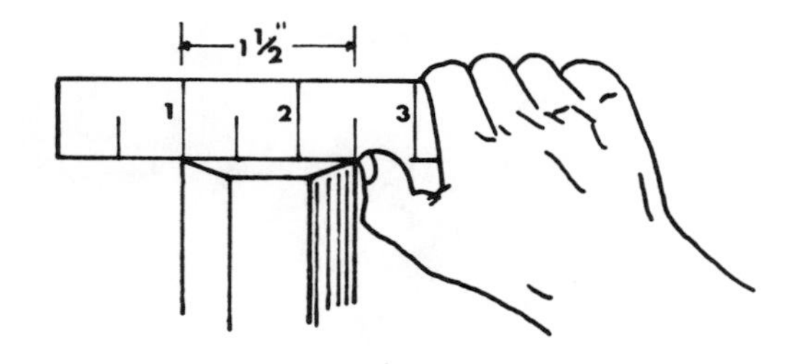

Figure 4. Measuring with the Steel Rule.

PROTRACTORS & COMBINATION SQUARES

The combination set, note Figure 5, is one of the more commonly used sets of measuring and layout tools in the metals shop. As shown, the combination set includes three separate measuring instruments, each using the steel rule or blade as the basic framework for measuring. The set includes a square head, bevel protractor and center head, any of which can be fastened quickly to the rule or blade.

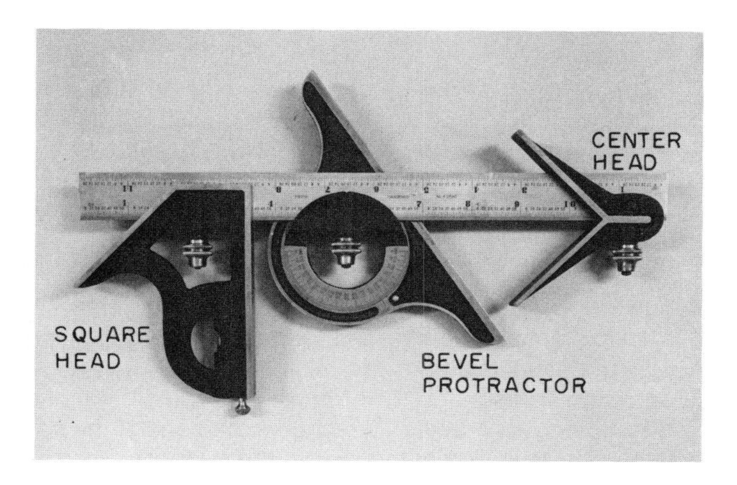

Figure 5. The Combination Set.

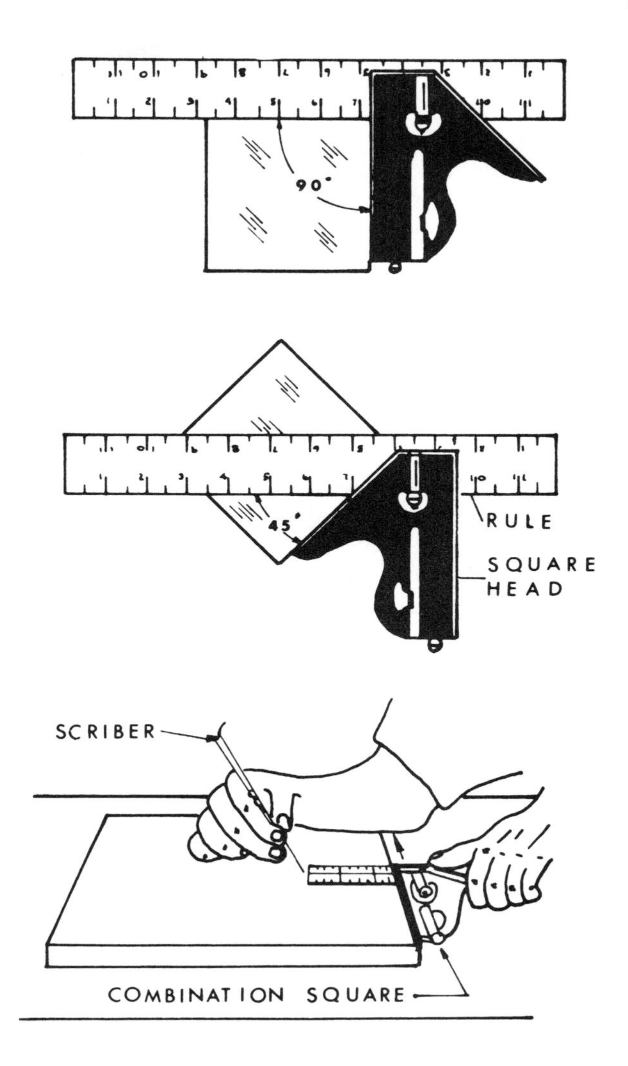

Figure 6. Common Uses of the Combination Square.

The square head commonly referred to as a combination square has many uses. As shown in Figure 6, it can be used to check the squareness of a steel plate or for laying out a 45° angle. Also shown is the common use of scribing a line. It can be used for checking the depth of a cylinder or hole or for checking the depth of a groove or keyway in a shaft. The combination square can also be used for setting a given measurement on the inside or outside caliper, setting the height of a surface gage or used as a rule for measuring sizes of various work materials. Two accessory parts to the combination square include a spirit level and scriber.

The bevel protractor can be used to scribe parallel lines at various angles as shown in Figure 7. It is divided into degrees and can be used to lay out or check any angle from 0 to 180°. It could also be used to check squareness of metals, to scribe a line, to check the depth of a hole or as a common measuring rule. Another use is to check the 59° angle commonly found on the tool gage used to check the cutting lip angle of the twist drill bit.

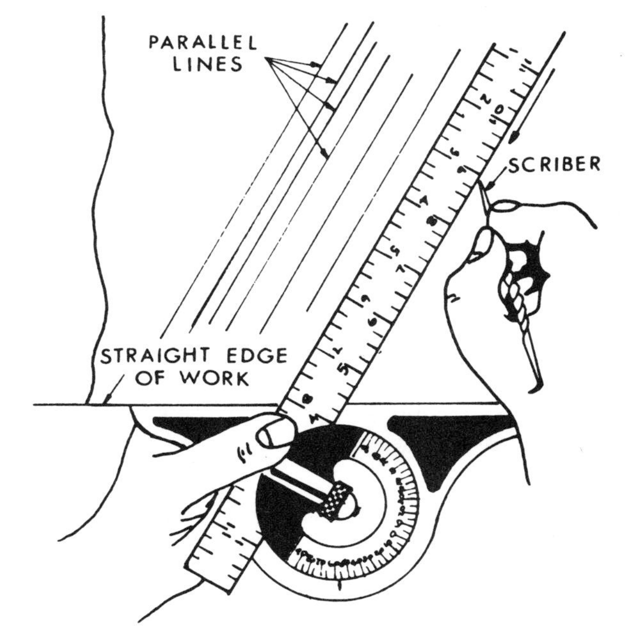

Figure 7. Using the Bevel Protractor to Scribe Parallel Lines.

The center head, the third main part of the combination set, has many important uses in metal work. With the steel rule fastened to it, it is called a center square. One of its common uses is finding the center of round stock, note Figure 8. Any two lines drawn across the end of a round peice of metal with the center square will cross at the center of the stock. The center head without the steel rule could be used for extending a line around a corner of metal stock as shown in Figure 8. The center square could also be used to lay out 45° angles or to check the squareness of 90° corners or edges of metal.

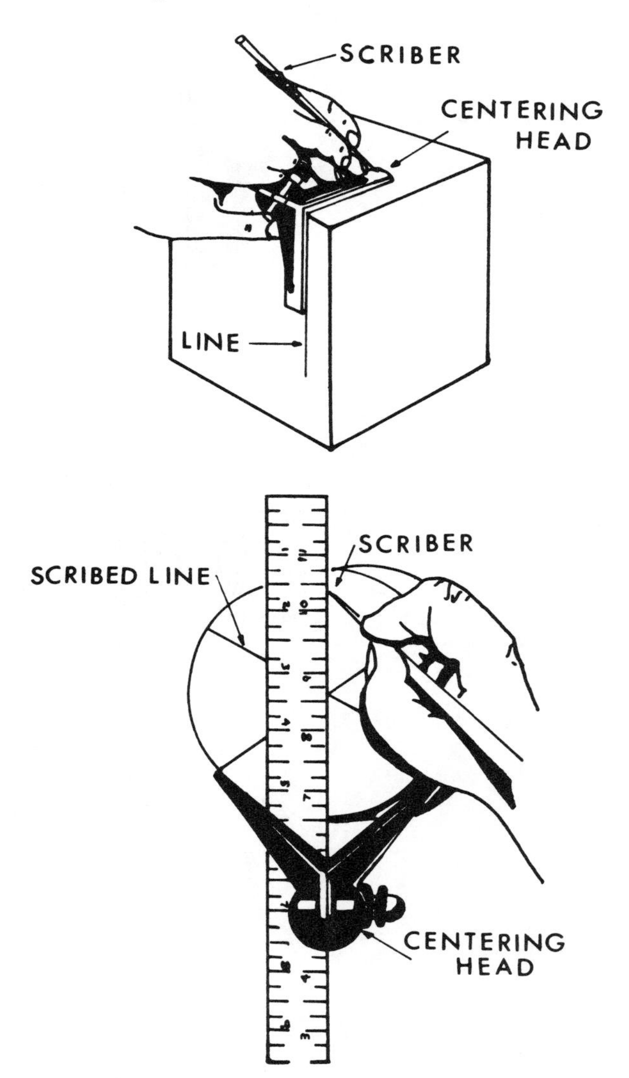

Figure 8. Using the Centering Head to Find the Center of Round Stock and for Transferring Lines Around a Corner.

THE MICROMETER CALIPER

The most common precision measuring instrument in the metal shop is the micrometer caliper. It is designed to measure in thousandths of an inch or in millimeters, Micro–meters with the thousandths measurements are said to have the English-measure whereas those with the gradu-ations in millimeters are said to have the metric-measure. In machine work where tolerances are critical, it is essential that measurements be accurate. The micrometer is an expensive but essential instrument. It is easy to use once a few basic skills are developed and principles are under-stood.

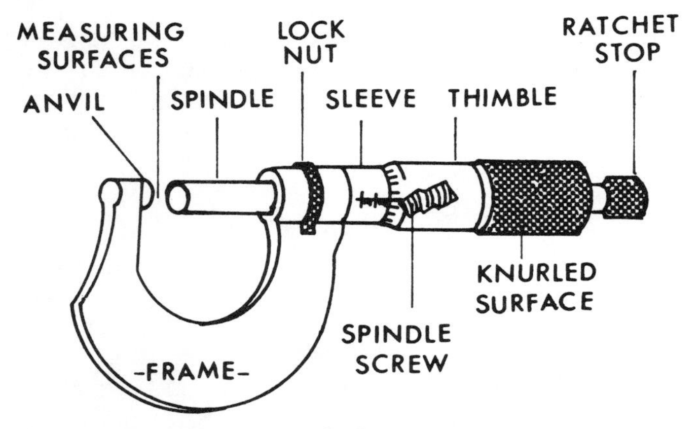

Figure 9. The Micrometer Caliper.

The micrometer caliper is illustrated in Figure 9. The micrometer has a number of important parts which must be fully understood before the micrometer can be used for precision measuring. The frame is the rigid body of the micrometer. The anvil is attached to one end of the frame and is the rigid side of the measuring surfaces. On some micrometers, the anvil is connected to the frame with a screw which is the adjustment point for the micrometer. The other measuring surface is on the end of the spindle. These surfaces contact the part to be measured. The spindle is the movable measuring surface. It moves back and forth as the thimble is turned. After taking a reading, the spindle can be secured with the lock nut and the reading will not be changed by accidently moving the thimble when the micrometer is removed from the work piece.

The sleeve of the micrometer consists of a horizontal line with forty vertical lines, each equal to 0.025 of an inch. On the sleeve, there are ten major graduations 0, 1, 2, 3, 4, 5, 6, 7, 8, 9, and 0 each equal to one hundred thousandths of an inch. The thimble rotates around the sleeve moving back and forth on the sleeve. There are 25 graduations, each equal to one thousandth of an inch, marked off around the thimble. The spindle screw fits inside the sleeve and moves on the threads inside the sleeve, note Figure 10.

The ratchet stop has three main functions: (1) it allows the operator to obtain consistent readings by ratcheting at a given tightness, (2) the ratchet diameter being smaller than the thimble allows the operator to rapidly move the spindle-back and forth, and (3) because the ratchet will slip when the spindle touches against the measuring surface, there is less chance of damaging the measuring surface by jamming it against the work piece.

The purpose of the knurled surface is to allow the operator to grip the thimble or ratchet stop. This also aids the operator in being able to obtain the same feel from job to job to develop accuracy in using the micrometer.

There are a number of micrometers manufactured for special uses. The outside micrometer is the most common type and is used to measure the outside diameter of round objects and the widths and thicknesses of flat pieces. The inside micrometer is used to measure the diameter of a hole such as the inside diameter or bore of an engine cylinder. The micrometer depth gage is used to measure the depth of holes, grooves or slots.

All micrometers are read alike. The end of the spindle inside the sleeve is referred to as the spindle screw. (Note Figure 10)

The thimble is fastened to the screw which has 40 threads to an inch. This means that the screw must make 40 turns or complete revolutions to move one inch. Since the micrometer is read in the thousandths of an inch instead of 40ths, 1/40 of an inch must be changed to thousandths of an inch by dividing the numerator by the denominator as:

$$40 \overline{\smash{\big)}\,1.000}^{\;0.025}$$

There are also forty lines to an inch on the sleeve, the same as the number of threads on the screw. These lines indicate the number of times the screw has turned. Each turn is 0.025". If one turn of the screw equals 0.025", then 1/25 of a turn equals 1/25 of .025", or 0.001" (one thousandths of an inch). Therefore, by dividing the edge of the thimble into 25 parts, it is possible to make exactly 1/25 of a turn which is 0.001" or 2/25 of a turn which would be equal to 0.002". This makes it possible to measure in thousandths of an inch.

To understand the marks on the sleeve and thimble, turn the thimble until the 0 mark on the sleeve and the 0 mark on the thimble come together; the micrometer is at the smallest measurement possible. On the 1" micrometer, the measurement is 0, on a 2" micrometer the measurement is exactly 1" etc. (Note Figure 11)

The marks on the thimble are 0.001" each. Turn the thimble to the next line on the thimble in the direction of the arrow, in Figure 12, and if this is a 1" micrometer, the spindle will be 0.001" from the anvil. Turn the thimble to the 5 mark on the thimble, the spindle will be 0.005" from the anvil.

Note that one complete revolution of the thimble is equal to 0.025" on the sleeve. For each turn of the thimble, the thimble moves one mark on the sleeve. This means that each mark on the sleeve is 0.025". Every fourth line on the sleeve is a little longer than the others and is stamped 1, 2, 3, etc., meaning 0.100", 0.200", 0.300", etc.

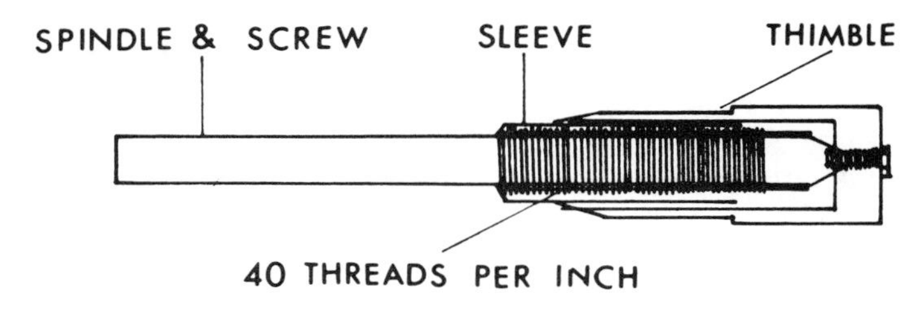

Figure 10. The Micrometer Thimble and Spindle Screw Assembly.

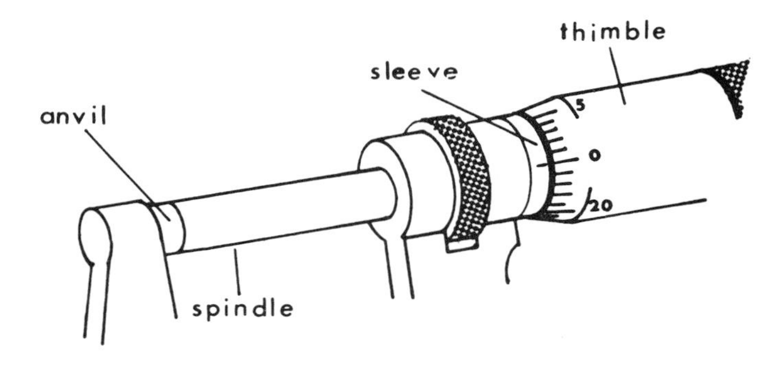

Figure 11. Testing the Exactness of a One Inch Micrometer.

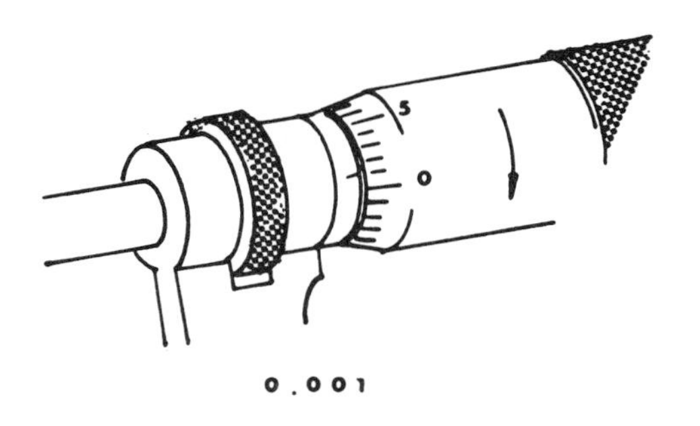

Figure 12. Reading One Thousandths of an Inch. (0.001)

To find out how much the micrometer is opened (the distance between the anvil and the spindle) the marks on the sleeve may be read like an ordinary rule remembering the numbers 1, 2, 3, 4, etc. mean 0.100", 0.200", 0.300", etc. To this add the number of twenty-five thousandths indicated by the small graduations between the numbers 1, 2, 3, etc. on the sleeve. Next add the number of thousandths that show on the thimble. The first reading is the number of hundred thousandths on the sleeve in twenty-five thousandths and the third reading is the number of thousandths on the thimble. For example, the readings on the micrometers in Figure 13 are:

	Reading	Thousandths
A.	1.	0.300 (dollars - numbers on sleeve)
	2.	0.000 (quarters - vertical lines on sleeve)
	3.	0.000 (pennies - divisions on the thimble)
		0.300
B.	1.	0.300 (dollars - numbers on sleeve)
	2.	0.025 (quarters - vertical lines on sheeve)
	3.	0.012 (pennies - divisions on the thimble)
		0.337

This method of making three readings has been called the dollars, quarters, and pennies system. Consider that you are making change from a ten dollar bill; the figures on the sleeve are dollars, the vertical lines on the sleeve are quarters and the divisions on the thimble are cents. Add up the change and place a decimal point in front of the figures rather than a dollar sign.

The accuracy of micrometer readings is dependent upon the operator's "sense of touch" if the micrometer does not have a ratchet stop. The ratchet stop is calibrated to insure a uniform pressure before the readings are taken and it will also slip before excessive pressure can be applied which might damage the measuring surfaces of the micrometer. Without the ratchet stop, a person must develop the correct "feel" for the exact tension before taking the reading. The correct feel is called the magnetic drag point.

Follow these steps when taking a reading:

1. Clean the anvil and spindle face with lens paper.
2. Hold the micrometer by the frame and not the thimble. (Note Figure 14)
3. Adjust thimble with thumb and first finger.
4. Do not force the item to be measured between anvil and spindle.
5. Adjust spindle pressure to the correct tension or until ratchet clicks twice.
6. Make an accurate reading before moving the micrometer from the work or secure thimble with lock nut, remove micrometer from work piece, then make reading.

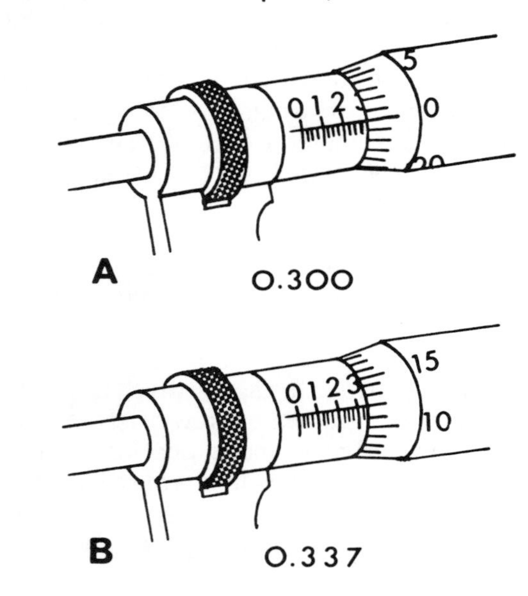

Figure 13. Reading the Micrometer.

When measuring a round shaft it is necessary to feel for the largest dimension and a "flat" or "out-of-round" shaft can be detected by taking four or more readings around the shaft. Normally it is recommended to take three readings around a shaft using the average or mean of the three readings as the actual shaft diameter.

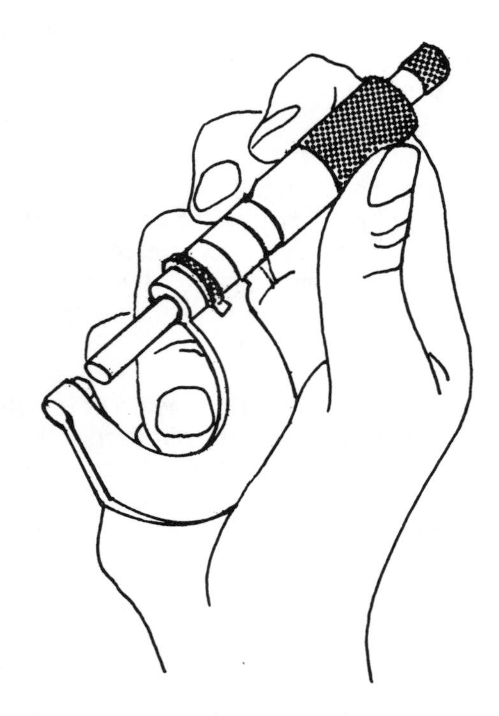

Figure 14. Holding the Micrometer.

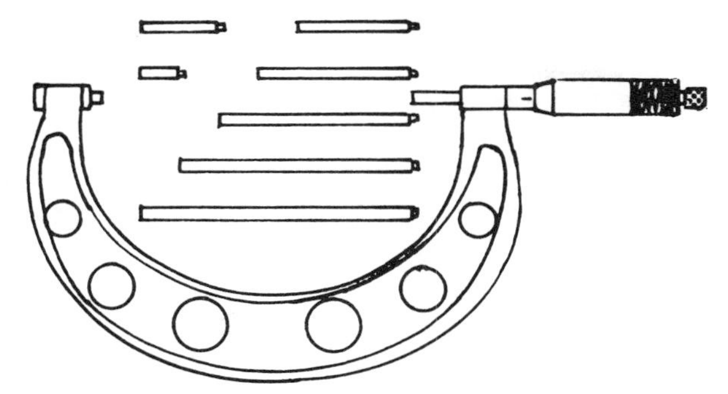

Figure 15. Micrometer for Measurements 0" to 8".

Micrometers are usually designed to measure within a one-inch range; for example, a 1" micrometer would measure between 0" and 1". To measure an engine piston that is approximately 2-1/4", a 3" micrometer, which would measure from 2" to 3" is needed. It is possible to purchase one micrometer, which could measure from 0" to 8". (Note Figure 15) This instrument has different length adaptor rods which attact to the spindle and make it possible to measure a steel rod less than 1" or an engine piston measuring over 6 inches in diameter with the same micrometer by changing the adaptor rods.

THE VERNIER MICROMETER

Some micrometers have what is called a vernier or vernier scale. A vernier micrometer is designed for use where accuracy to the nearest ten thousandths of an inch is required. As you have noted in reading the micrometer sometimes the marks on the thimble do not line up exactly with the horizontal line on the sleeve (Note Figure 16). If using a vernier micrometer, the numbers on the sleeve and the thimble are shown as on the normal micrometer, with each number on the thimble equal to 0.001".

The ten divisions on the back of the sleeve are the vernier; the lines are numbered 0, 1, 2, 3, 4, 5, 6, 7, 8, 9, 0 as in Figure 16. These lines have the same space as nine divisions on the thimble.

To read the vernier micrometer, first read the thousandths as on any ordinary micrometer. Assume this number to be between 0.275 and 0.276 as illustrated in Figure 16. To find the fourth decimal or the number of ten thousandths, select the line that matches exactly a line on the thimble. In this case, the line on the thimble matches line number 6 on the vernier.

This number indicates that the location on the thimble is 6 ten thousandths (0.0006) past the horizontal line on the sleeve. Thus the actual reading is 0.275"+ 0.0006" or 0.2756".

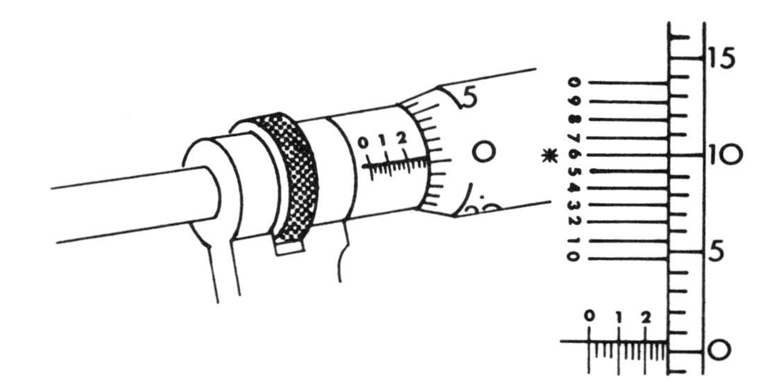

Figure 16. The Vernier Micrometer.

If a thimble mark does not line up with the sleeve line when using a micrometer without a vernier scale, you could interpolate for the number of ten thousandths or for most measurements the thimble could be turned and read to the nearest one-thousandth of an inch.

THE INSIDE MICROMETER

The inside micrometer illustrated in Figure 17 is used to measure the diameters of holes. Examples might be the cylinder bore of an engine or the inside of large pipe. The inside micrometer is calibrated in thousandths of an inch and read like the regular outside micrometer.

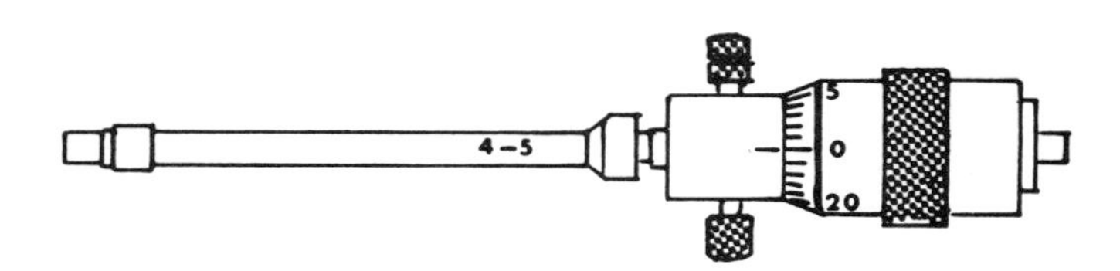

Figure 17. Inside Micrometer.

For the same micrometer to be used for a wide range of sizes, the inside micrometer is sold with a number of 1" adaptors ranging from 2" to 8".

For holes with small diameters, the instruments shown in Figure 18 can be used for inside measurements. Instrument A is referred to as a telescoping gage. The gage is sold in a variety of size ranges, one example is 1-1/4" to 2-1/8". The contact plungers are telescoped together, inserted into the hole, then allowed to expand. The small thumb screw on the handle is tightened holding the gage at the precise inside measurement of the hole. The gage is then removed from the hole and a regular micrometer is used to read the inside diameter of the cylinder or hole.

The telescoping gage is available in sizes up to 6 inches, therefore, it could be a substitute for the regular inside micrometer for many applications.

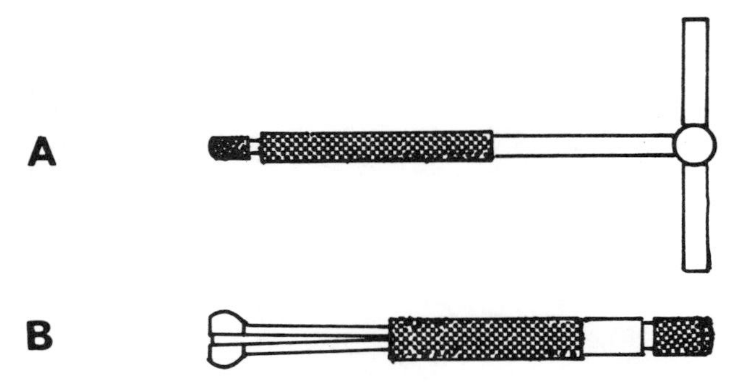

Figure 18. Gages Used for Inside Measurements.

Instrument B is called a small hole gage. The small hole gage is usually calibrated by tenths of an inch, for example, 4/10" to 5/10". The same procedure is followed as with the telescoping gage by inserting in the hole, expanding to the inside diameter and then reading the actual size with a 0-1" micrometer.

THE MICROMETER DEPTH GAGE

A micrometer depth gage as illustrated in Figure 19 is used to determine the depths of holes, grooves and slots. Common uses are measuring the depth of a keyway and measuring the stroke of a small engine.

The micrometer depth gage is read the same as the regular outside micrometer except for the fact that the numbers on the sleeve are just opposite. The highest numbers, 7, 8, 9, and 10, are to the left end of the scale on the sleeve rather than at the right end as with the micrometer caliper or inside micrometer.

When the spindle is out, this would indicate a greater depth or higher reading on the sleeve, note Figure 19. It is sold with adaptors to provide a number of ranges for different depth measurements.

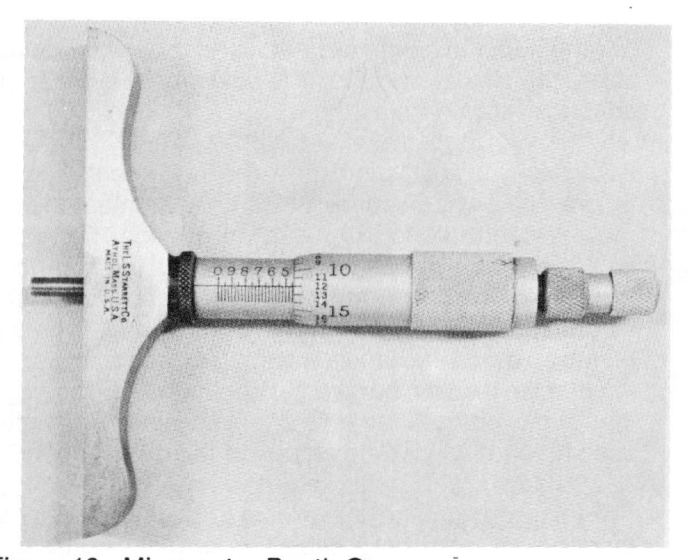

Figure 19. Micrometer Depth Gage.

THE VERNIER CALIPER

The caliper with a vernier scale as illustrated in Figure 20 is one of the most accurate precision measurement instruments used in the shop. The vernier caliper can be used for both inside and outside measurements. The graduations on this vernier scale are in decimals and in fractions. The vernier caliper consists of a beam (1), sliding jaw (2), slide release (3), outside nibs (4), jaws (5), inside nibs (6), vernier scale in fractions (7), and vernier scale in thousandths (8). The beam may range from 6 to 48 inches in length; however, a common size for most shop applications is 7 inches.

The graduations on the beam are called the "true scale". The beam shown in Figure 20 has two scales, a lower scale for measurements in one thousandths and an upper scale used for measuring in fractions; for this instrument the graduations are 1/128ths of an inch. The movable part of the caliper contains the vernier scale. This particular instrument has two vernier scales which correspond with the two scales found on the beam.

The caliper can be read in either fractions or thousandths of an inch. When reading in thousandths, use the lower scales on both the beam and the vernier scale. The vernier scale is divided in 25 divisions. The 25 divisions correspond to 49 divisions on the beam, Figure 21. The beam has inch 1, 2, 3, etc.; one-hundred thousandths of an inch, 100, 200, 300, etc. and twenty-five thousandths of an inch graduations. These divisions are the same calibrations as on the sleeve of the micrometer caliper. The procedure for reading the vernier caliper in thousandths is similar to reading a micrometer caliper calibrated in thousandths. The major difference is in using the vernier scale to determine the number of thousandths (.001") to add to the hundred and the twenty-five thousandths divisions on the beam. An example reading is given in Figure 21. Use the zero mark on the vernier scale to read on the beam. The first reading is the number of inches, 1.000". Next, note the number of hundred thousandths inches, .300". To determine the number of twenty-five thousandths count the number of small spaces on the beam that the zero mark has passed after the number 3 and multiply by .025", in this

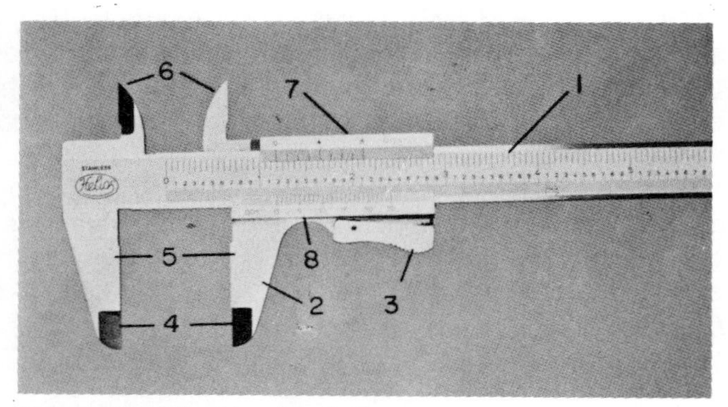

Figure 20. The Vernier Caliper.

instance it is 2 or .050". Note the scale marking on the vernier which lines with a line on the beam. The graduation number 18 corresponds with a line on the beam, this indicates .018". The total reading is:

Reading	Thousandths
1.	1.000"
2.	0.300"
3.	0.050"
4.	0.018"
Total	1.368"

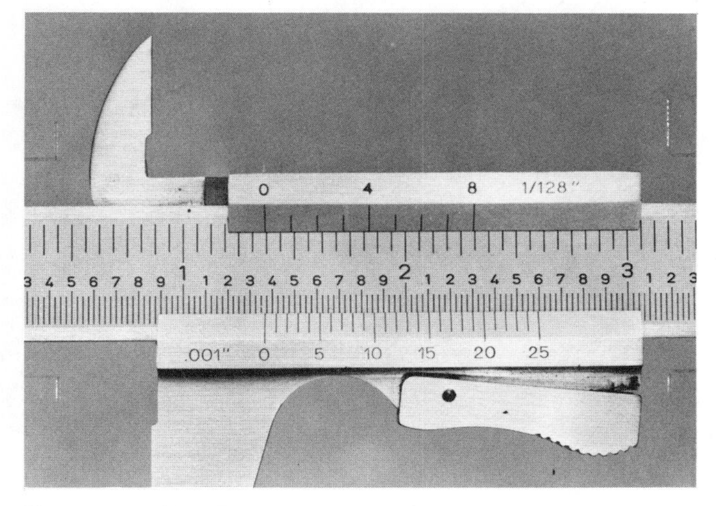

Figure 21. Reading the Vernier Caliper.

To make the same reading in fractions or 128ths use the upper part of the vernier scale. The scale is divided into 8 graduations each equal to 1/128ths of an inch. The beam is divided by inches and sixteenths of an inch (Note Figure 21). To make the reading in fractions, first note the zero point on the vernier scale and the number of inch divisions on the beam. Next count the number of sixteenths passed the one inch mark. In this reading, the zero point is passed 5/16" graduation but is not quite to the 6/16" graduation. To determine the number of 1/128ths that the point is passed the 5/16" mark, refer to the vernier scale. Again, note the line on the vernier scale that matches a lineon the beam. The number 7 scale marking matches exactly a line on the beam. This indicates 7/128ths of an inch. Note 16ths are changed to 128ths by multiplying the numerator by 8. The total reading is:

Reading	Fraction
1.	1"
2.	5/16" or 40/128"
3.	7/128"
Total	1 and 47/128"

This caliper can be used for both inside and outside measurement using the same scale on the beam and the vernier to determine the reading. A few simple rules as in reading any precision instrument should be followed:

1. The vernier caliper should not be forced on the work piece.
2. Carefully move the vernier slide against the part being measured.
3. Carefully remove the caliper from the part being measured.
4. Follow the procedure outline for accurate readings.

DIVIDERS AND CALIPERS

There are a number of dividers and calipers used in metal work. Those commonly used are illustrated in Figure 22. The divider, instrument A, is a two legged, steel instrument used to lay off distances, measure distances and to scribe circles. Size of the divider is determined by the greatest distance it can be opened between the two points. The divider is adjusted by turning the thumb screw to the desired distance between the points.

Instrument B is an outside caliper and is used to measure outside diameters of round objects and to measure widths and thicknesses of flat stock. The caliper should be held gently with the finger tips and moved back and forth over the work until both legs just touch the surface of the work piece. Accurate measurements with the outside caliper depends upon the sense of touch or feel in the operator's fingers. It is easy to force the legs over the work resulting in a wrong measurement.

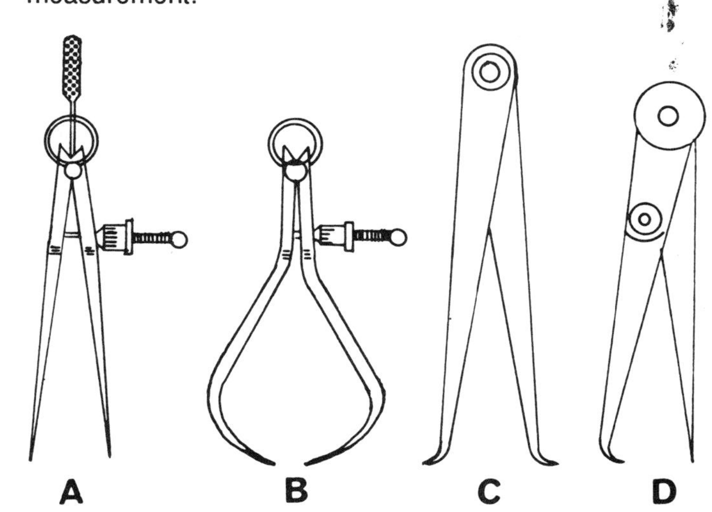

Figure 22. Dividers and Calipers.

Proper setting of the outside caliper is shown in illustration A, Figure 23. Hold a steel rule in the left hand and with the right hand place one leg of the caliper against the end of the steel rule. Place the finger behind the end of the steel rule to keep the leg from slipping off while setting the other leg to the desired width.

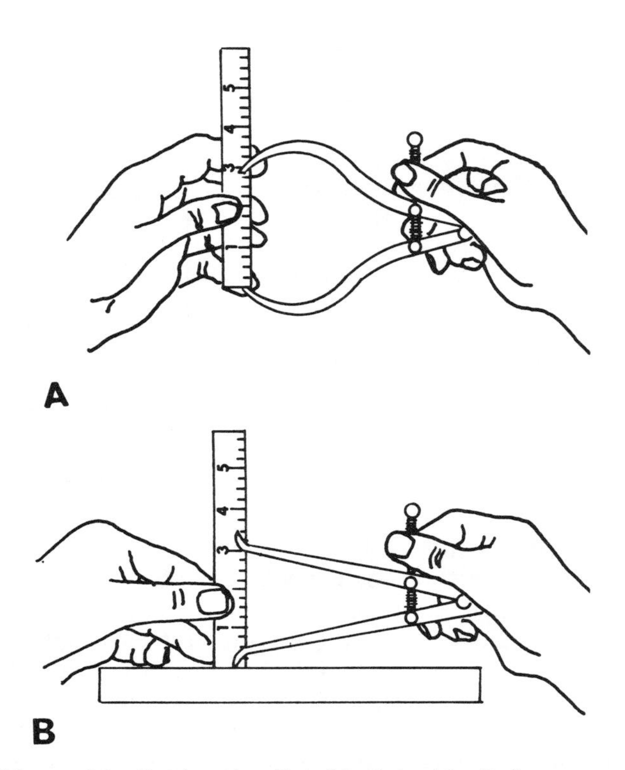

Figure 23. Setting the Outside & Inside Calipers.

An inside caliper, instrument C, Figure 22, is a two legged instrument with its leg ends bent outward. It is used to measure the diameter of holes or space between two objects such as the distance between two round shafts. Obtaining the correct sense of touch or feel is the same as described for the outside caliper. The inside caliper is set to a given size as shown in Figure 23, illustration B. Hold the end of a steel rule against a flat metal surface and set the caliper as shown.

Instrument D, Figure 22, is known as a hermaphrodite caliper. It has one pointed leg like the divider and one bent leg like the outside caliper. It could be used to find the center on the end of a round bar or to scribe lines parallel to the edge of a work piece.

Calipers are commonly used to transfer sizes from one piece to another such as when machining a round object on the lathe to duplicate the object. Measurements can be transferred from one caliper to another as from an inside caliper to an outside caliper. Measurements can also be transferred to other measuring instruments, such as, transferring a measurement from the inside caliper to a micrometer for exact reading of size.

Dividers and calipers are available in sizes according to the distance that can be measured between the two points. Common sizes are 4" and 6". They are sold with a thumb screw adjustment, instrument B, Figure 22, or with what is referred to as a "firm joint", instrument C. It is easier to maintain a given measurement with the adjustable thumb screw type than with the firm joint type. With the firm joint type, the measurement can be changed by bumping when transferring sizes. To achieve operator accuracy when using dividers and calipers, manual dexterity and practice is required.

SHEET METAL AND WIRE GAGES

A wire gage for measuring wire size as illustrated in Figure 24 is commonly found in the metalworking shop. Exact size of steel wire can quickly be determined with this gage.

Figure 24. Steel Wire Gage.

The gage consists of a number of slots in which the wire can be inserted for accurate size determination. Each slot is numbered on one side to indicate the decimal equivalent and on the other side to indicate the gage number. As noted in Figure 24, a piece of steel wire is inserted into the slot with a number 9 indicating that the wire is 9 gage. To determine the decimal equivalent, note the other side of the gage. In this case, the number 9 gage is equal to a decimal equivalent of 0.1483".

Steel wire such as fencing wire or welded wire fabric for concrete reinforcing would be measured with a U.S. Standard steel wire gage. The gage numbers on the gage are similar to the numbers on the American Standard Gage; however, the actual size decimal equivalent is slightly different. For example, a number 14 gage on the U.S. Standard gage has a decimal equivalent of 0.0800".

Non-ferrous sheet metal and wire such as aluminum, copper or brass are measured with the American Standard Wire Gage. This gage is frequently used by electricians to determine electrical conductor sizes. For example, a number 14 gage has a decimal equivalent of 0.0641.

The Manufacturer's Standard Gage is more commonly used for determining gage thickness of iron and steel sheets. At present, however, the actual measuring gage is not available. Therefore, for measuring and determining gage size using the manufacturer's gage sizes, one would use a micrometer as shown in Figure 25 to determine the thickness in thousandths, note the thickness of the steel plate in Figure 25. Its thickness is determined to be 0.165.

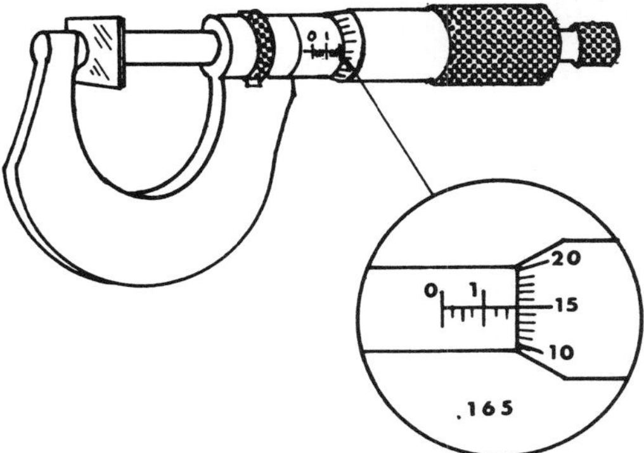

Figure 25. Using the Micrometer Caliper to Determine the Thickness of a Steel Plate.

Now refer to Table 1, page 15, and check the gage number under the column 1, Manufacturer's Standard Gage. The closest decimal to the reading is 0.1644 which is noted to be number 8 gage steel sheet. This is the most accurate method for determining gage thickness at least until the actual gage using the manufacturer's standard is available. A number of different gages are used for measuring different types of materials. The Manufacturer's Standard Gage when available, will be used to measure iron and steel sheets, whereas the American Standard Wire Gage is used to measure the thickness of all other metals such as aluminum or brass. Be sure that the gage which is used is stamped to correspond to the material being measured. Note the gages, gage numbers and decimal sizes in Tables 1. If in doubt, use the micrometer to check actual thickness, then refer to the table to check the actual gage number. Since several different kinds of gages are used for measuring different kinds of sheet metal and wire, confusion can result. When ordering, specify the decimal thickness of the material in inches, the name of the gage and the gage number.

OTHER GAGES

A number of other gages are used in the metal shop for special measuring applications. Use of the screw pitch gage shown in Figure 26 is the quickest and most accurate way to find the number of threads per inch on a bolt or a nut.

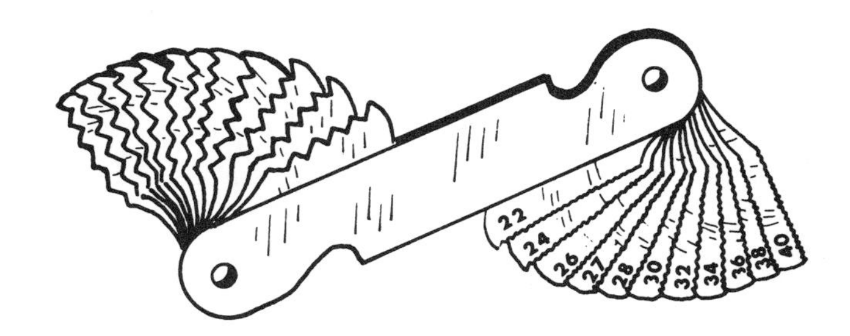

Figure 26. The Screw Pitch Gage.

As noted in Figure 27, the pitch gage can be used for determining the number of threads per inch on a bolt or the number of threads on the inside of a nut. The blades are stamped with numbers indicating the threads per inch. Matching the gage with the threads on the bolt quickly reveals the pitch of the bolt.

Other gages that have application in the metal shop are illustrated in Figure 28. Instrument A is a flat feeler gage. It is used to determine clearances between pieces of metal. It contains a number of thin metal blades usually in increments of .0001" with a range of 0.001" to 0.025" depending on the number of blades in the gage.

Instrument B is referred to as a taper gage and is used to measure the inside diameter of tubing, the widths of slots or as a standard in setting calipers. As shown, the taper gage is marked off in thousandths of an inch.

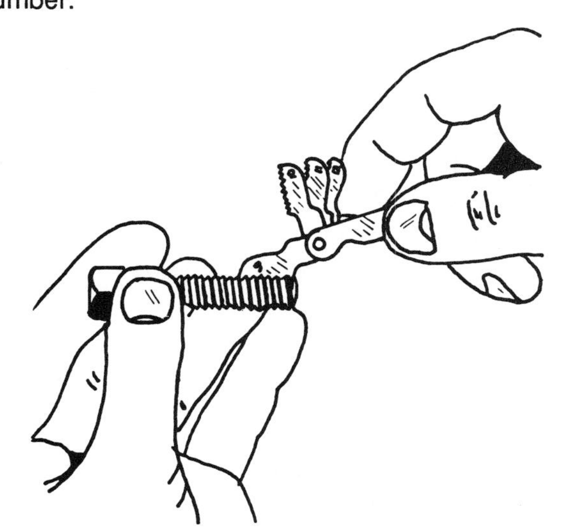

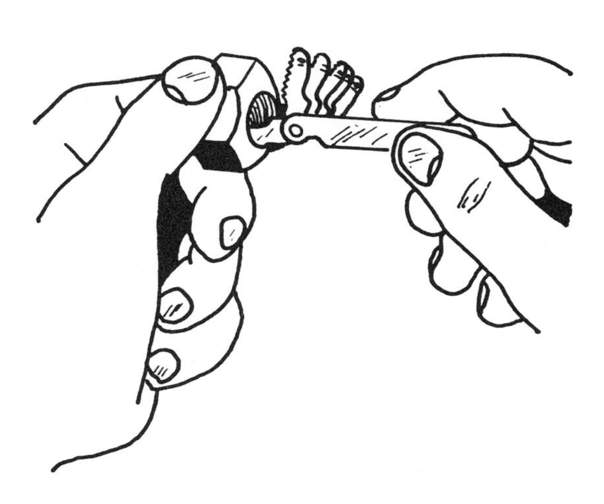

Figure 27. Using the Screw Pitch Gage to Determine the Number of Threads Per Inch.

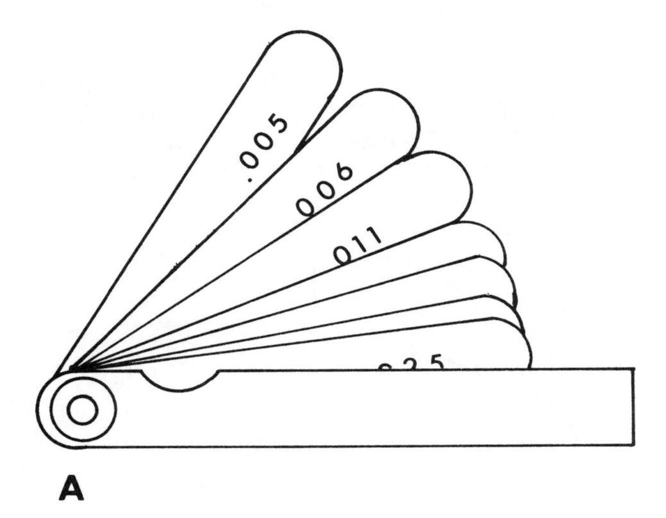

A

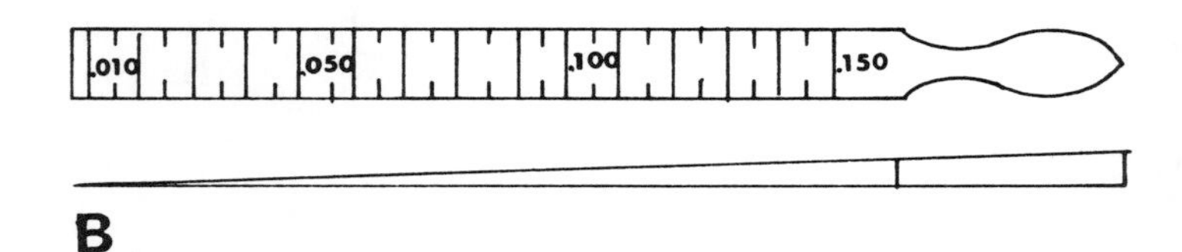

B

Figure 28. Thickness Gages.

METRIC MEASURING TOOLS

Metric measuring instruments used in a metal shop are illustrated in Figure 29. They include: (1) metric micrometer, (2) metric rule, (3) metric vernier caliper, (4) metric combination square, (5) metric tape and (6) metric thickness gage. The first three will be discussed in more detail and the procedure for reading and proper use will be covered.

The metric system of linear measurement originated in France in 1793 with the standard base unit of length as the meter. The units of linear measure are characterized by a relationship based upon the decimal system as illustrated in Figure 30.

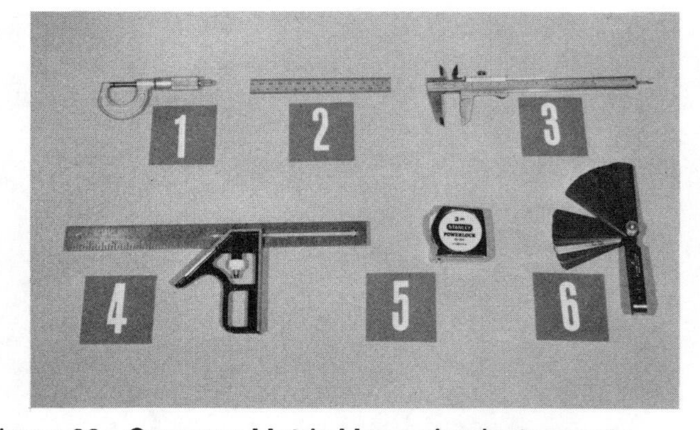

Figure 29. Common Metric Measuring Instruments.

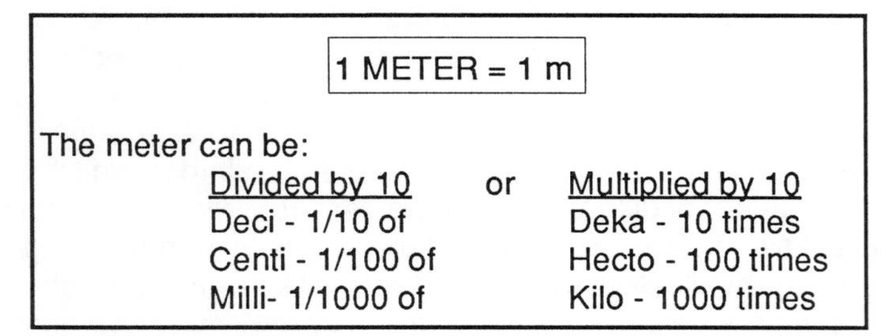

1 METER = 1 m

The meter can be:

Divided by 10	or	Multiplied by 10
Deci - 1/10 of		Deka - 10 times
Centi - 1/100 of		Hecto - 100 times
Milli- 1/1000 of		Kilo - 1000 times

Figure 30. The Meter is the Standard Base Unit of Length.

METRIC STEEL RULE

The metric steel rule is the simplest and most widely used of all metric measuring instruments. Metric rules are commonly subdivided into graduations or a scale of centimeters and millimeters. Some metric rules are further divided into 0.5 millimeter graduations. The procedure for reading the metric rule is shown in Figure 31.

Reading the measurement indicated at Point A.

1. Read the largest number possible and record in mm. 3 cm x 10. <u>30 mm</u>
2. Count the number of full mm's indicated past the major number and record in mm. <u>3 mm</u>
3. Read the number of 1/2 mm's and record. <u>1/2 mm</u>
4. Total all readings: <u>33 1/2 mm</u>
5. Written as: 33 1/2 mm or 33.5 mm or thirty three and one-half millimeters.

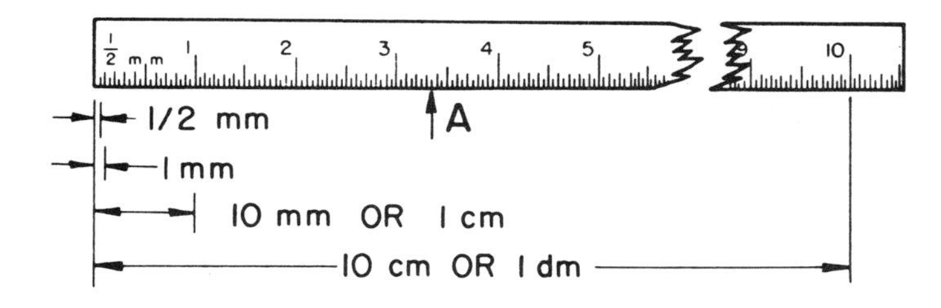

Figure 31. Reading the Metric Rule.

METRIC MICROMETER

Another commonly used metric instrument is the metric micrometer. Metric micrometers are scaled in millimeters with a range of 25 mm. Common sizes available include 0-25 mm, 25-50 mm and 50-75 mm. The parts of the metric micrometer are the same as the parts of the English micrometer illustrated in Figure 9, page 4; therefore only the procedure for reading will be discussed.

The sleeve on the 0-25 mm metric micrometer consists of a horizontal line with 50 vertical lines each equal to 0.5 mm. Therefore, each time the thimble is turned 1 full revolution as illustrated in Figure 32, the thimble moves 0.50 mm. There are 50 graduations marked around the thimble; therefore, turning the thimble from 0 to the first mark on the thimble would move the spindle 0.01 mm.

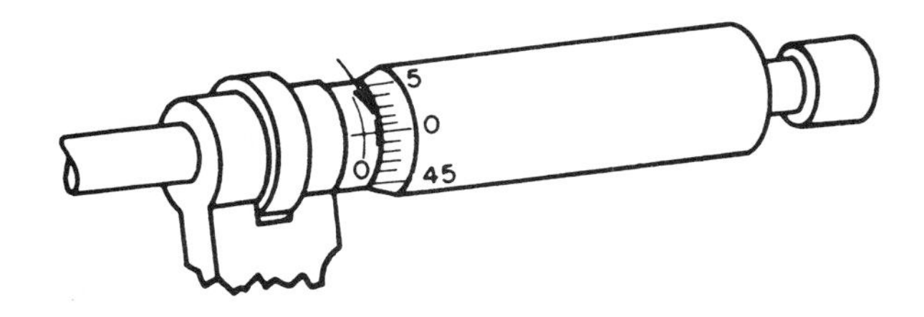

Figure 32. One Turn of the Thimble Equals 0.50 mm.

The metric micrometer may have a vernier scale as shown in Figure 33. The vernier allows for the instrument to be read in 0.001 mm rather than to 0.01 mm as is the case with the micrometer without the vernier scale.

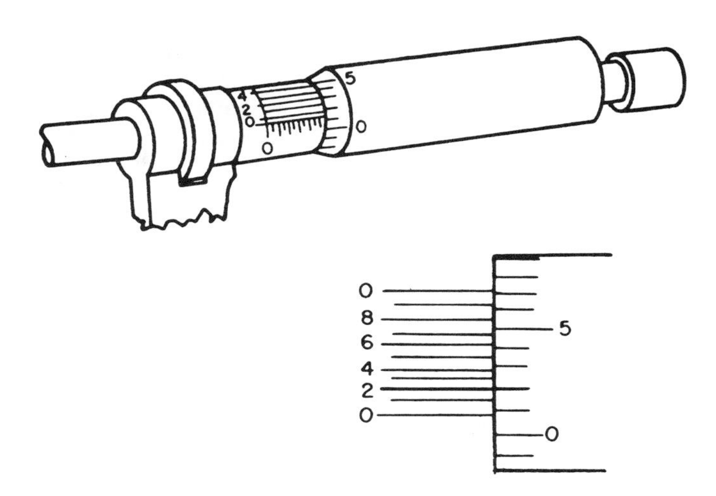

Figure 33. The Vernier Scale.

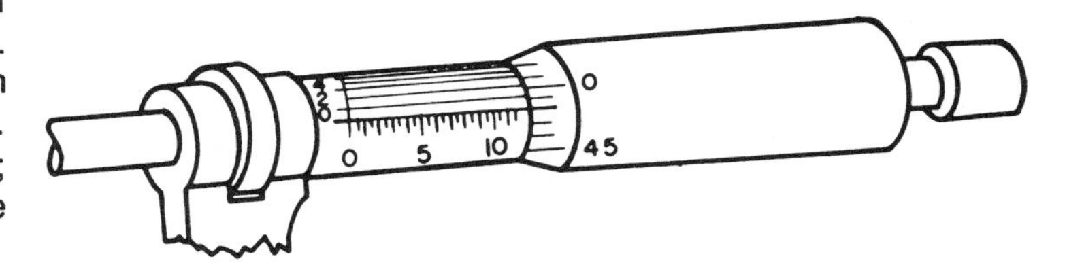

A 50 mm micrometer is selected

Figure 34. Reading the Metric Micrometer.

The procedure for reading the metric micrometer illustrated in Figure 34 is:

1. Record the smallest possible measurement: 25.000 mm
2. Read the largest possible whole number on the sleeve. 10.000 mm
3. Determine the number of graduations beyond the whole number. Each mark equals 0.500 mm therefore 3 x 0.500 equals. 1.500 mm
4. Read the number of full marks shown on the thimble and record. .480 mm
5. Record the mark on the vernier that matches a mark on the thimble. .004 mm
6. Total all readings. 36.984 mm

METRIC VERNIER CALIPER

Notes

The metric vernier caliper is commonly used in the metal shop for making inside, outside and depth measurements. The parts of the vernier caliper were covered earlier in this chapter, therefore, discussion will be confined to the procedure for reading the metric vernier caliper.

As noted in Figure 35 the beam or rule is divided into millimeters (smallest divisions) and centimeters (numbered divisions). The vernier scale has 20 divisions, each representing 1/20 of a mm or 0.05 mm. Each fourth division is numbered 0, 2, 4, 6, 8 and 10 to make the reading of the scale easier.

The procedure for reading the metric vernier caliper shown in Figure 35 is:

1. Record the largest possible whole number on the beam that the zero passed:
   1 cm x 10 — 10.00 mm

2. Determine the number of full marks between the zero point and the whole number.
   5.00 mm

3. Read the vernier. Record the number on the vernier that matches any line on the beam. Remember each vernier mark is equal to 0.05 mm.
   13 x 0.05 — 0.65 mm

4. Total all readings. — 15.65 mm

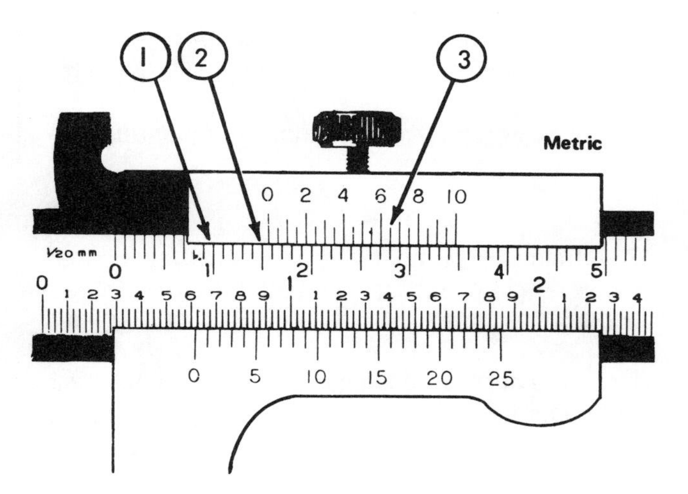

Figure 35. Reading the Metric Vernier Caliper.

TABLE 1. GAGES, GAGE NUMBERS AND DECIMAL SIZES

	Manufac-turer's Standard Gage	Galvanized Sheet Gage	American Standard Wire Gage or Brown & Sharpe Gage	Steel Wire Gage or American Steel & Wire Co.
	1	2	3	4
Gage number	For Iron and Steel Sheets	For Galvanized Steel Sheets	For Wire & Sheet Metal Except Iron Steel & Zinc	For Steel Wire (Not Music Wire or Drill Rod)
0			0.3249	0.3065
1			0.2893	0.2830
2			0.2576	0.2625
3	0.2391		0.2294	0.2437
4	0.2242		0.2043	0.2253
5	0.2092		0.1819	0.2070
6	0.1943		0.1620	0.1920
7	0.1793		0.1443	0.1770
8	0.1644	0.1681	0.1285	0.1620
9	0.1495	0.1532	0.1144	0.1483
10	0.1345	0.1382	0.1019	0.1350
11	0.1196	0.1233	0.0907	0.1205
12	0.1046	0.1084	0.0808	0.1055
13	0.0897	0.0934	0.0720	0.0915
14	0.0747	0.0785	0.0641	0.0800
15	0.0673	0.0710	0.0571	0.0720
16	0.0598	0.0635	0.0508	0.0625
17	0.0538	0.0575	0.0453	0.0540
18	0.0478	0.0516	0.0403	0.0475
19	0.0418	0.0456	0.0359	0.0410
20	0.0359	0.0396	0.0320	0.0348
21	0.0329	0.0366	0.0285	0.0317
22	0.0299	0.0336	0.0253	0.0286
23	0.0269	0.0306	0.0226	0.0258
24	0.0239	0.0276	0.0201	0.0230
25	0.0209	0.0247	0.0179	0.0204
26	0.0179	0.0217	0.0159	0.0181
27	0.0164	0.0202	0.0142	0.0173
28	0.0149	0.0187	0.0126	0.0162
29	0.0135	0.0172	0.0113	0.0150
30	0.0120	0.0157	0.0100	0.0140

CLASSROOM EXERCISE MEASURING INSTRUMENTS

1. Identify the three basic parts of the combination set and give three specific uses for each.

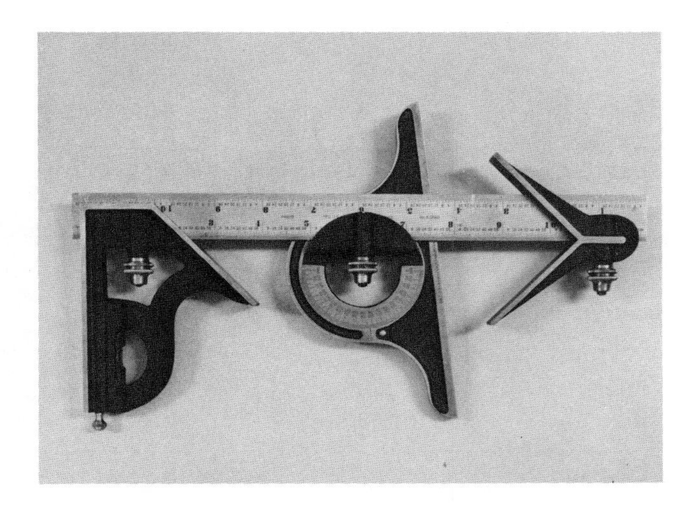

Name: ______________	Name: ______________	Name: ______________
Uses: 1. ______________	Uses: 1. ______________	Uses: 1. ______________
2. ______________	2. ______________	2. ______________
3. ______________	3. ______________	3. ______________

2. Identify the parts of the micrometer caliper shown below:

1 2 3 4 5 6 7

9 10

8

1. ______________
2. ______________
3. ______________
4. ______________
5. ______________
6. ______________
7. ______________
8. ______________
9. ______________
10. ______________

3. What are the decimal equivalents of the fractions:

   7/8", ______________; 1-3/10", ______________; 21/64", ______________.

4. Write the following numbers: seven hundred forty-two thousandths, ______________
   and three ten-thousandths, ______________.

5. Write out the following:

   a. .288" ______________ b. .0004" ______________

6. List the purposes of the ratchet stop on the micrometer:
   1 ______________ 2. ______________ 3. ______________

7. How many threads per inch are there on the spindle screw? ____________________
   How many divisions around the thimble? ____________________
   Each graduation on the sleeve is equal to ____________________".
   Each individual division on the thimble of the micrometer is equal to____________________".

8. One turn of the micrometer thimble is equal to what in a fraction?____________________".
   In a decimal?____________________".

9. What is the shaft-bearing clearance in thousandths of an inch if micrometer A below represents the shaft diameter and micrometer B below represents the inside bearing diameter? (Micrometer A is a 3" micrometer while B is a 4" micrometer) ____________________".

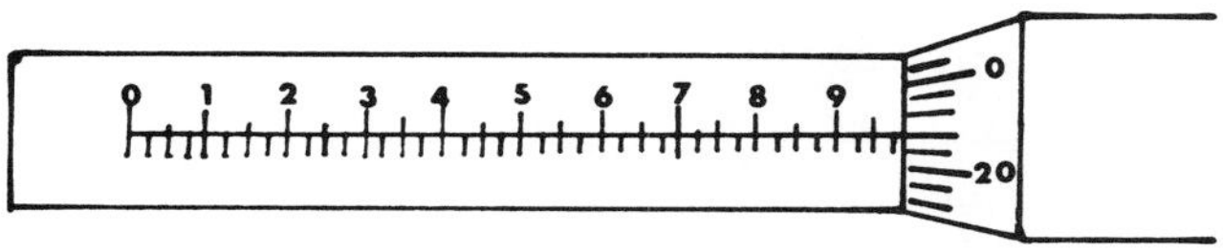

A

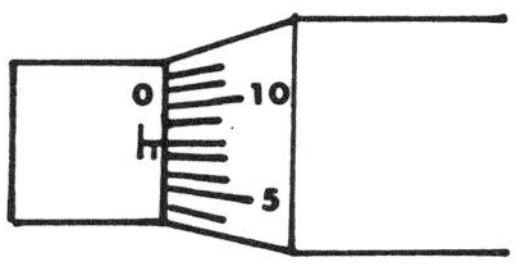

B

10. Determine the micrometer readings.

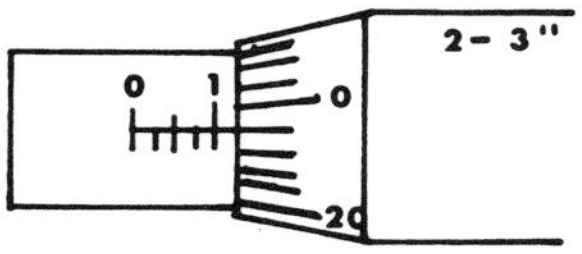

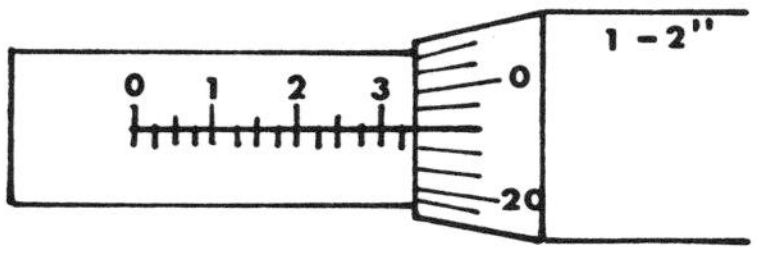

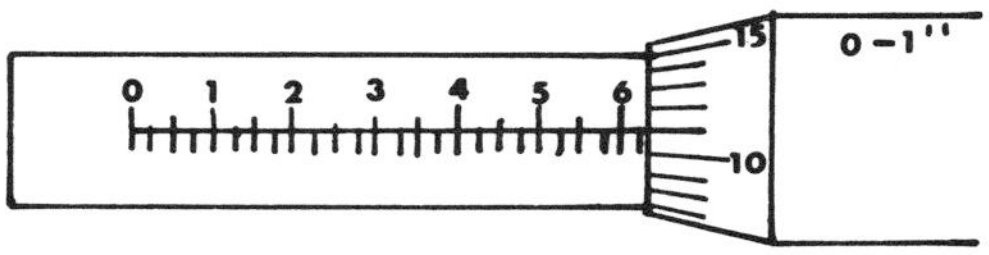

1.____________________ 2.____________________ 3.____________________

11. What is the purpose of the vernier scale on the vernier micrometer? ____________________

____

12. Determine the reading of the following vernier micrometer.

____________________"

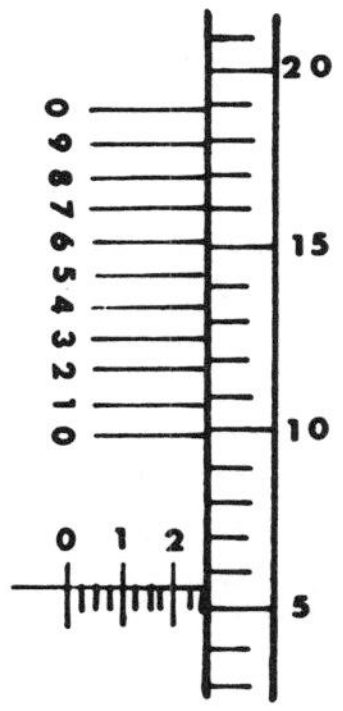

Micrometer 1" - 2"

13. What instruments studied in this unit could be used to measure the inside diameter of a hole or cylinder?

____

14. Read the following vernier caliper in decimals (thousandths) and fractions (1/128ths) of an inch.

a.______________________"

b.______________________"

15. A screw pitch gage blade with a number 16 on the blade could be used for:

____

The number 16 would indicate: ______________________________

16. List common uses for the flat feeler gage.

1.______________ 2.______________ 3.______________

17. List the gage number or decimal equivalent for the following materials.

Material	Gage Number	Decimal Equivalent
a. Sheet steel	8	____________"
b. Copper elec. conductor	____________	0.0808"
c. Galvanized zinc steel	22	____________"
d. Steel wire	____________	0.1483"

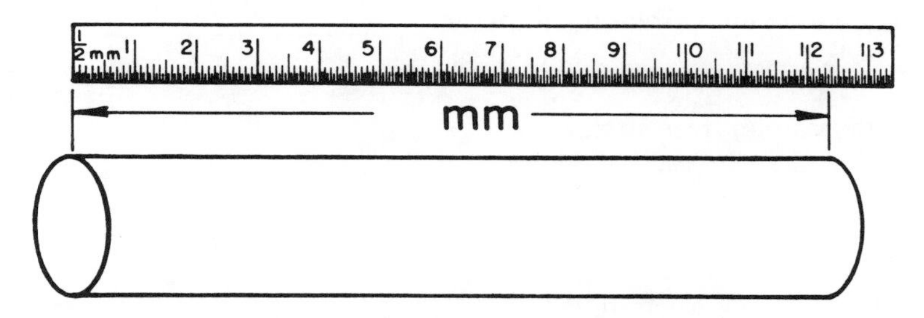

18. Read the metric rule for measuring this round stock.______________mm

19. Determine the reading of this 0-25 mm metric micrometer. ______________mm

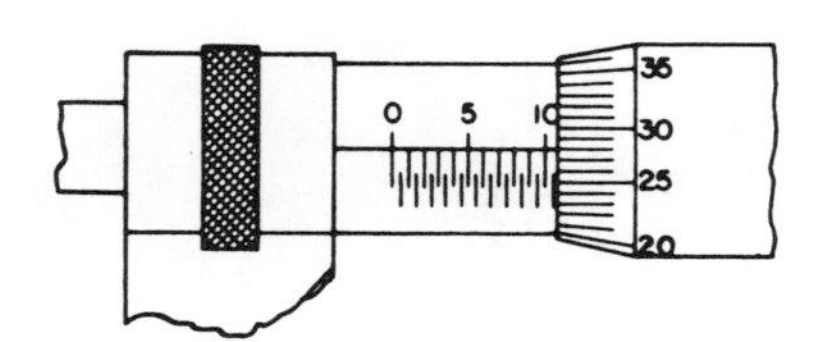

20. Read the vernier caliper in metrics.______________mm

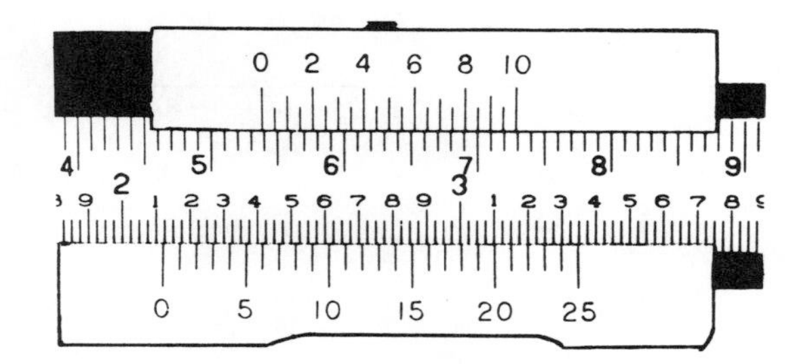

LABORATORY EXERCISES MEASURING INSTRUMENTS

LAB EX - 1 USING THE MICROMETER CALIPERS

1" - 2" Micrometer

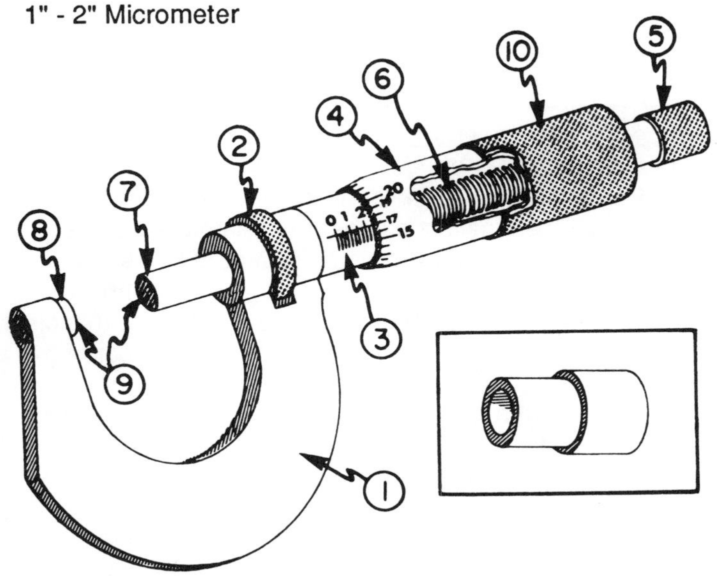

Part Identification:

1.______________ 6.______________
2.______________ 7.______________
3.______________ 8.______________
4.______________ 9.______________
5.______________ 10.______________

Materials:

1 - 1" - 2" micrometer.
1 - two-step practice cylinder (top step between 1.75" and 2.00" lower step 1.25" to 1.75").

A. 0"-1" Micrometer ________"

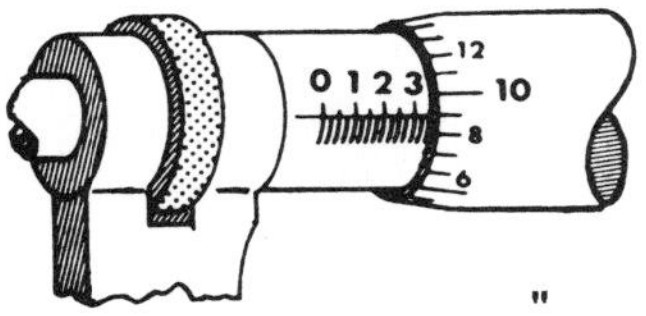

B. 2"-3" Micrometer ________"

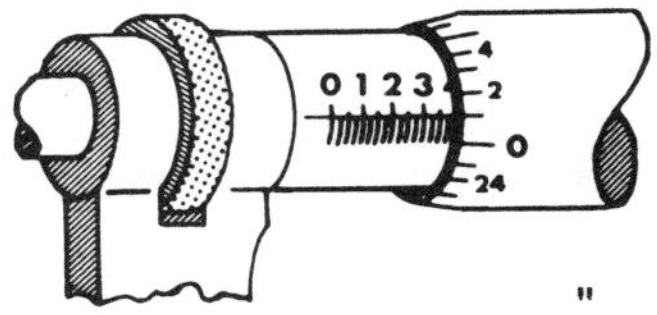

Operational Procedure:

1. Complete the part identification section.
2. Reading the micrometer:
   a. List the smaller number of inches that can be read with the micrometer illustrated above ________"
   b. Denote the number on the sleeve that the thimble edge just passed, this indicates the number of hundred thousandths ________"
   c. Count the number of full spaces that is between the last numbered line (Step b) and the thimble edge and multiply by .025" ________"
   d. Locate the line on the thimble that matches the horizontal line on the sleeve and list this number in thousandths ________"
   e. Total the values (a+b+c+d) ________"
3. Determine the measurements of the two-step machined, practice cylinder:
   a. Using a 1" - 2" micrometer list the measurement of the top step in thousandths. Proper measurement should be the average of measurements made at 3 points around the cylinder ________"
   b. Determine the measurement of the lower step ________"
   c. Subtract the reading in Step (b) from reading in Step (a) to determine the difference in thousandths of an inch ________"
4. Determine the readings of the micrometer inserts A and B.

Operation Teaches: (Ability to...

1. Identify the parts of the micrometer.
2. Understand the function of the various parts.
3. Understand the use of decimals and fractions in measurements.
4. Convert fractions to decimals and decimals to fractions.
5. Properly hold the micrometer.
6. Feel the correct "drag" or tension.
7. Read the micrometer to the nearest one thousandths of an inch.
8. Use the micrometer to measure flat, round or square stock.

Evaluation Score Sheet:

Item	Points Possible	Points Earned
1. Parts identification	20	________
2. Reading the micrometer	25	________
3. Determining the difference in the measurements of the practice cylinder (3-c) (plus or minus .001" = 30 pts., + or - .002" = 20 pts., + or- .003" = 10 pts., greater than .004" off correct reading = 0 pts.)	30	________
4. Determine the reading of the inserts A and B. (5 pts. each)	10	________
5. Handling the micrometer	10	________
6. Attitude and work habits	5	________
Total Points	100	________

Name:______________________ Date:____________ Grade:____________

LAB EX-2 USING THE METRIC MICROMETER

0-25 Millimeter Micrometer

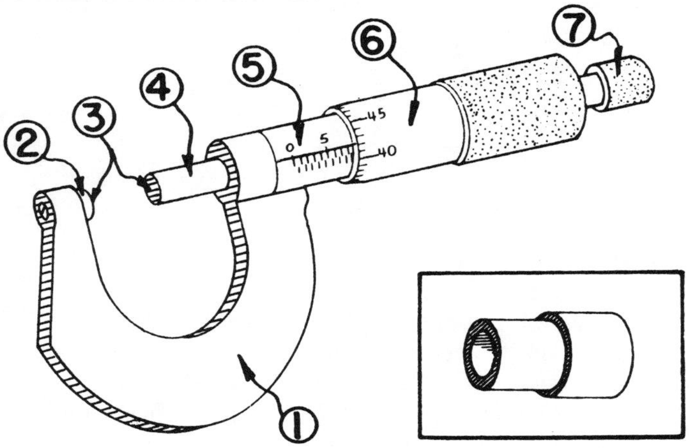

Operation Teaches: (Ability to. . . .

1. Identify the parts of the metric micrometer.
2. Describe the function of the various parts.
3. Properly hold the micrometer.
4. Feel an accurate reading.
5. Read the micrometer to the nearest hundredths of a millimeter.
6. Use the micrometer to measure flat, round or square stock.

Evaluation Score Sheet:

Item	Points Possible	Earned
1. Micrometer part identification (3 pts. per correct item)	21	______
2. Reading the micrometer (2a, b, c, d & e	24	______
3. Determining the difference in the measurements of the practice cylinder (3-c) (plus or minus .01 mm = 30 pts., + or - .02 mm = 20 pts., + or - .03 mm = 10 pts., greater than .04 mm off correct reading = 0 pts.)	30	______
4. Determining the reading of the micrometer inserts (5 pts. each)	10	______
5. Handling the micrometer	10	______
6. Attitude and work habits	5	______
Total	100	______

Name: ______________________________

Date: ________________ Grade: ________________

Part Identification:

1. ________________ 5. ________________
2. ________________ 6. ________________
3. ________________ 7. ________________
4. ________________

Operational Procedure:

1. Complete the part identification section.
2. Reading the micrometer: (Picture on worksheet)
   a. List the smallest number of millimeters that can be read with the micrometer illustrated to the left. __________mm
   b. Denote the number on the sleeve that the thimble edge just passed; this indicates the number of millimeters. __________mm
   c. Count the number of full spaces that are between the last numbered line (Step b) and the thimble edge and multiply by 0.50 __________mm
   d. Locate the line on the thimble that matches the horizontal line on the sleeve and list this number in hundredths of millimeters. __________mm
   e. Total the values(a+b+c+d) __________mm
3. Determining the measurements of the two-step practice cylinder:
   a. Using a 0-25 mm micrometer, list the measurement of the top step in millimeters. Proper measurement should be the average of measurements made at 3 points around the cylinder. __________mm
   b. Determine the measurement of the lower step. __________mm
   c. Subtract the reading in Step (b) from the reading in Step (a) to determine the difference in millimeters __________mm
4. Determine the readings of the micrometers shown below:

A. 25-50 mm Micrometer

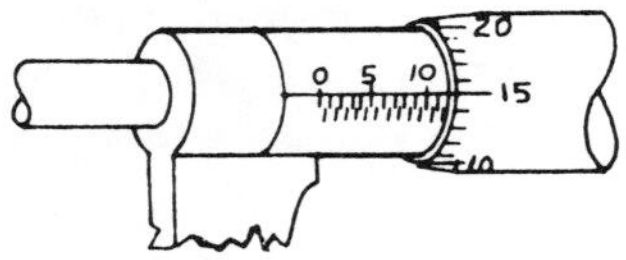

__________mm

B. 50-75 mm Micrometer

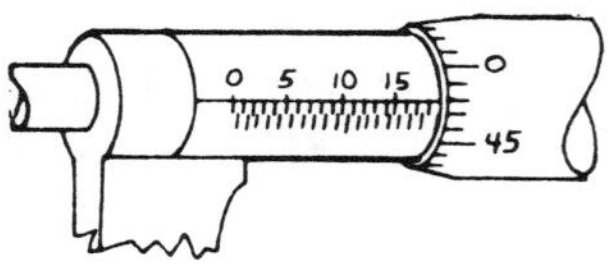

__________mm

Materials:

1 - 0-25 mm micrometer

1 - two-step practice cylinder (top step between 15 and 25 mm and lower step between 5 and 12 mm)

LAB EX - 3 USING THE MICROMETER DEPTH GAGE

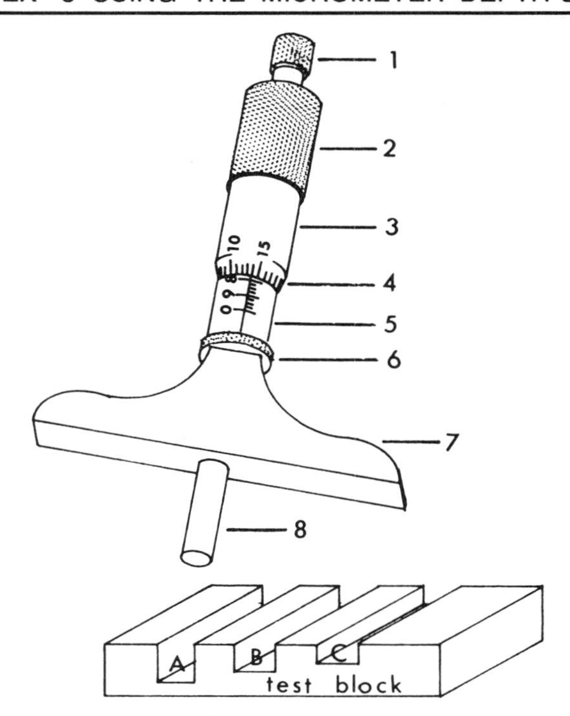

Part Identification:

1.____________________ 5.____________________
2.____________________ 6.____________________
3.____________________ 7.____________________
4.____________________ 8.____________________

Materials:

1. Micrometer Depth Gage
2. Test Block with Pre-determined Dimensions

Operation Teaches: (Ability to. . . .

1. Identify depth gage parts.
2. Understand fractions and decimals.
3. Understand micrometer scale graduations.
4. Read the scale.
5. Take accurate readings.

Name:____

Date:______________ Grade:____________________

Operational Procedure:

1. Identify parts of depth gage.
2. Turn thimble and observe change in its position to lines on sleeve.
3. Reading gage as pictured:
   a. Lines on sleeve are graduated to measure inches. Each line represents .025 inch. Every fourth line is longer and designates hundredths of thousands. Line "1" represents .100 inch and line "2" .200 inch.

   b. Beveled edge of thimble is divided into 25 equal parts. Each line represents .001 inch, each line is numbered consecutively. Rotate thimble from one thimble line to the next. Spindle will move 1/25 of .025 or .001 inch. Moving it two lines is .002 inch.

   c. Reading the micrometer depth gage:

      (1) Determine the number of hundred thousandths on the sleeve. Select largest number thimble has passed.______

      (2) Multiply the number of vertical lines on the sleeve passed the whole number by .025 in. (_______ x .025). ________

      (3) Add the number of thousandths indicated by the lines on the thimble which coincides with the horizontal line on the sleeve. _______

      (4) Total 1+2+3 _________"

4. Hold base firmly against surface of test block.

5. Turn thimble until rod contacts bottom of hole or recess.

6. Tighten lock nut.

7. Remove instrument and determine reading.

Evaluation Score Sheet:

Item		Points Possible	Points Earned
1. Parts Identification		24	______
2. Test block readings:	A.	15	______
(+or-.001" = 15 pts)	B.	15	______
(+or-.002" = 10 pts)	C.	15	______
(+or-.003" = 5 pts)			
(+or-.004" = 0 pts)			
3. Handling the instrument		16	______
4. Attitude and work habits		15	______
	Total	100	______

LAB EX - 4 USING THE VERNIER CALIPER

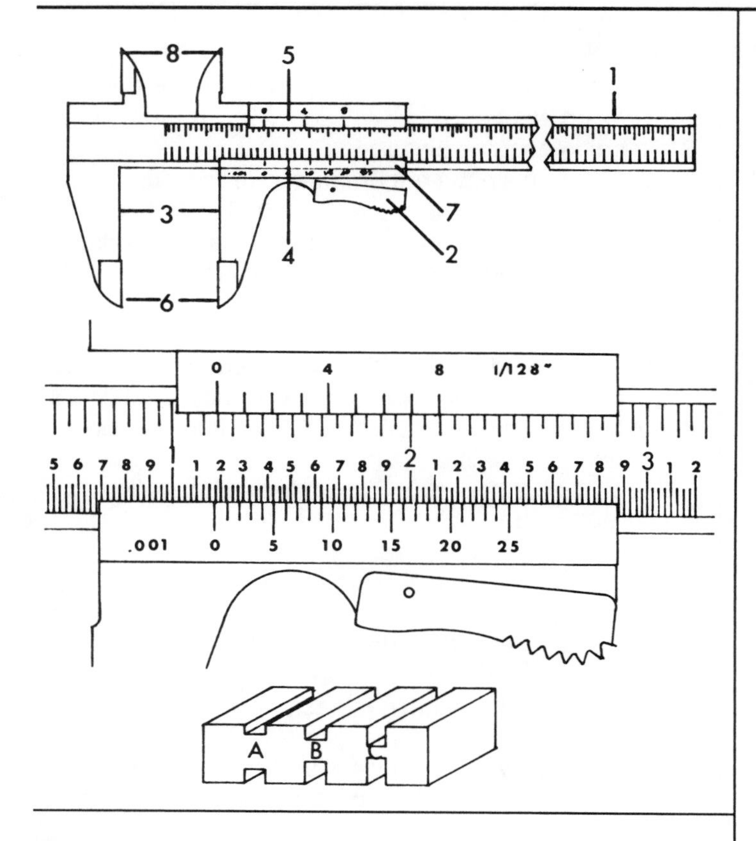

Part Identification:

1.________________	5.________________
2.________________	6.________________
3.________________	7.________________
4.________________	8.________________

Materials:

1. Vernier caliper, 1/1000, and 1/128ths or metric measure.
2. Test block with pre-determined dimensions.

Operation Teaches: Ability to...

1. Identify the parts of a vernier caliper.
2. Understand function of the various parts.
3. Understand decimals & fractions to decimals.
4. Convert fractions to decimals.
5. Properly hold the caliper.
6. Understand how the vernier scale is read.
7. Read the vernier caliper.
8. Take inside readings.
9. Take outside readings.

Operational Procedure:

1. Complete the parts identification section.
2. Reading the vernier caliper in thousandths.
   a. Read the vernier calipers illustrated to the left. List the major inch marking on the beam that the zero mark on the vernier scale just passed
   ____
   b. Note the number of one-hundred thousandths (this refers to the small numbers between the inch graduations________________
   c. Read the number of twenty-five thousandths marking. Multiply by .025________________
   d. Note the graduation on the vernier scale that lines up with a graduation on the beam. This is the number of one-thousandths. ____________
   e. Total values (a+b+c+d) ____________
3. Measuring the test block in thousandths.
   a. ________________________________
   b. ________________________________
   c. ________________________________
4. Make the same above readings if the caliper has fraction graduations.

Evaluation Score Sheet:

Item		Points Possible	Earned
1. Parts identification		16	______
2. Reading the vernier calipers		20	______
3. Test block readings:	A.	15	______
(+or-.001" = 15 pts)	B.	15	______
(+or-.002" = 10 pts)	C.	15	______
(+or-.003" = 5 pts)			
(+or-.004" = 0 pts)			
4. Handling of the instrument		9	______
5. Attitude and work habits		10	______
Total		100	______

Name:________________________________

Date:______________ Grade:________________

LAB EX - 5 MEASUREMENT LABORATORY

Name:________________________ Date:______________ Grade:__________

This lab shall consist of 7 drawers with measurements to be taken at 8 jobs. Please read each job carefully before filling in the requested measurement.

<u>Drawer 1, Job 1</u> - Cold Finished Round and Brass Wire; Measurement of Length and Diameter

A. <u>Cold Finished Round</u>	<u>Length</u>	<u>Diameter</u>
1. Instrument used - Steel Rule, measure in 64ths of an inch.	______	________
2. Instrument used - Vernier Caliper, measure in 128ths of an inch.	______	________

B. <u>Brass Wire</u>	<u>Diameter</u>
1. Instrument used - Vernier Caliper, measure in 128ths of an inch.	________
2. Instrument used - Vernier Caliper, measure in 1000ths of an inch.	________

<u>Drawer 2, Job 2</u> - Cold Finished Bar, Metric Measurement;
Measurement of Length, Width and Thickness in Millimeters

	<u>Length</u>	<u>Width</u>	<u>Thickness</u>
A. Metric Rule	______	_____	_________
B. Metric Micrometer			_________
C. Metric Vernier Caliper		_____	_________

<u>Drawer 2, Job 3</u> - Measurement of Angles

A. Measure angles of aluminum pieces with a protractor head and combination square.

X__________ V__________ W__________

S__________ T__________ U__________

<u>Drawer 3, Job 4</u> - Measurement of steel sheets to determine thickness in inches and gage number. Use the micrometer to determine thickness and then locate the gage number from Table 1, page 15 of the Metals and Welding Manual.

	<u>Inches</u> (thickness)	<u>Gage</u>
A. Roofing (galvanized)	______	____
B. Steel Plate	______	____
C. Steel Plate	______	____
D. Steel Plate	______	____

Drawer 4, Job 5 - Steel Wire; Measure gage from the Standard Wire Gage No. 287.
NOTE - Flip the gage to find the size in inches.

	Gage	Inches
A. Barbed Wire (Measure 1 strand only.)	____	______
B. Livestock Fence		
1. Bottom Strand (largest)	____	______
2. Mesh Wire (Line or stay)	____	______
C. Steel Wire (galvanized)	____	______
D. Steel Wire (not galvanized)	____	______

Drawer 5, Job 6 - Measurement of Electrical Conductors and Non-Ferrous Sheet Metal

A. Copper and Aluminum Conductors - Measure with American Standard Wire Gage No. 688.

	Gage	Inches
1. Yellow Wire	____	______
2. Pink Wire	____	______
3. Black Wire	____	______
4. Aluminum Cable	____	______

B. Non-Ferrous Sheet Metal - Measure with Standard Wire Gage No. 688 for non-ferrous metal.

	Gage	Inches
1. Sheet K - Brass	____	______
2. Sheet L - Copper	____	______

Drawer 6, Job 7 - Measurement of holes, sleeves and pins with small-hole gage, telescoping gage and micrometer

	Decimal	Fraction
A. Determine the inside diameter of the hole.	______	________
B. Determine the inside diameter of the sleeve.	______	________
C. Determine the outside diameter of center pin.	______	________
D. Determine the clearance between the pin and sleeve.	______	________

Drawer 7, Job 8 - Measurement of Pipe Wall Thickness in Fractions.
Use inside and outside calipers and combination square.

	Inside Diameter	Outside Diameter	Wall Thickness	
			Cal.	Mea.
A. Pipe Y	__________	__________	______	______
B. Pipe Z	__________	__________	______	______

UNIT II
PROJECT PLANNING AND SHOP DRAWINGS

Before discussing shop projects and using shop drawings, a question must be answered. Why are shop drawings and project planning sheets needed? Most everyone at sometime has seen a drawing or picture of an object or the actual object and at a later time decided that they would like to construct something similar. However, the problem oftentime faced is the fact that they do not have the drawing and can only picture the object in their minds from memory. A number of solutions are possible. The best, of course, would be to find the drawing once seen. If this is not possible, one might find a picture of a similar or look-alike object. Or they might sit down and produce by sketching, their own plan or drawing from memory.

In the latter two alternatives, one still only has a picture or rough sketch of the proposed object. Examples might be the picture or sketch of the battery terminal puller in Figures 1 and 2. To do an effective and accurate job of construction of this object, more information is needed than the sketch or picture. First, a complete working drawing including dimensions, types and kinds of materials, and the various views are needed. Also, a complete step by step procedure indicating the various tools and equipment used in construction would be essential. In addition, one must have a good understanding and working knowledge of how to read and follow the complete plan or blueprint. These factors should point out the importance of having and using complete and accurate working drawings before attempting to construct shop projects.

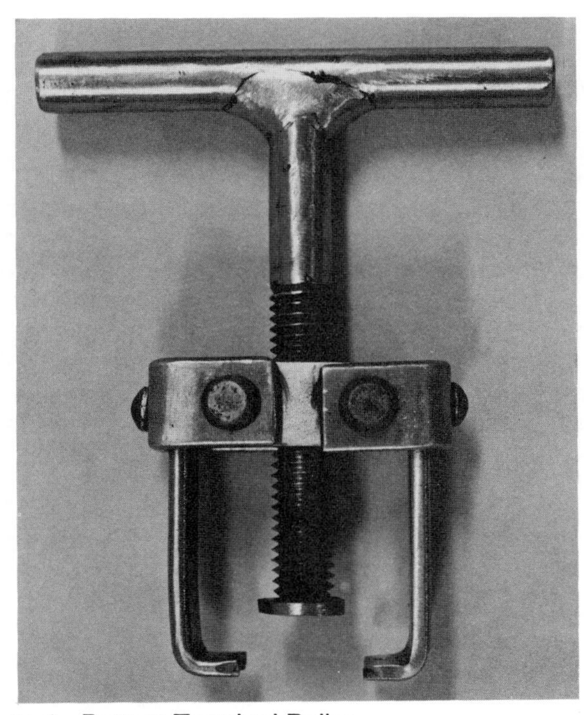

Figure 1. Battery Terminal Puller.

There is another important reason for shop drawings; that is to have a plan to pass on to someone who might be interested in constructing the same object. Shop drawings and blueprints could be considered a language in that they are a method of communication from the draftsman to the shop worker. Normally, the symbols, views, and description are quite standard on shop drawings. This makes it possible for other people to take a plan and assuming they have sufficient ability, they should be able to construct a near equal to the original project. Now that we realize the importance of shop drawings, we need to understand the various types and uses of shop drawings.

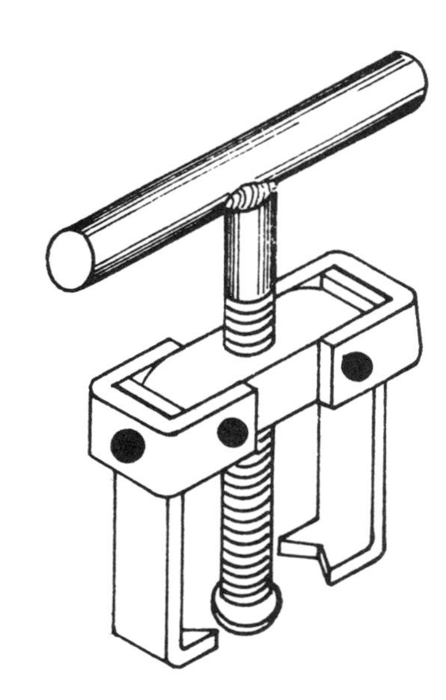

Figure 2. Sketch of Battery Terminal Puller.

WORKING DRAWINGS

Projects in the shop are normally constructed from working drawings. A working drawing is a complete description of an object with all the information needed for construction or fabrication. In other words, it is the drawing from which you work in the shop. As discussed earlier, when a working drawing is given to different individuals in different shops, exactly the same object can be made. The working drawing must show:

1. Shape of each part of the object.
2. Exact size of all parts.
3. Kinds of materials.
4. Kind of finish.
5. Number of each part needed.
6. A view of the complete project as well as views of specific parts.

An example of a working drawing is shown in Figure 3. The metal worker in the metal shop should be able to take this drawing and fabricate the battery terminal puller. One must agree that it should be easier and the final project more accurate in using this working drawing rather than attempting to fabricate from either the picture or sketch shown in Figures 1 and 2.

MECHANICAL DRAWINGS

A mechanical drawing is a working drawing made with the aid of drawing instruments. It differs from the artist's drawing in that no attempt is usually made to give the drawing perspective or to present the object as the eyes see it with portions near the viewer appearing larger than portions farther away. In mechanical drawing, the object is drawn in such a way as to leave no doubt about its exact size and shape, even though the drawing may seem to be out of proportion. The working drawing in Figure 3 could also be considered a mechanical drawing in that it was made by a draftsman using mechanical drawing instruments.

BLUEPRINTS

Mechanical drawings, as used in the shop, are usually not the original drawing made by the draftsman, but a reproduction referred to as a blueprint. An original drawing for each part, component, and assembly, or for each repair or fabrication job, would be far too expensive. Although they are called blueprints, copies of drawings depending upon the reproduction process may or may not be blue in color. Technically, a blueprint is any working drawing produced by any process, not just the familiar reproduction with a blue background and white lines.

The more common reproduction process today is the use of the diazo printer. The diazo printer makes an exposure through a tracing to light sensitive paper using a bright light. Diazo copies are on white paper rather than blue as blueprints.

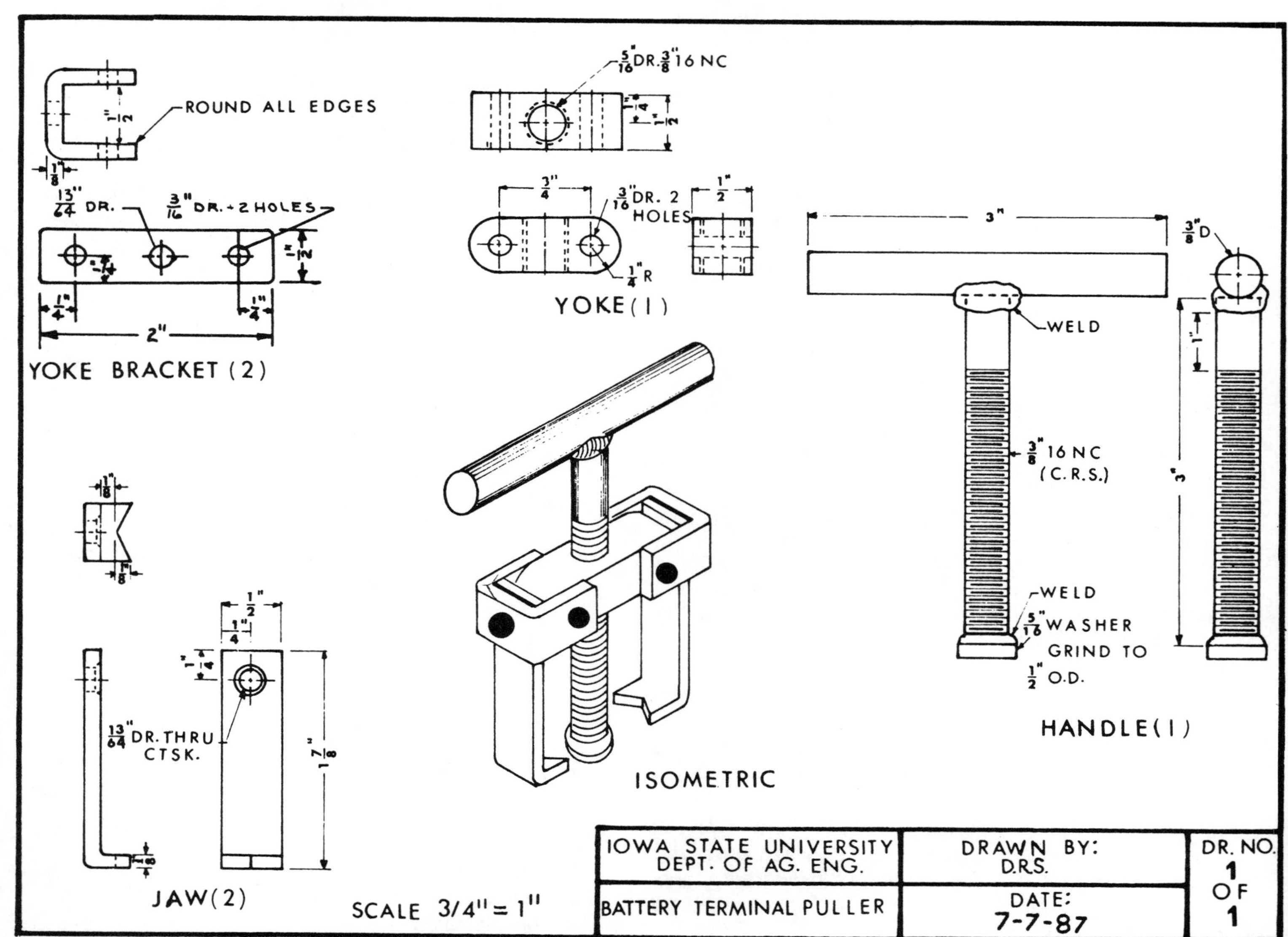

Figure 3. Working Drawing of Battery Terminal Puller.

MAKING WORKING DRAWINGS

Since the shop drawing is the language of the shop worker, before one can fully understand how to use shop drawings one must have some idea of how to make drawings for use in the shop. Oftentimes the first drawing of an object is just a rough sketch made by the shop worker. In order for the draftsman to take the sketch and make a complete working or shop drawing, the sketch must include certain important parts.

SCALE

In most cases the object to be drawn is too large to draw full-size. Therefore, one must understand what is meant by the scale of a drawing. When an object is drawn to scale, it means that it is drawn so that a given size on the drawing would be equal to the size of the actual object. A common scale would be half-size which could also be stated as 1/2 inch equals 1 inch, meaning that 1/2 inch on the drawing would be equal to one inch on the actual object.

Note the scale of 3/4" = 1" on the drawing in Figure 3. One must keep in mind, however, that the sizes given on the drawing are the actual sizes and not the scale sizes. For example, the length of the handle shown in Figure 3 is 3 inches as given on the drawing while the actual length on the drawing is only 2-1/4 inches. Other common scales are 1/4" equals 1", 1/8" equals 1", 1/4" equals 1 foot, etc. If a drawing is other than full size, the scale will always be given on the drawing as: "Scale 1/2" = 1 foot".

DIMENSIONS

Other important parts of working drawings are the dimensions. Rather than discuss the size of an object, one should properly refer to the size as the dimension. Dimensions are the most valuable part of the drawing because they give to the fabricator the actual size of the object.

The most important dimensions are the overall dimensions as the total length, height, width, and thickness of the object. Dimension lines and dimensions are shown in Figure 3. Draw long, narrow lines with arrowheads that touch the extension lines extending from the object to show the actual distances given by the dimensions. Note the 3" length of the handle in Figure 3. Dimensions for small spaces are shown as the 1/8" thickness of the jaw, lower left corner in Figure 3.

The dimension of a circle should be given as the diameter rather than the radius and written as 3/8" D as the diameter of the handle of the battery terminal puller. Dimensions between holes should be given center to center as the holes in the yoke in Figure 3. Fractions must be made with a horizontal line as $\frac{3}{4}$" or $\frac{3}{8}$".

LINES

One must understand the lines on a working drawing to be able to fully understand just how the object is to be constructed from the drawing. The important lines for a working drawing are shown in Figure 4. The most common line is the **visible line** or object line. The visible line is a thick solid line used to show edges of the object that can be seen.

The **hidden line** consists of short, evenly spread dashes 1/8", long 1/16" apart to show hidden features of an object. They are sometimes referred to as an invisible edge line. Examples of hidden lines are shown in Figure 3 on the yoke and other parts of the battery terminal puller.

The **dimension line** terminates with arrowheads at each end. It is broken where the dimension is inserted, note Figure 3. Dimension lines are thin lines drawn approximately 1/4" from the object. They extend to the extension or witness line.

Extension lines are used to indicate the extent of a dimension. They do not touch the outline of the object, but should be 1/16" from the object and extend 1/8" past the arrowhead.

Centerlines consist of long and short dashes alternately and evenly spaced with a long dash at each end. Centerlines locate the center of a circle or arc. A centerline is a thin line made of a 5/8" line, then a 1/16" dash, a 5/8" line, etc. Note Figures 3 and 4.

Leader lines are used to indicate a part or area to which a number, note or other reference applies and usually terminate with an arrowhead as the 3/8" D for the diameter of the handle in Figure 3.

When an object is uniform in shape, such as a pipe or shaft and the length prevents its being shown completely on the drawing, a portion is often removed from the midsection and shown by **break lines**, note Figure 4.

Border lines are extra thick lines used to frame or enclose the entire drawing, giving the drawing a finished appearance as shown in Figure 3.

Line	Weight
Outline or Visible Line	Thick
Hidden	Medium
Dimension	Thin
Sectioning and Extension Line	Thin
Center line	Thin
Leader	Thin
Break Line	Thin
Border Line	Extra Thick

Figure 4. Lines Commonly Used in Working Drawings.

VIEWS

A drawing normally contains a number of views of the object to be constructed. Some objects can be described with only one view; however, to show an object completely, two or more views are usually necessary. An example of an object requiring only two views is a cylinder as shown on Figure 5. In this case, only the top and front views are necessary to show all dimensions for constructing the simple cylinder.

Many objects require three views, a front, top and side view such as the rectangular block shown in Figure 6. To understand this 3-view type of projection, imagine that the object is resting inside a rectangular-shaped box having transparent or glass sides as shown in Figure 7. If you look straight down at the top of the transparent box, the object seen would be the rectangle such as you see in Figure 6. Note how each of the four corners of this rectangle matches up with the corners of the object. Also note how the front and side views appear on the transparent box of what you would see if you were looking directly at the object from those particular angles. To fully understand view alignment, imagine swinging the hinged sides of the transparent box around, flattening it into a single plane as though you were laying them flat on drawing paper. This flat plane is represented by the top, front and side views as seen in Figure 6.

LETTERING

One other important part of the drawing is the lettering. It is often necessary to add short statements, words, or abbreviations of words to give all necessary information on a working drawing. All lettering must be freehand, neat and plain so that it can be easily read by anyone planning to use the drawing. Abbreviations should not be used unless necessary because of space. Note examples of lettering in Figure 3.

DRAWING SYMBOLS

A number of symbols and abbreviations are common to the working drawing. As noted, symbols and abbreviations are used to conserve space on a drawing. The following symbols and abbreviations are common to working drawings found in the metal shop.

'	Feet. or Minutes
"	Inches, or Seconds
°	Degree
±	Plus or minus; more or less
1/2RD	Half-round
1/4RD	Quarter-round
℄	Centerline
~~	Finish

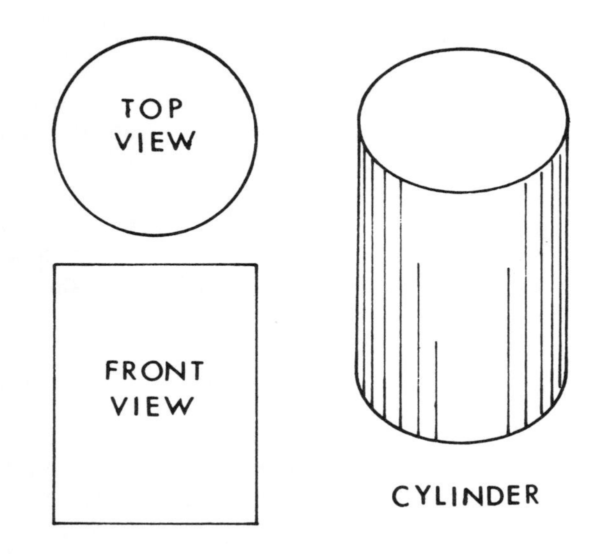

Figure 5. Front and Top Views of a Cylinder.

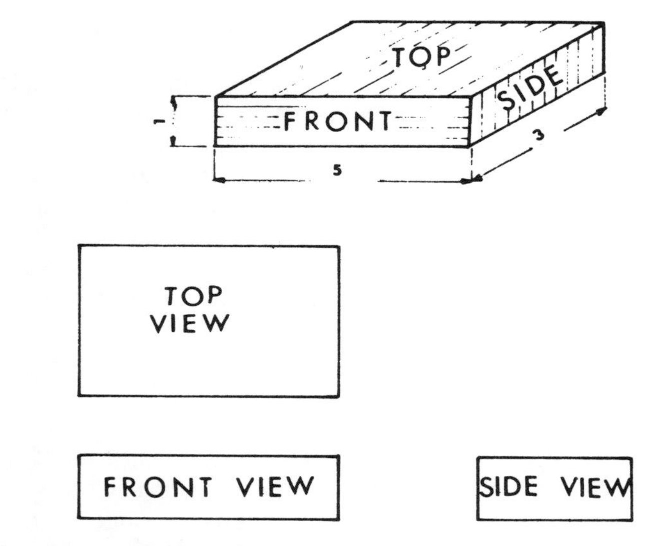

Figure 6. Rectangular Block Showing Top, Front and Side Views.

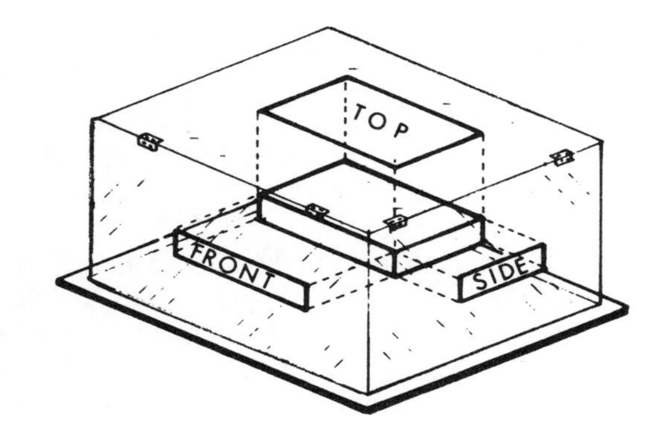

Figure 7. Three Projected Views of an Object.

Arc/w	Arc Weld
B/M	Bill of Material
Csk	Countersink
Dr	Drill
H	Hard
Stl	Steel
Ir	Iron
CI	Cast Iron
GALV	Galvanize
GI	Galvanized Iron
AL	Aluminum
BRS	Brass
BRZ	Bronze
SST	Stainless Steel
WI	Wrought Iron
HGT	Height
LG	Length
THK	Thickness
W	Width
D or Dia	Diameter
R or Rad	Radius
OD	Outside Diameter
ID	Inside Diameter
X HVY	Extra Heavy
Hex	Hexagon
Hor	Horizontal
Vert	Vertical
RH	Right Hand
LH	Left Hand
Thds.	Threads
Det	Detail
Dim	Dimensions
BS	Both Sides
CTR	Center
C to C	Center to Center
Cir	Circle
Dwg	Drawing
Jt	Joint
Sk	Sketch
NTS	Not to Scale
P	Pitch
Ga	Gage
BH	Brinnell Hardness
RH	Rockwell Hardness
Hyd	Hydraulic
NC	National Coarse
NF	National Fine
USS	United States Standard
SAE	Society of Automotive Engineers
CS	Carbon Steel
HRS	Hot Rolled Steel
CRS	Cold Rolled Steel
HSS	High-speed Steel
HCS	High Carbon Steel
Flor	Fluorescent
B & S	Brown and Sharp
AWS	American Welding Society
NEMA	National Electrical Manufacturers Association
AISI	American Iron and Steel Institute
ASTM	American Society for Testing Materials

WELDING SYMBOLS

Welding symbols on the working drawing or blueprint provide complete and concise information for the shop worker as to the type, position, dimension and amount of weld for fabrication of a particular object. It is important that one understands and has the ability to read and use welding symbols if working in the metal shop as most metal projects do involve some type of welding in fabrication. A set of weld symbols has been standardized by the A.W.S., American Welding Society.

Shown in Figure 8 is a common weld symbol and the standard locations of elements of welding information. The reference line of the weld symbol designates the welding process to be used, its location, dimension, extent, contour, and other supplementary information. If necessary to provide specific notations, a tail is attached to the reference line.

There is a definite distinction between the terms "weld symbol" and "welding symbol". The weld symbols as shown in Figure 9, such as details A and B indicate the desired type of weld. The assembled welding symbol, Figure 8, consists of the reference line, arrow, dimensions, tail, specifications, process, the basic weld symbols and other data. The basic weld symbols indicate the welding processes used in metal joining operations such as whether the weld is localized or all around or a shop or field weld as well as the contour of the specified weld. The basic weld symbols are summarized as follows:

1. As noted in part A of Figure 9, the basic arc and gas weld symbols include: bead; fillet; plug or slot; arc spot or arc seam; square, V, bevel, U and J groove welds; and edge and corner flange welds.

2. Resistance weld symbols, as shown in Part B of Figure 9, include spot, projection, seam, and flash or upset welds.

3. Supplementary weld symbols are illustrated in part C of Figure 9. These symbols designate requirements that are common in many processes and include symbols for: (1) weld-all-around, (2) field weld, and (3) flush and convex contour welds.

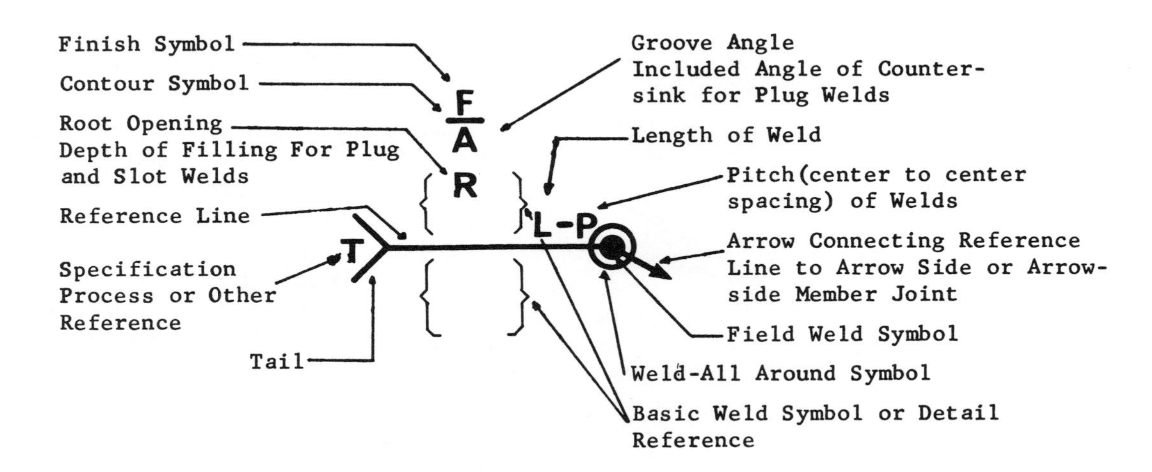

Figure 8. Common Welding Symbol.

A

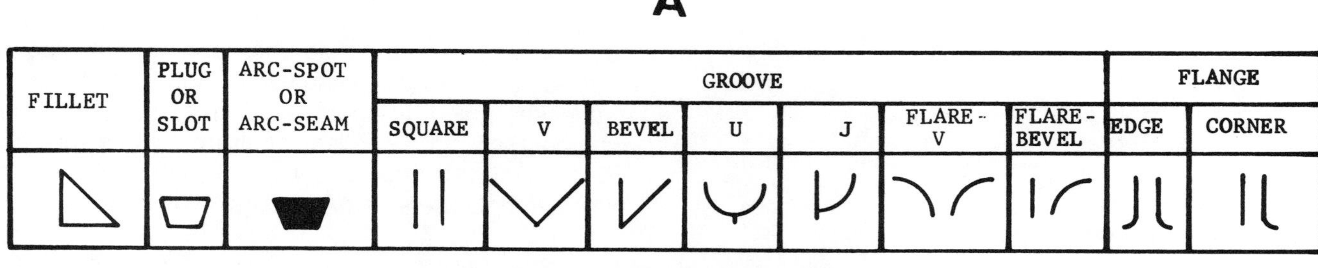

FILLET	PLUG OR SLOT	ARC-SPOT OR ARC-SEAM	GROOVE							FLANGE	
			SQUARE	V	BEVEL	U	J	FLARE-V	FLARE-BEVEL	EDGE	CORNER

Basic Arc and Gas Weld Symbols

B

TYPE OF WELD			
RESISTANCE-SPOT	PROJECTION	RESISTANCE-SEAM	FLASH OR UPSET

Basic Resistance Weld Symbols

C

WELD ALL AROUND	FIELD WELD	CONTOUR	
		FLUSH	CONVEX

Supplementary Symbols

Figure 9. Weld Symbols.

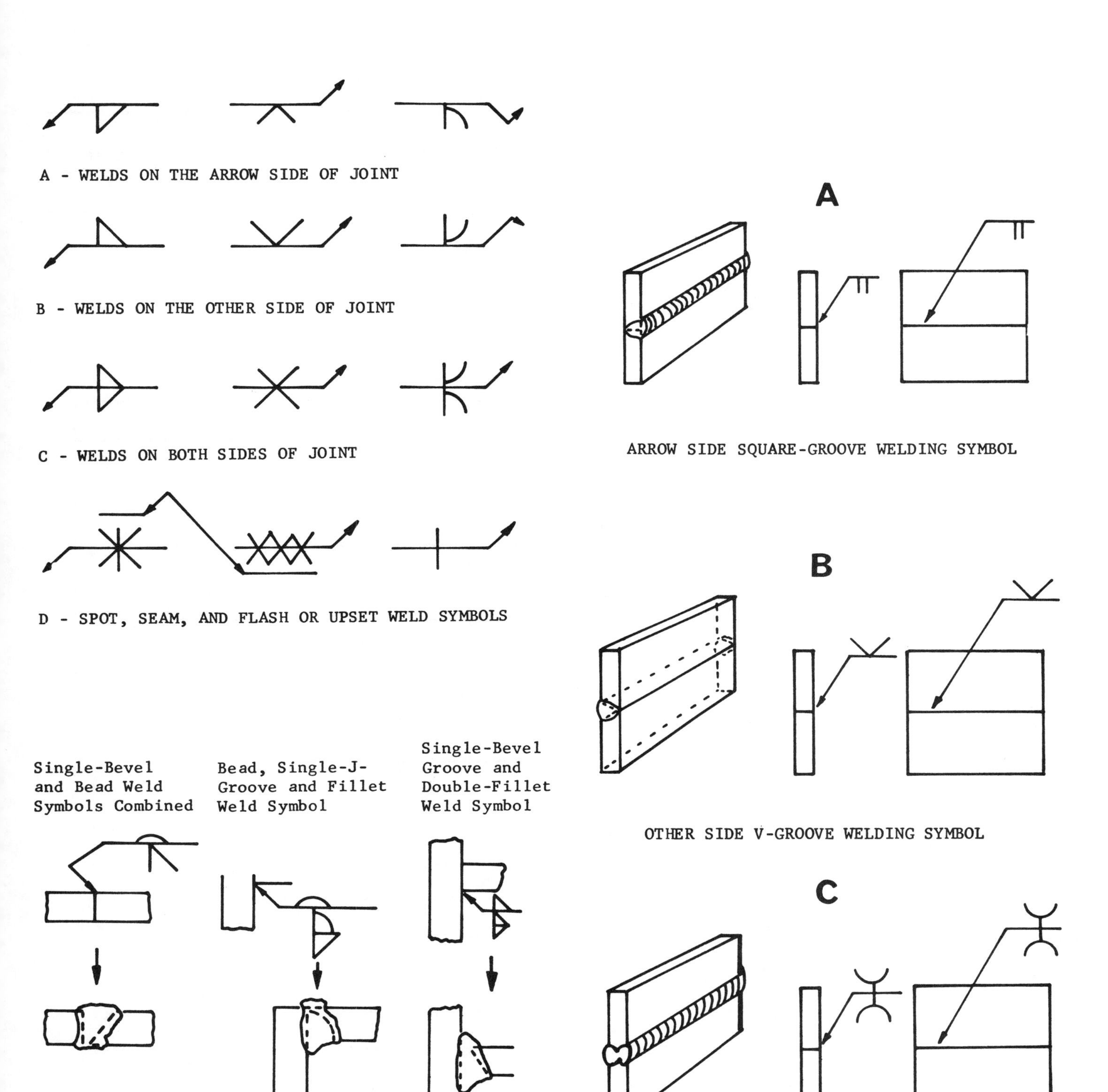

Figure 10. Typical Welding Symbols.

Figure 11. Typical Groove Welds and Welding Symbols.

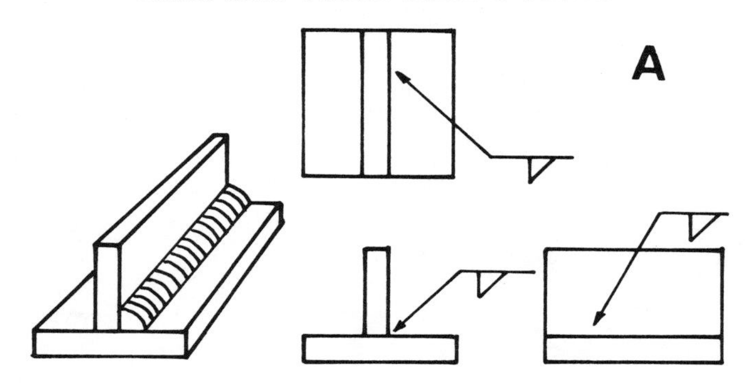

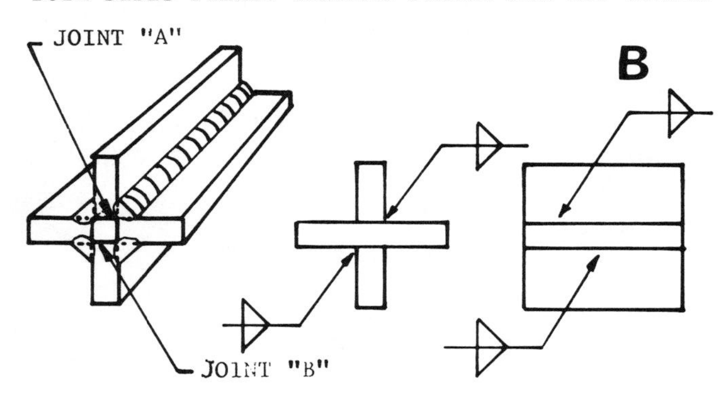

Figure 12. Typical Fillet Welds and Welding Symbols.

LOCATION SIGNIFICANCE	FILLET	PLUG OR SLOT	SPOT OR PROJECTION	SEAM	FLANGE	
					EDGE	CORNER
ARROW SIDE						
OTHER SIDE						
BOTH SIDES		NOT USED	NOT USED	NOT USED	NOT USED	NOT USED
NO ARROW SIDE OR OTHER SIDE SIGNIFICANCE	NOT USED	NOT USED			NOT USED	NOT USED

LOCATION SIGNIFICANCE	GROOVE						
	SQUARE (FLASH OR UPSET)	V	BEVEL	U	J	FLARE-V	FLARE-BEVEL
ARROW SIDE							
OTHER SIDE							
BOTH SIDES							
NO ARROW SIDE OR OTHER SIDE SIGNIFICANCE	NOT USED EXCEPT FOR FLASH OR UPSET WELDS	NOT USED	NOT USED	NOT USED	NOT USED	NOT USED	NOT USED

Figure 13. Basic Welding Symbols by Location.

Welds on the arrow side of the joint are indicated by the weld symbol that is on the side of the reference line toward the reader as shown in Figure 10, detail A. Welds on the other side of the joint are indicated by the weld symbol that is on the side of the reference line away from the reader, note detail B, Figure 10. Welds on both sides of the joint are indicated by the weld symbol on both sides of the reference line, as in detail C, Figure 10. Weld symbols illustrated in detail D of Figure 10 reveal the symbols for spot, seam, flash and upset lines when shown on the reference line. Also shown in Figure 10, part E are combinations of weld symbols and the desired weld for the finished product.

To give you an idea as to the use of weld symbols, study the groove welds and weld symbols illustrated in Figure 11. Part A, Figure 11 illustrates the arrow side square-groove weld whereas in part B, another side V-groove weld is shown. A both sides U-groove welding symbol is shown in part C of Figure 11.

Applications of fillet welding symbols are illustrated in Figure 12. Detail A indicates the proper symbols for the arrow-side fillet welding symbol whereas the both sides fillet welding symbol for two joints is shown in Part B of Figure 12.

The basic welding symbols for arc and gas welding symbols by location are illustrated in Figure 13. These basic welding symbols and specific applications should be valuable to you as you prepare working drawings and plan for shop projects.

JOB PLAN

Now that we have a basic understanding of why we use working drawings, the various types of drawings, how to make a working drawing and of symbols and abbreviations, we have one basic step to complete before attempting to construct the actual project. That is to prepare a job plan. A job plan should include the following items:

1. A working drawing of the project.
2. A bill of material with estimated cost.
3. A list of tools and equipment needed.
4. A step-by-step procedure for fabrication of the project.
5. Source of materials.
6. An estimate of time required.
7. Approval by instructor, if required.
8. Proposed completion date.

A job plan form sheet is illustrated in Figure 14. Completion of a form sheet, such as this one or others provided by your instructor, will make your job much easier. The working drawing should be attached to the form sheet making the job plan complete; thus, you are ready to go to work on your project.

BILL OF MATERIALS

As shown in Figure 14, space is provided for the complete bill of materials. A bill of material is a must for any project as most any project requires metal, hardware, standard parts, and some type of finishing material such as paint. The bill of materials should include:

1. The parts identified by number.
2. The number of pieces for each part.
3. The size of the materials.
4. A description of the material, shape and kind.
5. The source of the material or parts.
6. The per unit cost of the material.
7. The total cost of each part.
8. The total cost of the project.

A completed job plan form sheet is illustrated in Figure 15. This completed job plan is for the Battery Terminal Puller shown in Figure 3 page 26. By combining the working drawing and the job plan, you could be ready to construct a battery terminal puller.

The procedure of planning this project as described in this unit on project planning and working drawings is very similar to the procedure used in industry each day in the manufacture of most of the materials around us from the pencil you use for writing to the welder you use in lab to the automobile you drive. Working drawings must be prepared for each part; the proper materials must be selected; the proper tools, machinery, and equipment must be available; the procedures must be analyzed for most economical production and last, the cost must be analyzed before the item can be produced. The job plan that you will prepare, therefore, involves many of the same activities performed in our manufacturing plants. From this discussion you should be able to realized the importance of complete working drawings and job plans for constructing any project.

JOB PLAN FORM SHEET

Name:______________________ Grade:______________ Date:______________

Project Name: ____

Source of Project Idea:______________________________________

Est. Time:__________ Actual Time:__________ Est. Cost:__________ Actual Cost:__________

Proposed Completion Date:______________Completion Date:______________

Instructor:______________________ Approved:______________________

Bill of Materials

Part No.	No. of Pieces	Size T	Size W	Size L	Description of Material	Source	Unit Cost	Total Cost

Total Cost__________

Tools and Equipment:

Construction Procedure:

Figure 14. Job Plan Form Sheet.

COMPLETED JOB PLAN FORM SHEET

Name: Marvin Marks Grade: 10 Date: Sept. 18, 1987

Project Name: Battery Terminal Puller

Source of Project Idea: Found in suggested plan list

Est. Time: 5 hrs. Actual Time: 5 1/2 hrs. Est. Cost: $ .42 Actual Cost: $ .50

Proposed Completion Date: Sept. 24, 1987 Completion Date: Sept. 23, 1987

Instructor: Mr. Emory Approved: JFE

Bill of Materials

Part No.	No. of Pieces	Size T W L	Description of Material	Source	Unit Cost	Total Cost
1	2	1/8" 1/2" 2"	Yolk Bracket-flat iron	Steel Inc.	30¢/lb.	.03
2	2	1/8" 1/2" 2 1/8"	Jaw-flat iron	Steel Inc.	30¢/lb.	.04
3	1	1/2" 1/2" 1 1/4"	Yoke-cold rolled steel	Steel Inc.	35¢/lb.	.05
4	1	3/8" Dia. 6"	Handle-cold rolled steel	Steel Inc.	30¢/lb.	.06
5	2	1/2" x No. 6	Rivets-countersink head	Bill's Hdwe.	4¢ each	.08
6	2	1" 3/16" Rd Hd	Rivets-soft iron	Bill's Hdwe.	4¢ each	.08
7	1	5/16" Dia.	Flat Washer	Bill's Hdwe.	4¢ each	.04
8	1	3/32" Dia.	E6013 Arc Welding Elect.	Bill's Hdwe.	4¢ each	.04

Total Cost $.42

Tools and Equipment:

Drill press, grinder, forge, arc welder, ballpeen hammer, anvil, steel rule, center punch, 3/16", 5/16" and 13/64" twist drill, 3/8-16 NC die, die holder, 3/8-16 NC tap, tap wrench, hacksaw, combination square, scratch awl, file and countersink.

Construction Procedure:

1. Cut yoke brackets and jaws to length from 1/8" x 1/2" flat iron.
2. Drill 3/16" and 13/64" holes in jaws and yoke brackets.
3. Countersink for rivet heads in jaws.
4. Cut and file groove in ends of jaws.
5. Bend yoke brackets and jaws.
6. Round edges on yoke brackets.
7. Cut yoke stock to length from 1/2" x 1/2" cold rolled steel.
8. Drill 5/16" and 3/16" holes.
9. Tap 3/8"-16 NC threads in yoke.
10. Round edges of yoke.
11. Cut handle and screw from 3/8" round stock.
12. Cut 3/8"-16 NC threads up 2" on screw part of handle.
13. Grind 5/16" washer to 1/2" outside diameter.
14. Rivet jaws to yoke brackets using 1/2" x No. 6 flat head rivets.
15. Rivet jaw-yoke brackets to yoke using 1" x 3/16" round head rivets.
16. Weld handle and threaded screw.
17. Turn handle into yoke.
18. Weld washer to end of screw.
19. Polish and clean arc welds on handle.
20. Finish by painting or light coating of oil to avoid rust.

Figure 15. Completed Job Plan Form Sheet.

CLASSROOM EXERCISE

PROJECT PLANNING AND SHOP DRAWINGS

1. Describe a working drawing. ____
____

2. A working drawing should show:

   a.______________________ d.______________________

   b.______________________ e.______________________

   c.______________________ f.______________________

3. What is a blueprint? ____
____

4. What is the actual size of a part if on a drawing having a 1/4" = 1" scale, the part is 2 1/2" long on the drawing? ______________________, if the drawing scale is 1" = 1'0"?______________________

5. Explain where a hidden line might be used on a drawing? ____
____

6. What line would be used to indicate the center of a circle or arc? ____
____

7. How many views would be required to show a rectangular box? ______________________

8. Identify the following symbols or abbreviations by either giving the symbol or abbreviation or the written meaning:

   a. Csk ______________________ f. AISI ______________________

   b. Degree ______________________ g. American Welding Society ______________________

   c. B/M ______________________ h. NC ______________________

   d. D ______________________ i. Ga ______________________

   e. Bronze ______________________ j. Cold rolled steel ______________________

9. Identify the following weld symbols or draw in the appropriate symbol.

   a. 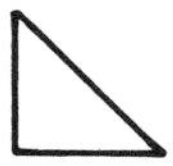 ______________________ d. convex ______________________

   b. ______________________ e. V-groove ______________________

   c. ______________________ f. Weld all around ______________________

10. Identify the following welding symbols and type of joint as found on a reference line.

   a. ______________________

   b. ______________________

   c. ______________________

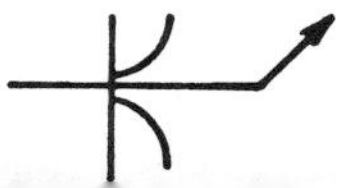

11. Draw in the welding symbols for the weld joints below:

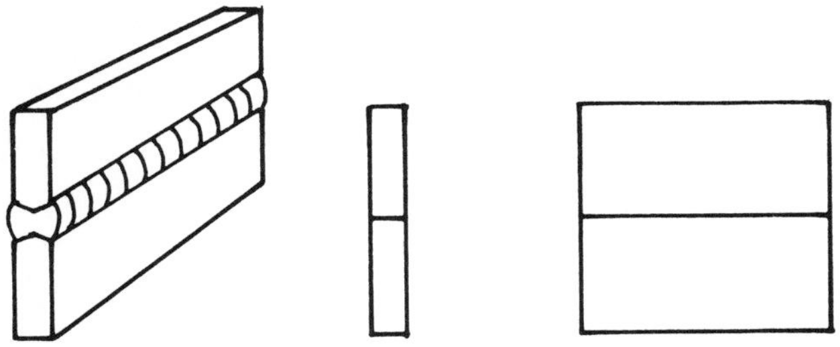

Both sides V-groove weld

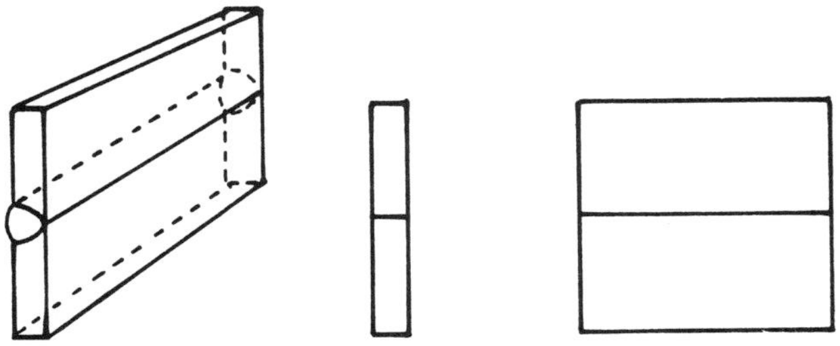

Other side V-groove weld

12. Describe the job plan and list 8 major items that should be included in a complete job plan.

a.______________________ e.______________________

b.______________________ f.______________________

c.______________________ g.______________________

d.______________________ h.______________________

13. Use the job plan form page 34 to figure complete bill of materials and construction procedures for one of the metal construction projects in the Project Construction Unit of this manual.

Notes

UNIT III: METALS

CHARACTERISTICS OF METALS

METALLURGY

The art and science of extracting metals from their ores, refining them and then processing them for use, has had much to do with our industrial expansion and the raising of our standard of living. **Process metallurgy** deals with the extracting of the metal from the ore, refining to obtain relatively pure metals and forming into useable forms, such as plates, castings and sheets for industry. **Physical metallurgy** is concerned with the behavior of the structure and composition of the metal during the shaping, fabricating, heat treating, and welding.

Properties of Metals - The properties of metals are the characteristic qualities by which we recognize the metal and determine its usefulness. In evaluating the usefulness of metal, both the mechanical and the physical properties must be considered. A knowledge of the mechanical properties- tensile strength, compression strength, ductility, is very important in selecting the right metal for the job. These properties can be determined by consulting metals handbooks.

Mechanical Properties - Tensile strength is the mathematical expression of the resistance of a metal to a force that is acting to pull it apart. A common specimen size for tensile testing is called a 505 bar measuring 0.505 inch in diameter. The overall length is 2 1/4 inches with 3/4 inch threaded ends. Two center punched gage marks are made 2 inches apart. The specimen is mounted in the testing machine where the pulling load is gradually applied at a uniform rate. The testing machine has a motor-driven strip chart where an automatic record is made. Shown in Figure 1 is a stress-strain diagram. The following are terms relative to a tensile strength test:

Stress - is the load measured in 1000 psi, pounds per square inch. It is the vertical axis of Figure 1.

Strain - is the deformation that occurs as elongation during loading. It is the horizontal axis.

Yield Point - is the stress point beyond which the metal stretches without an increase in load usually considered yield strength.

Elastic Range or Limit - is the maximum stress the metal will support without permanent deformation.

Proportional Limit - is that point where the increase in strain (elongation) and the increase in stress (load) are no longer proportionate. Up to this point the curve is a straight line.

Ultimate Strength - is the maximum load the metal will support in tension.

Modulus of Elasticity - is the ratio of stress to strain used to compare elasticity of metals.

Percent of Elongation - is the measurement of increase in length of the specimen tensile tested. It is an indication of ductility considered with reduction of cross sectional area.

In addition to the mechanical properties, metals have physical properties which are extremely important in working with metals but are sometimes difficult to measure. The following are important physical properties of metals:

Hardness - is the characteristic of metal which resists scratching, abrasion or indentation.

Toughness - is the ability of a metal to absorb repeated abuse before failing. No standard test of this quality has been developed because of difficulty in describing the load.

Brittleness - is the ability of the material to absorb shock or impact. If metal is brittle, it lacks toughness.

Corrosion Resistance - is the ability of a metal to resist chemical action.

Electrical Resistance - is the opposition that an electric current encounters when it flows through the metal. Conductivity is the opposite of resistance.

Fusability - is a measure of the ease of melting the metal.

Thermal Expansion - is the increase in measurements of a metal due to changes in temperature.

Cost - is not a physical property but it must be considered in the choice of a metal or alloy.

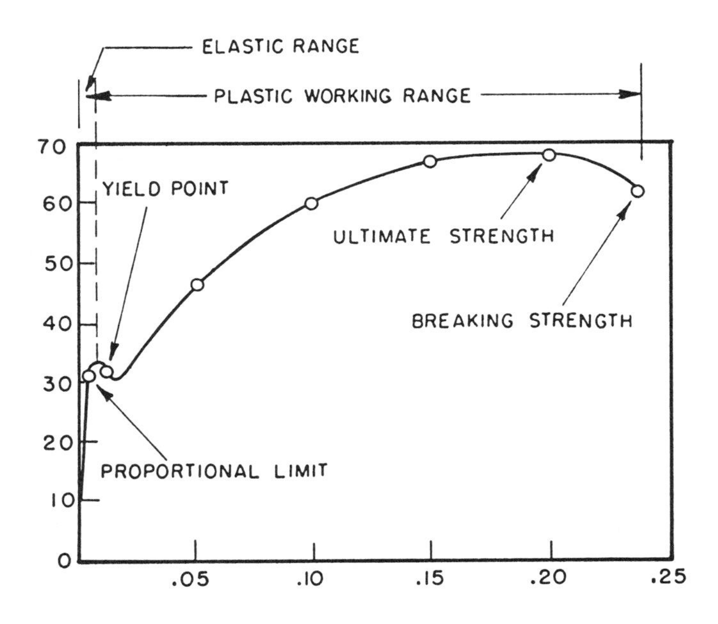

Figure 1. Stress-Strain Diagram.

STEEL-CAST IRONS (FERROUS METALS)

The ferrous metals are made up principally of iron and carbon. Those having less than 1.7 percent carbon are classified as steels and those with more than 1.7 percent are cast irons. Steel with .83 percent carbon is called eutectoid steel. Steel with more than eutectoid is called hypereutectoid. Steel with less than eutectoid is called hypoeutectoid steel.

Hot rolled and cold drawn steels are the simplest of all steels. They consist primarily of iron with small measured amounts of carbon and other elements as sulphur and phosphorus, silicon and manganese as impurities that in most cases do not affect the quality of the steel.

The following are only a few of the more common steels and cast irons:

1. **Carbon Steels - Low Carbon Steels** - have a carbon range between .05 and .30 percent carbon. Steel in the low carbon class is usually tough, ductile, easily formed, machined and welded.
Medium Carbon Steels - have a carbon range from .30 to .45 percent. They are strong and hard but are not as easily forged or welded as low carbon steels.
High and Very High Carbon Steels - have a carbon range from .45 to 1.7 percent (High .45 to .75 percent)(Very High .75 to 1.7 percent). Both of these steels respond well to heat treating so that almost any degree of hardness, temper or strength may be ob tained. In an annealed state they may be machined Special techniques and special welding rods are required if they are to be successfully welded.

2. **Alloy Steels** - Alloy steels are those having special physical or chemical properties which come about as a result of the addition of small amounts of metallic elements as nickel, molybdenum, chromium, vanadium, tungsten, manganese and silicon. Each of these elements give certain qualities to the steel to which they have been added. Stainless steels are common examples of alloy steels. Their outstanding characteristic is their ability to resist corrosion. This group of steels contains measured amounts of nickel and chromium. One type of stainless steel also contains from 7 to 20 percent of nickel. Low carbon steels and most cast irons are relatively inexpensive while some alloy steels and many of the non-ferrous metals may be quite expensive.

3. **Cast Irons -Gray cast Iron** - is the most common type of cast iron used in machinery castings. Carbon and silicon are the alloying elements. The carbon, due to slow cooling of the molten iron, separates into graphite flakes, giving a gray color to the fracture of the metal.
White Cast Iron - contains carbon in the form of iron carbide. It is produced by cooling the molten metal so quickly that the carbon does not separate from the carbon carbide compound. It is distinguished by being very hard and by its white or silvery color when broken. White Cast because of its brittleness, has few applications in its original state, it is used extensively in processing into other forms.
Malleable Cast Iron - is produced by subjecting white cast iron to a long period of annealing at from 1500 to 1700° F. This treatment tends to produce a "skin" of steel about the casting. It will stand shock and will tend to bend before it breaks.
Ductile Cast Iron - a relatively new cast iron that has been markedly improved by a change in its internal structure. The addition of magnesium at the time of melting causes the carbon, which forms in graphite flakes in gray cast, to form into nodular or spheroidal form. Ductile iron has high strength and will bend much like steel before fracturing.

Shown in Table 1 is a comparison of carbon content, characteristics and common uses of iron and steel. Data in Table 2 reveal representative alloy steels and stainless steel compared with carbon steel. In the table, carbon steel has an index of 100, the alloy steels are compared on various mechanical properties based on the index of 100 for carbon steel. With the data given, a ready comparison of properties can be made.

NON-FERROUS METALS

While much of present day metal fabrication involves ferrous metals, there are many metals that do not contain appreciable amounts of iron. The more common nonferrous metals are copper, aluminum, magnesium, nickel, zinc and lead or their alloys.

Brass - is an alloy of copper and zinc. Some types of brass may be spun or cold drawn; other types are made into screws and rivets.
Bronzes - are copper alloys in which the alloying elements are other than zinc. There are a large number of bronzes the most common of which is a copper-tin alloy.
Aluminum - and its many alloys are used where there is need for light weight, good strength, high thermal conductivity, high electrical conductivity or corrosion resistance. It is rolled, drawn, or cast into many shapes.
Magnesium - is the lightest of all of the commercial metals. However, in its pure form it has low strength. Alloys are stronger, some responding to heat treatment.
Zinc - is used extensively for die-castings and for galvanizing. It has very good corrosion resistance and alloying properties, being used extensively with copper.
Nickel - has unusual corrosion and oxidation resistance qualities. It is used extensively as an alloy in steel and copper. Money makes up a group of the more important alloys of nickel.

TABLE 1. CARBON CONTENT, CHARACTERISTICS AND COMMON USES OF IRON AND STEEL

Designation	% Carbon	Characteristics	Common Uses
Wrought Iron	.02-.03	High ductility, easily, worked cold and welded. High corrosion resistance.	Waterpipes, rivets
Low Carbon	.05-.30	Easily worked refined iron.	Bolts, better grades of water pipe, rivets, nails, sheet metals
Medium Carbon	.30-.45	Malleable steel, easily worked.	Agricultural machinery, bars, plates, rails, structural shapes.
High Carbon Steel	.45-.75	High quality steel, which may be hardened by cooling suddenly, may contain some manganese or other elements.	Screw drivers, pliers, drive shafts, crankshafts
Very High Carbon	.75-1.00		Chisels, punches, tools, knives, springs
	1.0-1.2		Lathe tools, dies, taps
	1.2-1.7		Files, reamers, metal cutting saws
High Speed Steel	.55-.85	Contains carbon with varying amounts of cobalt, molybdenum, and tungsten. Withstands heat resulting from cutting at high speeds. Tools cost from 2 to 3 times carbon steel.	Drills, taps, dies, lathe tools, milling cutters and reamers.
Gray Cast Iron	1.7-4.5	Alloy of iron with free carbon(graphite) and silicon, brittle, resists rust.	Agricultural machinery, casting stoves, engine blocks
Chilled (white) Cast Iron	1.7-3.75	Combined carbon, very hard and brittle. High wear resistance	Disc and rotary hoe bearings
Malleable Cast Iron	1.7	Combined carbon of white cast iron changed to graphite carbon by 50 hour heat treatment. Resists shock and has higher tensile strength than ordinary cast. May be considered as gray or white cast with a steel-like layer on the outside, thus bending without breaking.	High strength - agricultural machinery parts, some tools, castings
Ductile Cast Iron (Nodular Spheroidal)	3.44-3.88 (Nickel 1.5% Magnesium or cerium .1 - .15%)	High ductility, 8 times that of gray cast, good machinability, shock resistant.	Replaces gray and malleable cast and unusual steel shapes

TABLE 2. REPRESENTATIVE ALLOYS AND STAINLESS STEELS COMPARED WITH CARBON STEEL INDEX OF MECHANICAL PROPERTIES (COMPARED WITH CARBON STEEL)

	Representative Types of Steel	Breaking Strength	Relative Elasticity	Ductility	Hardness	Distinguishing Characteristics	Typical Uses
A.	CARBON STEEL 1. (C-O.40%)	100	100	100	100	Good strength and workability	Railroad track, bolts, auto axles, bicycle pedals
B.	ALLOY STEELS HEAT TREATED						
	1. MEDIUM MANGANESE (Mn 1.75%)	145	155	58	138	Good strength and workability	Logging, road and agricultural machinery
	2. 3 1/2% NICKEL-MOLYBDENUM (Ni 3.5, Mo 0.25%)	202	224	63	192	Good strength and toughness	Rock-drill and air-hammer parts, high-duty gears
	3. CARBON-VANADIUM (V-0.18%)	158	179	68	153	Resists impact	Locomotive parts, hand tools
	4. CARBON MOLYBDENUM (Mo 0.50%)	149	162	53	164	Resists heat	Oil refining and high-pressure steam equipment
	5. SILICON-MANGANESE (Si 2.00, Mn 0.75%)	198	224	42	180	Springiness	Auto and railroad car springs
	6. PLAIN CHROMIUM (Cr 0.95%)	157	177	63	147	Good strength & workability	Springs, shear blades, wood cutting tools
	7. CHROMIUM-NICKEL (Cr 0.60, Ni 1.25%)	115	125	94	120	Surface easily hardened	Auto rings gears, pinions, piston pins, transmission
	8. CHROMUIM-VANADIUM (Cr 0.95, V-0.18%)	202	229	52	225	High strength & hardened	Automobile gears propeller shafts, connecting rods
	9. CHROMIUM-MOLYBDENUM (Cr 0.95, Mo 0.20%)	130	135	94	125	Resists impact fatigue and heat	Aircraft structures, missile parts
	10. NICKEL-MOLYBDENUM (Ni 1.75, Mo 0.35%)	155	177	68	153	Resists fatigue surface easily hardened	Railroad rollers bearings, auto transmission gears
	11. MANGANESE-MOLYBDENUM (Mn 1.30, Mo 0.30%)	158	177	68	151	Good strength & toughness	High-pressure tanks
	12. NICKEL-CHROMIUM-MOLYBDENUM (Ni 1.75, Cr 0.65, Mo 0.35%)	158	203	63	161	High strength	Aircraft engine and landing gear parts
C.	STAINLESS STEEL 1. 18-8 STAINLESS (Cr 18, Ni 8%) (Cold worked)	207	219	53	165	Resists corrosion	Food machinery, kitchenware, railroad cars, building panels
D.	ALLOY SHEET STEEL 1. HIGH SILICON SHEETS (Si 4.00%)	Electrical Properties of Prime Importance				High electrical efficiency	Transformers, motors, generators
E.	ALLOY TOOL STEEL 1. HIGH SPEED STEEL (Tungsten 18, Cr 4, V-1.0%)	Cutting Properties of Prime Importance				Stays hard at high temperature	High-speed metal-cutting tools
F.	ALLOY MAGNET STEEL 1. COBALT MAGNET STEELS (Co 35.0%)	Magnetic Properties of Prime Importance				High magnetic strength	Permanent magnets in electrical apparatus

Lead - one of the oldest known metals has many industrial applications and is widely used in paints and finishes.
Tin - very high corrosion resistance and adheres to iron. Has a low melting point 232°C (450°F). It is used extensively in solder, brass, bronze and pewter. So-called tin cans are made from thinly rolled sheet steel coated with tin. In the molten state, tin dissolves and alloys readily with many metals.
Copper - has wide use as wire, tubes, sheets and plates. It has excellent workability either hot or cold and has the highest electrical and heat conductivity of all commerical metals. It is used as an alloy with other metals.
Tungsten - a very heavy and high melting point metal. Because of its high melting point 3370°C (6098°F), it is widely used in incandescent lamp filaments. Tungsten Carbide, near diamonds in hardness, is used extensively in cutting and piercing tools.
Silver - is usually thought of as a precious metal but it is used widely in metal alloys. Pure silver is quite soft. Before 1964, silver coins were made from nine parts silver and one part copper. Sterling silver is 92.5 percent silver and 7.5% copper. Silver brazing alloy may contain up to about 45 percent silver. It is an excellent conductor of electricity.
Gold - its purity is measured in karats, 24 karats is pure gold. It is widely used as plating and ornamental jewelry.

THERMOPLASTAIC (NON-METAL)

Nearly 50% of piping in the home, chemical, industrial and agricultural uses involves thermoplastics. Thermoplastics have many desirable qualities they are light, flexible, tough and corrosion resistant. Polyvinyl chloride (PVC) is used extensively in home construction and agricultural uses.

IRON & STEEL AND THEIR ALLOYS

HOW STEEL IS MADE

1. **Production of Pig Iron** - Iron ore, limestone, coking coal and compressed air are the basic raw materials used in the production of pig iron. The first three materials are "charged" in alternate layers through the top of a blast furnace, which may be as much as 100 feet high. Preheated compressed air is then forced into the base, which, in combination with the burning coke, produces temperatures of about 3500° F. at the base of the furnace. Large quantities of carbon monoxide produced when oxygen in the air comes in contact with highly heated coke, act as a reducing agent resulting in the separation of the molten iron. Limestone combines with the impurities of the ore forming slag, which, being lighter than molten iron, floats to the top. The slag is drawn off at a higher "cinder notch" leaving relatively pure iron to be drawn off at the lower "iron notch". The hot metal cast into molds is called pig iron. Most pig iron is moved hot for further processing. The following equation shows the approximate tonnage of materials involved in producing 1 ton of pig iron:

Iron Ore + Coke + Limestone + Air =
(1.93) (0.96) (0.48) (3.93)

Pig Iron + Slag + Gas + Flue Dust
(1.0) (.55) (5.68) (0.09)

Steel is produced from the pig iron by open hearth, Bessemer, electric furnace or basic oxygen processes. Note Figure 2, flow-chart illustrating the production of ferrous metals from pig iron to complete products.

2. The **Open Hearth Process** - At one time this was the major process used in the production of steel in the U.S. The furnace consists of a large brick-lined, dish-shaped hearth into which a total of approximately 200 to 400 tons of molten pig iron and steel scrap are placed. The hearth is heated by a gas flame sweeping across the top of the molten metal. The undesirable materials such as carbon, manganese, silicon, phosphorous and sulfur are almost completely removed by oxidation or by combination with slag.

There are two types of processes; "Basic" and "Acid". In the "Acid" method, most of the silicon, manganese, and carbon are removed; but it does not remove phosphorous and sulfur. The "Basic" process removes all but a trace of the impurities including phosphorous and most of the sulfur.

3. The **Bessemer Process** - In this process, approximately 25 tons of pig iron (90%) and steel scrap (10%) are placed in the converter, a brick lined steel cylinder closed at the bottom. Two types of Bessemer process "Acid" and "Basic" are used. The kind of lining of the converter and flux (sand or limestone) added determines the type of process.

4. The **Electric Furnace Process** - This furnace may be charged with up to 200 tons of entirely scrap steel or a mixture of pig iron and scrap with the usual lime and other fluxes. Heat is applied by huge carbon electrodes touching the slag through the top of the furnace. This method is used for production of stainless steel, carbon tool steel, high alloy and other special types of steel.

5. The **Basic Oxygen Process** - The basic oxygen process, developed in Europe, now produces the majority of the U.S. output. Instead of six to ten hours in an open-hearth furnace, only about 45 minutes is needed to cook a batch of quality steel in all carbon ranges.

Computer duties are vital in this process. To meet customers' steel analysis requirements, amounts of additives must be accurate to the pound. Determined by computer, required amounts of ore pellets, burnt lime, fluorspar, limestone and alloy materials are weighed automatically.

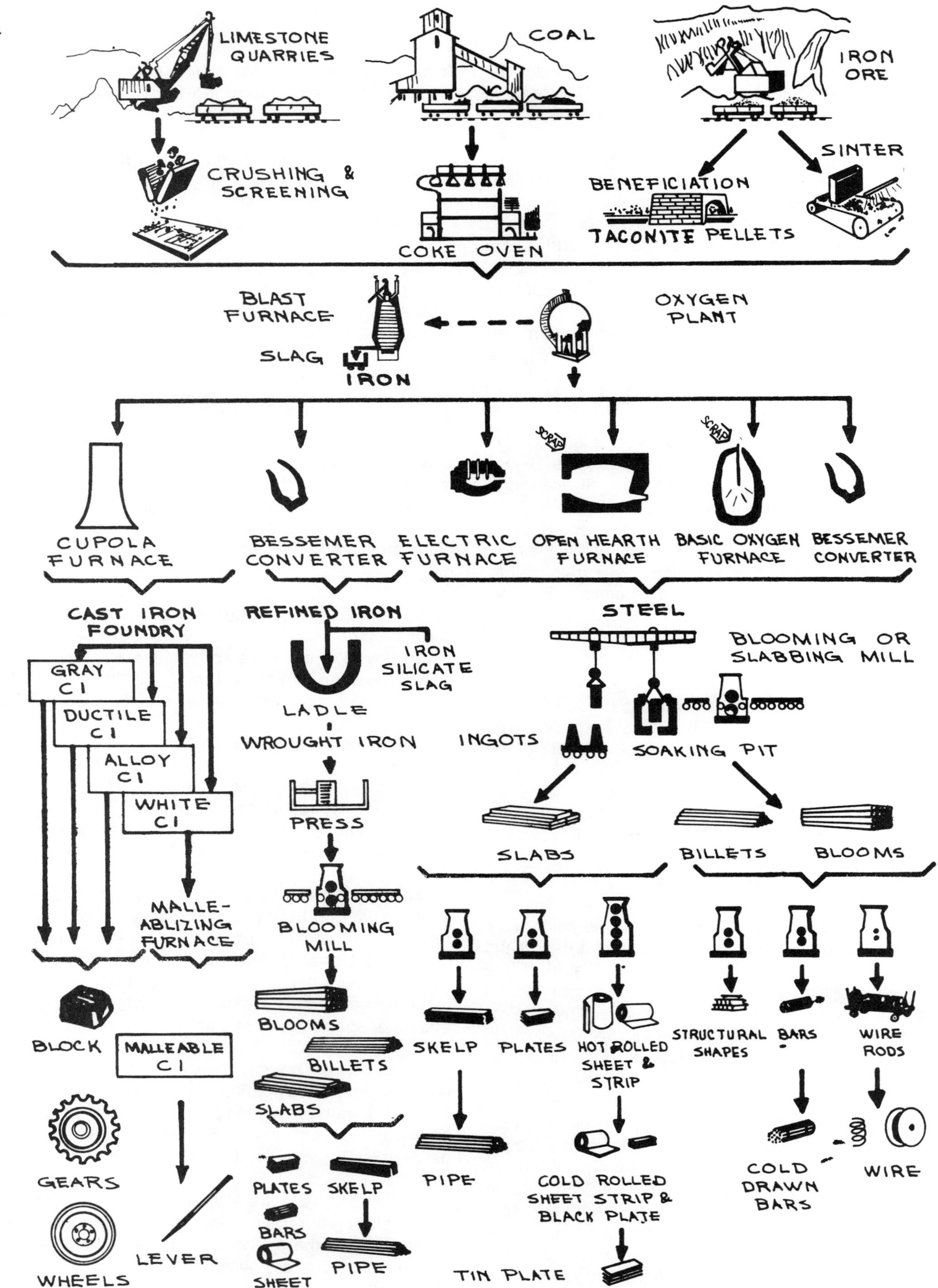

Figure 2. Production of Ferrous Metals.

The computer also reads Spectrovar (automatic analysis equipment) findings and calculates the analysis of hot metal and steel samples. In a minute or two, the presence of a dozen elements is analyzed. Measured to the thousandths of a percent, proportions of the elements are typed on a teletype in the lab and in the operator's pulpit.

The Steel Desk of the plant's Order Department sends specifications for preparing each heat to the computer room. Data on ten heats can be handled at one time. Tape-fed to the computer, the data are stored for up to 30 heats and called for by the melter as needed.

The computer system is operated from a control desk in the furnace pulpit.

Oxygen steelmaking requires no external fuel to create heat as does the open hearth process. An exothermic reaction occurs, the pure oxygen combining with carbon in the bath. Carbon burns off as carbon monoxide and dioxide.

The high-purity oxygen is supplied by four oxygen-producing units totaling 1410 tons per day. A water-cooled lance 60 ft. long and 10 inches in diameter stabs the molten metal in the furnace with a Mach-2 jet of high-pressure (125 to 150 psi) oxygen.

At the end of the oxygen process, a scrubber system achieves the same degree of cleaning as electrostatic precipitators accomplish in Bethlehem's open hearth shops. Waste gases discharged through a 200 feet, high stack are cleaned of more than 98% of the dust and iron oxide fume.

RIMMED STEEL

Molten steel is poured into cast iron ingot molds. Nearly pure iron solidifies on the walls and bottom of the mold. Oxygen becomes carbon monoxide moving toward the core. Some of it escapes to the atmosphere. This increases the carbon content of the remainder of the steel ingot. This cooling reaction continues until a rim about 3 inches thick is formed. Rimmed steel in its various finished shapes has a skin of almost pure iron and a core of steel containing carbon, phosphorus and sulfur. This may contribute toward welding difficulties known as crack sensitive conditions. Welding electrodes and other forms of wire, plate and sheet used in deep drawing are made from rimmed steel.

KILLED STEEL

Killed steel is completely deoxidized by adding aluminum, silicon, titanium, calcium or zirconium in varous combinations and quantities. The result is a premium grade steel. Semi-killed steel becomes the major tonnage product of the industry. A comparison of rimmed, capped, semi-killed and killed steel is given in Table 3.

CONTINUOUS CASTING

In the interest of accuracy, this covers a cast rather than a rolled product but it is discussed here because the end product is classed as semi-finished. While there are somewhat over 60 continuous casting installations in existence throughout the world, only a few are located on this continent. Most of the existing installations cover low volume output and have as their main purposes (1) improved yield of the order of 10 to 15 percent and (2) freedom from the expense of slabbing or billet mills.

A Canadian steel company installation which has been operating since 1954 is described. This unit handles slabs primarily but also casts square billets two strands at a time. Various refinements of the process are planned elsewhere but the following description covers the operation presently in use in Figure 3.

TABLE 3. COMPARISON OF RIMMED, CAPPED, SEMI-KILLED AND KILLED STEELS

	Rimmed	**Capped**	**Semi-killed**	**Killed**
Carbon content	Up to 0.3%	Up to 0.2%	Up to 1.0%	Up to 1.5%
Silicon content	Less than 0.005%	Less than 0.01%	0.010/.10%	Over 0.10%*
Primary piping (shrinkage)	None	None	None	None
Porosity in ingot	Considerable	Considerable	Some	None
Segregation	Pronounced	Little	Little	Very Little

* Killed steel may have a very low silicon content if another deoxidizer, like aluminum, is employed.

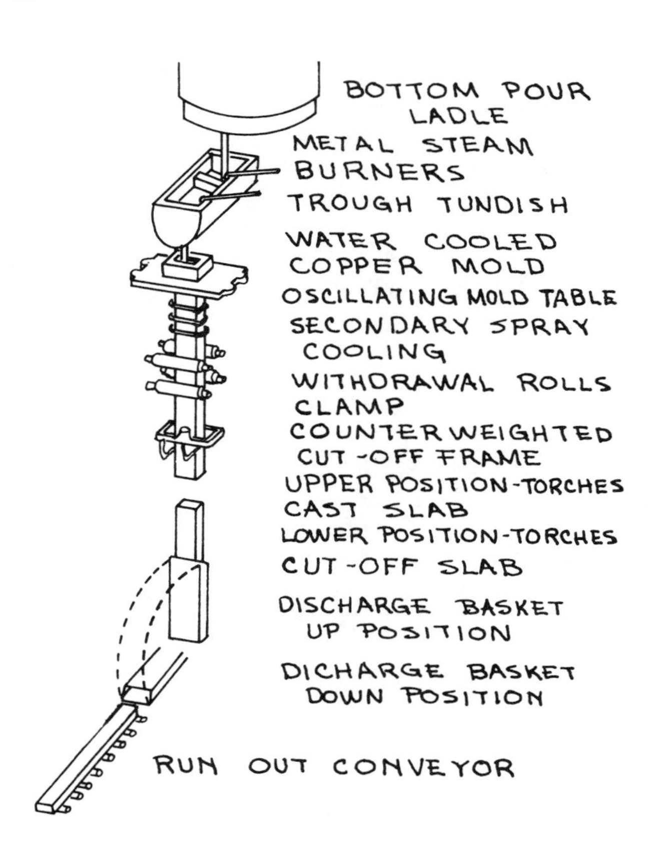

Figure 3. Continuous Casting of Steel.

Machine Description - based on the Junghans-Rossi Principle, this machine is capable of casting any size in the general range of 5 to 8-1/2 in. squares and up to 6-5/8 X 24 in. slabs. It is a vertical unit, with the casting floor 31 ft. above ground and discharge pit 20 ft. deep.

Steel to be cast through this machine is tapped into bottom-pour ladles of 10, 30 or 50 ton capacity. They have 1 in. of insulating brick under 11 in. of standard ladle brick lining and have refractory-lined steel covers.

The preheated ladle is filled with steel from an electric arc furnace, and is then moved by transfer car to the casting machine. An overhead crane lifts the ladle to the casting floor. As the ladle nozzle is opened, steel flows into a refractory lined tundish, which acts as a reservoir, immediately above the mold. The liquid metal is at all times protected from oxidation with a propane atmosphere.

At the start, a starting or dummy bar approximately 20 ft. long, of the same configuration as the section to be cast, is inserted into the bottom of the mold. As molten steel flows from the tundish into the mold, it is chilled, forming a shell 1/4 to 1/2 in. thick.

Pinch rolls act to pull down the starting bar, followed by the cast slab or billet. Synchronized machinery oscillates the mold vertically through 3/4 in., lowering at the same speed as the billet movement and rising at three times this speed. A vegetable oil is constantly injected to provide lubrication between mold and billet.

As the billet moves lower, it passes through a series of small rollers which contain and support all sides. At the same time, water sprays at 20 to 50 psi pressure impinge on the bar, speeding up cooling.

Below this cooling chamber are two oxyacetylene torches mounted so they clamp on the billet or slab and travel downward at the same speed as the billet while cutting is in progress. At the desired intervals, the torches go into action to cut the cast billet into desired lengths up to 16 ft.

The cut lengths are lowered onto a carrier which brings them back to a cooling bed at ground level. From the pickup cradle, the billets or slabs, now about 1500° F, may be moved to reheating furnaces at the rolling mills.

Casting speed can be set at any desired point up to 175 in. per minute. The 6-5/8 X 24 in. section has been cast at a rate of more than 50 in. per minute, or 60 tons per hour. Billets 6 in. square can be cast at 30 tons per hour.

Water consumption for mold cooling varies between 100 and 235 gpm. In addition, the water sprays may use between 30 and 100 gpm, depending on the section being cast, the rate of casting and the steel composition. For each cut made on 6-5/8 X 24 in. stainless slabs, 2-1/2 to 3 cu.ft. of acetylene, 60 to 75 cu. ft. of oxygen and 2-1/2 to 3 lb. of iron powder are used.

This particular facility is casting largely stainless steel.

SHAPING OF STEEL

Steel may be found as castings, forgings or as rolled shapes. The mass tonnage of steel is in the rolled condition such as angles, bars, flats, etc. Forgings are made by pressing or pounding red-hot steel either by gigantic drop-forge machines or by hand. The result is much the same, that of increasing the strength of the product by hot working. Table 4 shows sizes of various hotrolled products. Standard steel shapes are illustrated in Figure 4.

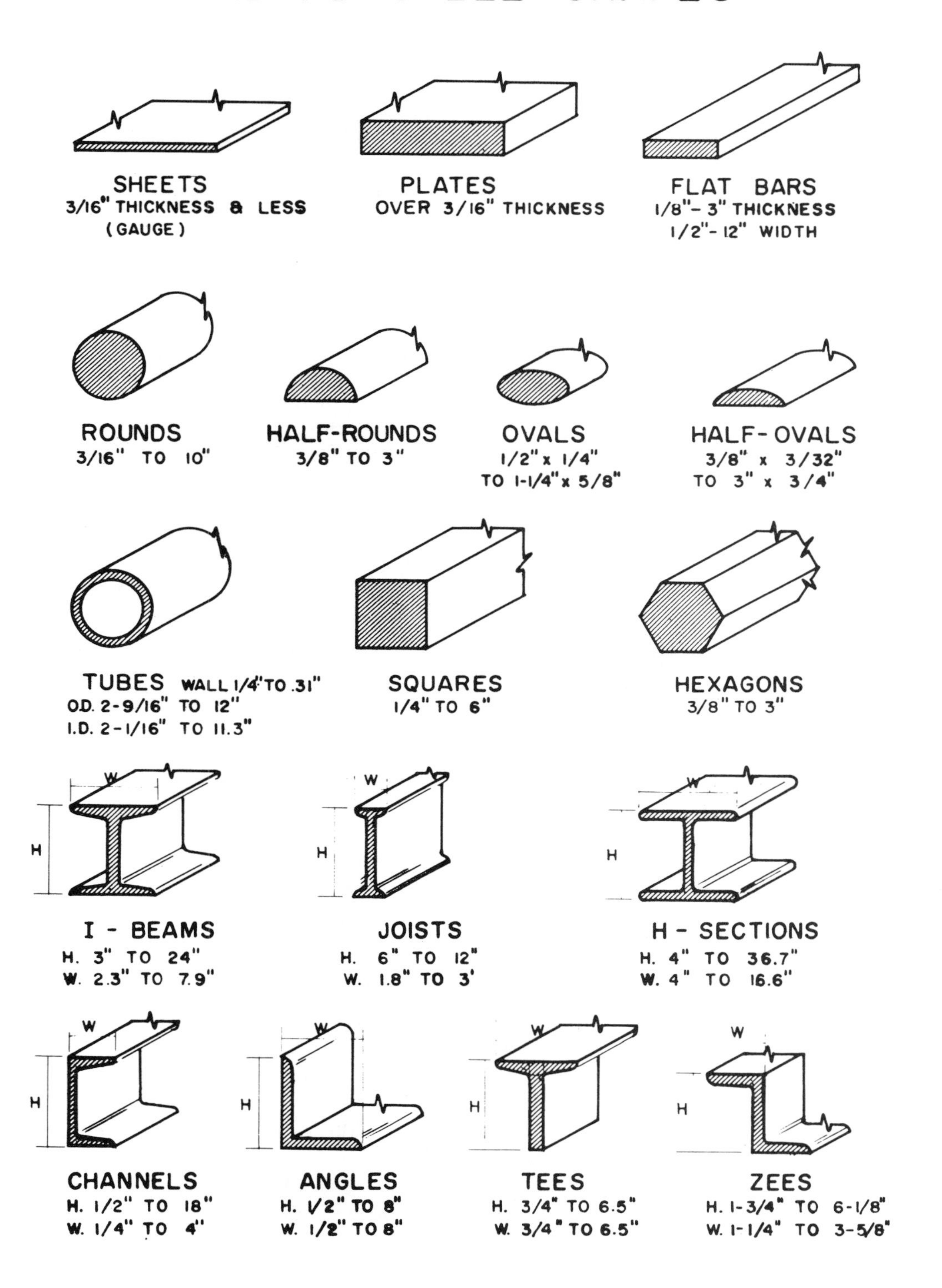

Figure 4. Standard Steel Shapes.

TABLE 4.
CLASSIFICATION OF HOT-ROLLED MILLPRODUCTS

Ingots	Any shape, as cast
Blooms	Cross-sectional area 36 sq. in. min., usually square or rectangular
Billets	Cross-sectional area 36 sq. in.max., 1-1/2 in min. thickness width at least twice thickness
Structural Shapes	Channels, I-beams, angles, etc.
Bars	Rounds, squares, hexagons and flats
Rods	About 3/8 in. diameter round and smaller, hot-rolled
Plates	Thickness 3/16 in. min. and of substantial width
Sheet	Thickness 3/16 in. max. and of substantial width
Strip	Up to 1/4 in. thickness and 12 in. width

Most steel is rolled in hot condition at high speeds. The finished hot-rolled shapes are air-cooled where they acquire an oxide scale. Cold drawn steel sometimes called cold rolled or cold finished (CR or CF) is shaped when cold under pressures. Cold drawn steel is outstanding for precise size (accuracy), high torsion strength and smooth, shiny appearance. Hot rolled steel is not known for these characteristics.

CLASSIFICATION OF STEELS

A numerical index, sponsored by the Society of Automotive Engineers (SAE) and the American Iron and Steel Institute (AISI), is used to identify the chemical compositions of the steels. In this system, a four-numeral series is used to designate the plain carbon and alloy steels; five numerals are used to designate certain types of alloy steels. The first two digits indicate the type of steel, the second digit also generally (but not always) gives the approximate amount of the major alloying element and the last two (or three) digits are intended to indicate the approximate middle of the carbon range. However, a deviation from the rule of indicating the carbon range is sometimes necessary.

The series designation and types are summarized in Table 5.

TABLE 5. INDEX FOR AISI-SAE STEEL SPECIFICATION NUMBERS

(Letter "X" in Specification No. indicates digit which changes with different types)

Classifications		Specification Numbers
Carbon Steels		10XX
Carbon Steels, Resulfurized		11XX
Carbon Steels, Rephosphorized & Resulfurized		12XX
Manganese Steels		13XX
Nickel Steels		2XXX
Nickel-Chromium Steels		31XX
High Nickel-Chromium Steels		33XX
Carbon-Molybdenum Steels		40XX
Chromium-Molybdenum Steels		41XX
Chromium-Nickel-Molybdenum Steels		43XX
Nickel-Molybdenum Steels		46XX
High Nickel-Molybdenum Steels		48XX
Low Chromium Steels		50XX
Chromium Steels		51XX
Carbon-Chromium Steels		52XXX
Chromium-Vanadium Steels		61XX
Low Nickel-Chromium-Molybdenum Steels		86XX
Low Nickel-Chromium-Molybdenum Steels		87XX
Silicon-Manganese Spring Steels		92XX
Silicon-Manganese-Chromium Spring Steels		92XX
Nickel-Chromium-Molybdenum Steels		93XX
Nickel-Chromium-Molybdenum Steels		98XX
Boron Containing Steels		XXBXX
Boron-Vanadium Containing Steels		XXBVXX
Water-Hardening Tool Steels		WX
Shock-Resisting Tool Steels		SX
Oil-Hardening Tool Steels		OX
Air-Hardening Tool Steels	Cold	AX
High-Carbon, High-Chromium Tool Steels	Work	DX
Hot-Work Tool Steels		HXX
High-Speed (Tungsten Base) Tool Steels		TX
High-Speed (Molybdenum Base) Tool Steels		MX
Special Purpose Tool Steels		LX
Carbon-Tungsten Tool Steels		FX
Mold Steels		PX
Chromium-Nickel-Manganese Stainless Steels		2XX
Chromium-Nickel Stainless Steels		3XX
Chromium Stainless Steels		4XX
Low Chromium Heat-Resisting Steels		5XX
AISI (Only) High-Temperature, High-Strength Alloys		
Martensitic Low Alloy Steels		60X
Martensitic Secondary Hardening Steels		61X
Martensitic Chromium Steels		61X
Semiaustenitic Precipitation and Transformation Hardening Steels		63X
Austenitic Steels Strengthened by "Hot-Cold" Work		65X
Austenitic Alloys, Iron Base		66X
Austenitic Alloys, Cobalt Base		67X
Austenitic Alloys, Nickel Base		68X, 69X

The American Iron and Steel Institute steel quality classifications system is shown in Table 6.

TABLE 6. AISI STEEL DESIGNATION SYSTEM

Prefix and Suffix Letters

Prefix Letter	Meaning
A	Basic Open-Hearth Alloy Steel
B	Acid Bessemer Carbon Steel
C	Basic Open-Hearth Carbon Steel
D	Acid Open-Hearth Carbon Steel
E	Basic Electric Furnace Steel
TS	Tentative Standard Steel
Q	Forging Quality, or Special Requirements
R	Re-rolling Quality Billets

Suffix Letter	Meaning
A	Restricted Chemical Composition
B	Bearing Steel Quality
C	Guaranteed Segregation Limits
D	Specified Discard
E	Macro-Etch Tests
F	Rifle Barrel Quality
G	Limited Austenitic Grain Size
H	Guaranteed Hardenability
I	Non-Metallic Inclusions Requirements
J	Fracture Test
T	Extensometer Test
V	Aircraft Quality or Magnaflux Testing

Identification of steel for warehouse purposes is done by means of paint applied usually by spray can on the ends of steel bars. Unfortunately, all colors are not standardized in the industry.

Common Steel Color Designations:

Color desigantion is not completely standardized by the steel industry or by manufacturers. The color is usually sprayed on the ends of the steel bars. The following are the color designations used by a steel supplier. (Ryerson)

Green - C 1015 to C 1025
Blue - C 1035
Yellow - C 1040 to C 1045
Pink - C 1095
Brown - C 1140
Gold - C 1117
White - B 1112
Orange - B 1113

WROUGHT IRON

Most wrought iron is produced by the Byers process developed in 1920. Wrought iron is essentially a refined low carbon steel from a Bessemer converter where most of the silicon, manganese, and carbon have been removed. Tiny droplets of almost pure iron mixed with silicates (slag) form when the molten iron is poured into molten slag, there being a considerable difference in the melting points of the two materials. The resulting spongy mass is mechanically worked and rolled, the result being wrought iron shapes with the original slag particles rolled with the iron into extremely fine long strands. This makes wrought iron have a good grain-like structure. Ductility and corrosion resistance are well-known qualities of wrought iron.

Wrought iron is weldable; however being a very ductile ferrous metal, it is seldom used where high strength is required. Best grades of water pipe are made of wrought iron. A typical percentage analysis of wrought iron is shown below.

Ingredient	**Percentage**
Carbon	0.02
Manganese	0.03
Phosphorus	0.12
Sulfur	0.02
Silicon	0.15
Slag	3.00

FORGING AND HEAT TREATING METALS

FORGING METALS

A. OPERATION OF THE GAS FURNACE AND FORGE

1. **Lighting Furnace for Heat Treating**
   a. Start the furnace blower.
   b. Open one door of the furnace.
   c. Turn gas valve to 2/3 open position to ignite burners.
   d. Light furnace according to manufacturer's instructions.
   e. **Warning**: Do not stand in front of open door while lighting furnace. A mild gas explosion may occur, blowing heat and flame from the door.
   f. Close the door.
   g. Adjust gas valve until slight flame is seen coming from upper exhaust ports while doors are closed.

2. **Lighting Furnace for Forging** (See Figure 5)
   a. Start the blower.
   b. Swing refractory lid off center.
   c. Turn on gas.
   d. Light furnace according to manufacturers' instructions.
   e. Swing lid to center position.
   f. Adjust gas flame with air or gas control.

3. **Operating precautions**
   a. Do not have more fire than necessary to obtain temperature and speed of heating desired.
   b. While putting in or removing metal from furnace, avoid exposure to flame which may extend out from the furnace hearth.
   c. The ceramic interior of the furnace is easily damaged. Avoid bumping the ceramic lining or dropping the metal. Wear protective gloves.

4. **Extinguishing the furnace flame**
   a. Close the valve.
   b. Turn off gas furnace blower.
   c. Allow to cool with doors closed.

B. BUILDING AND MAINTAINING A FORGE FIRE USING BLACKSMITHING COAL

1. **A good fire should be**
   a. Clean - free from clinkers, cinders, and other impurities.
   b. Deep - core of live burning coke depending on size and type of material being forged. Too large a fire wastes coal and overheats working area.
   c. Compact - well banked with dampened coal. Smithing coal is a special grade of coal free of impurities. When dampened and packed about the fire, it changes to coke which burns with a hot, clean flame.

2. **Starting the fire**
   a. Clean fire pot with forge shovel or hands, push coke back on the hearth and discard all clinkers. Empty the ash pit. Be sure the clean-out door is closed. Turn blower and check the blast of air through the tuyere.
   b. Secure a large, single handful of coarse shavings to which has been previously added a small amount of kerosene. Place in extreme bottom of fire pot. Light, then quickly cover outside of burning shavings with coke and coal, being careful to keep center open for good draft. Turn blower at a moderate speed, continue adding fuel to outside of fire and force toward the center as shavings are burned.

3. **Maintaining the forge fire**
   a. Coke all green coal by adding previously moistened coal to other edge of fire. As coke is formed, force toward the center. Keep fire compact.
   b. Coke burns with a blue flame. Since burning coke has greater affinity for oxygen, a hotter and more nearly neutral flame may be maintained than with ordinary green coal.

Figure 5. Gas Fired Furnace for Heating and Hot Shaping of Metals.

C. USING THE ANVIL

1. Make sure anvil is securely mounted preferably on wood.
2. Locate anvil near forge in uncrowded area with horn of anvil to left for right-handed operator. Anvil should be knuckle high for ease of working.
3. Avoid hammer marking the face or chipping the corners of the anvil.

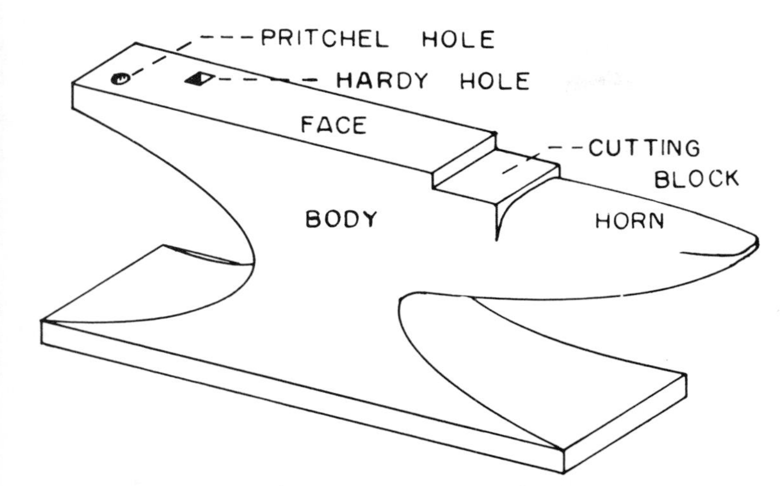

Figure 6. Anvil Commonly Used in Working Hot Metal.

D. HOLDING THE WORK

1. Tongs used must be of type and size to hold the material firmly. Note Figure 11. Since they are made of mild steel, slight changes in shape may be made by carefully reforging the jaws over the stock.
2. Repeated cooling of the outer end of stock and the use of gloves make it possible to do a considerable amount of work without the use of tongs.

E. HEATING THE METAL

1. Place the metal in a horizontal position with the pointed or thinner portions toward the outer or cooler parts of the fire.
2. Heat mild steel to a bright cherry-red, orange-yellow or near white color, depending upon the size of the piece and the stage of the forging process. "Hit while the iron is hot".
3. Tool Steel should never be heated above a cherry-red or forged below a dull-red. The quality of the forged tool is dependent upon the maintenance of the carbon content of the steel.
4. Check temperature of the working stock from time to time. Correct working temperature is very important.

F. CUTTING

1. Heat metal and cut on hot-cutting hardy. Do not cut completely off by striking directly over the hardy as this will dull the hardy and cause flying metal that may endanger other workers. Note Figure 8.
2. A hot cutter placed opposite the hardy will hasten cutting.

FORGING TOOLS

ANVIL - The anvil as shown in Figure 6 is a basic forging tool. It has the following characteristics: Weight about 100 to 150 pounds - cast steel body with 1/2 inch steel face welded to the body. The parts of the anvil are cast body, hardened steel face, pritchel hole, cutting block, hardy hole, and horn. It is usually best to mount the anvil on a built-up wooden base with the height of the face near the knuckles with the arms hanging fully extended, and with the horn to the left of the worker.

The following is a list of hand tools frequently used in forging and hot metal work:

Hardy 7/8" shank, 2" cutter
Cold Cutter 1 3/8", 16" handle
Hot Cutter 1 3/4", 16" handle
Hot Punches 3/8", 1/2", 3/4"
Tongs, straight hipped 3/8", 1/2", 3/4"
Tongs, round lipped 1/4", 3/8", 1/2", 5/8"
Machinist's Vise 4 1/2" jaws, stationary base, weight 60 pounds -
Leg vise, a device for very heavy work
Heavy Duty Hammer 2 1/2 lb. cross peen, New England pattern, rim tempered face
Ball Peen Hammers 12 oz. and 20 oz. rim tempered faces
Sledge - 6 to 12 lbs. 36" handle

These hand tools are discussed in more detail in the following illustrations:

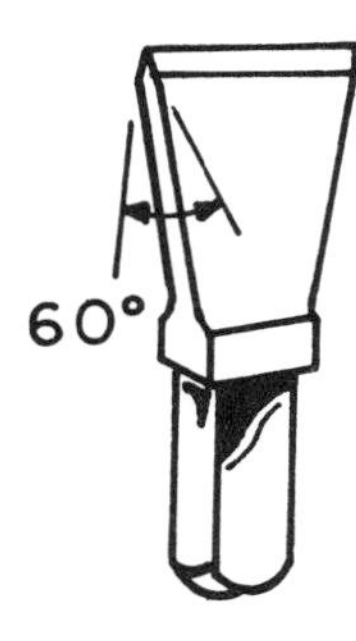

Figure 7. The Hardy.

The hardy shank fits into the hardy hole on the anvil. The cutter of the hardy is sharpened to 60°, the same as a chisel. As shown in Figure 8, the hardy is being used to cut cherry red metal using a blacksmith's chisel. In Figure 9, metal is being cut with a hardy and blacksmith hammer. The metal has been heated to cherry red. Cutting is usually accomplished by nicking the metal and breaking as it cools.

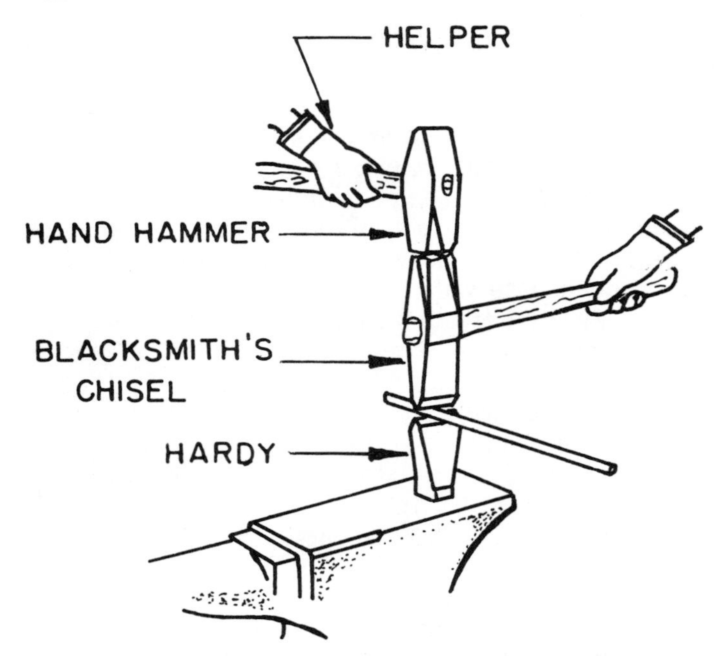

Figure 8. Cutting Metal with a Blacksmith Chisel and Hardy.

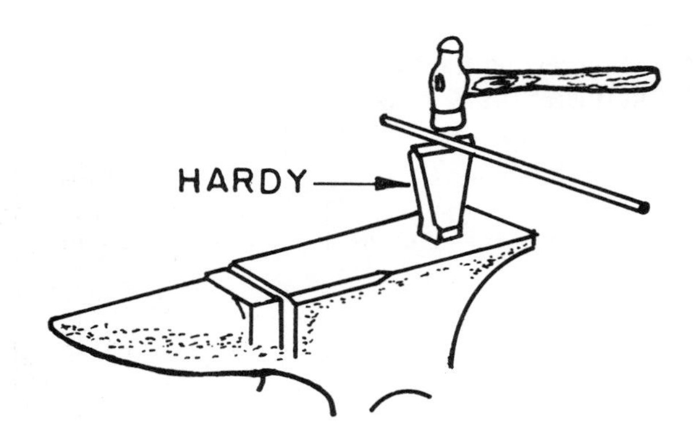

Figure 9. Cutting Metal with a Hardy.

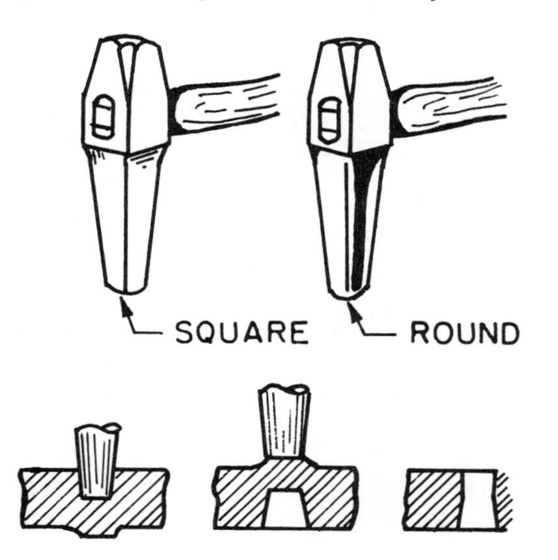

Figure 10. Blacksmith Hot Punches.

Shown in Figure 10 are hot punches and the steps in using to punch a hole in metal. Blacksmith hot punches are made in many sizes. The metal is heated to a bright red color and a hole is punched as shown. The punch must be cooled repeatedly.

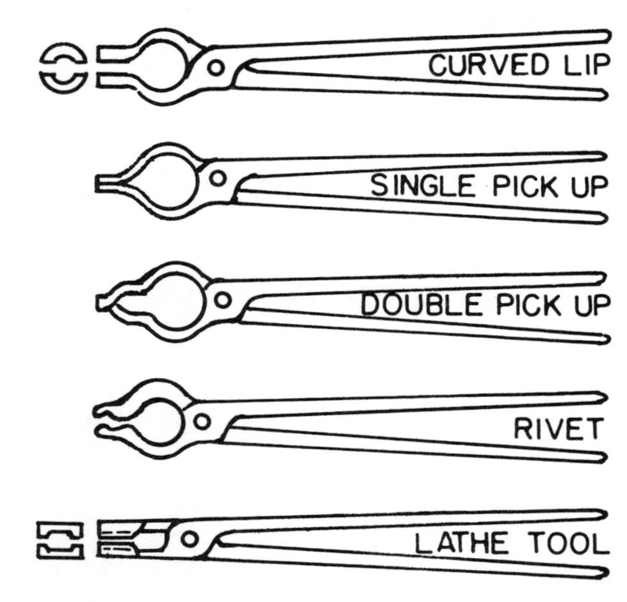

Figure 11. Metal Forging Tongs.

A few of the commonly shaped tongs are illustrated in Figure 11. They are available in many sizes and lip shapes for different forging applications. The lips and jaws of the tongs must be kept cool to avoid distortion of the lips. Never hold a piece of metal in the tongs while heating in the forge.

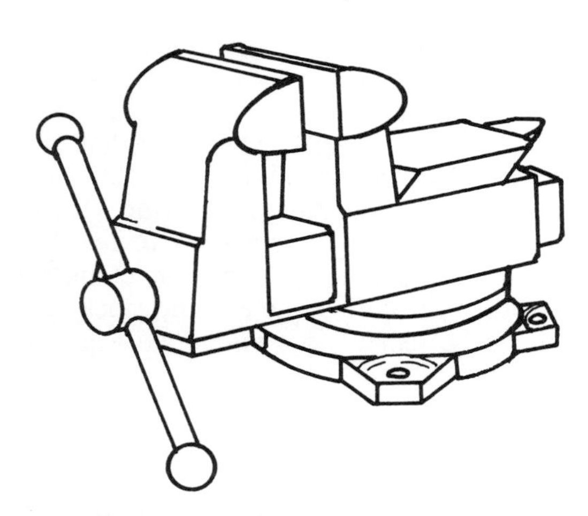

Figure 12. The Machinist's Vise.

Metal vises, as illustrated in Figure 12, vary widely in size, weight, jaw width and opening. The vise shown has a back jaw anvil and a swivel base. Vises often have replaceable jaw faces and pipe jaws for holding pipe. For heavy work, a vise with a leg would help transmit the pounding pressure to the floor.

BENCH VISE USE AND CARE

1. **Mount Vise Firmly** - keep it tight on bench. A loose vise is dangerous and inefficient.

2. **Lock Swivel Base Securely** - tapered-gear lock bolt prevents slippage.

3. **Never Hammer the Handle** - too much pressure may damage the work.

4. **Don't Use Handle Extension** - normal leverage will hold work securely in place.

5. **Don't Hammer the Beam** - the vise will give almost unlimited use. But it won't stand continued abuse. The beam is not an anvil.

6. **Oil the Screw** - remove front jaw. Use oil or, preferably, light grease. Do this frequently. It will prevent screw wear.

7. **Keep Jaw Faces Clean** - use wire brush or file card to remove chips and dust.

8. **Convert to Swivel Base** - if you wish to change the stationary base vise to a swivel base type, install a Swivel Kit.

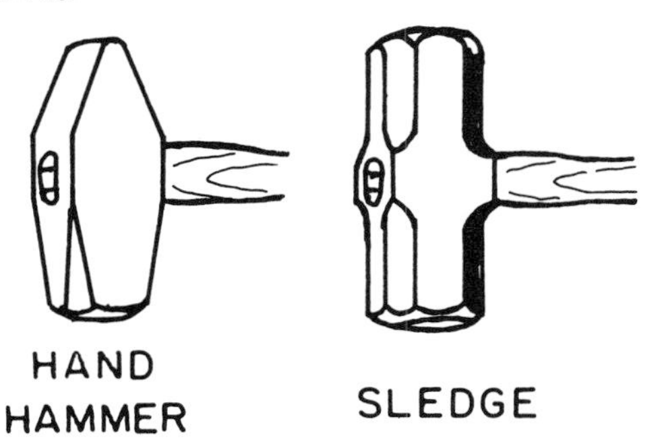

Figure 13. The Blacksmith Hammer and Sledge.

The blacksmith hammer or sledge as shown in Figure 13 is a basic forging tool. Hand hammers weigh up to 2 1/2 pounds with a 16 inch handle while sledges weigh 6 to 12 pounds and have a 36 inch handle. The lighter duty ball peen hammer used for lighter forging work will weigh 16 to 20 ounces. All hammers or sledges should have rim tempered faces to avoid chipping of the hammer face.

HEAT TREATMENT

Most low carbon steels are used as rolled and the properties desired in the finished product are so predicated. However, in most cases involving higher carbon alloy steels, the ultimate property capabilitiy of a steel cannot be realized without heat treating in some manner. Heat treatment may be defined as an operation, or series of operations, involving the heating and cooling of steel in the solid state to develop the required properties. A tempering furnace is illustrated in Figure 14. There are in general four different forms of heat treatment employed and these modify the mechanical properties of the steel to suit the end use.

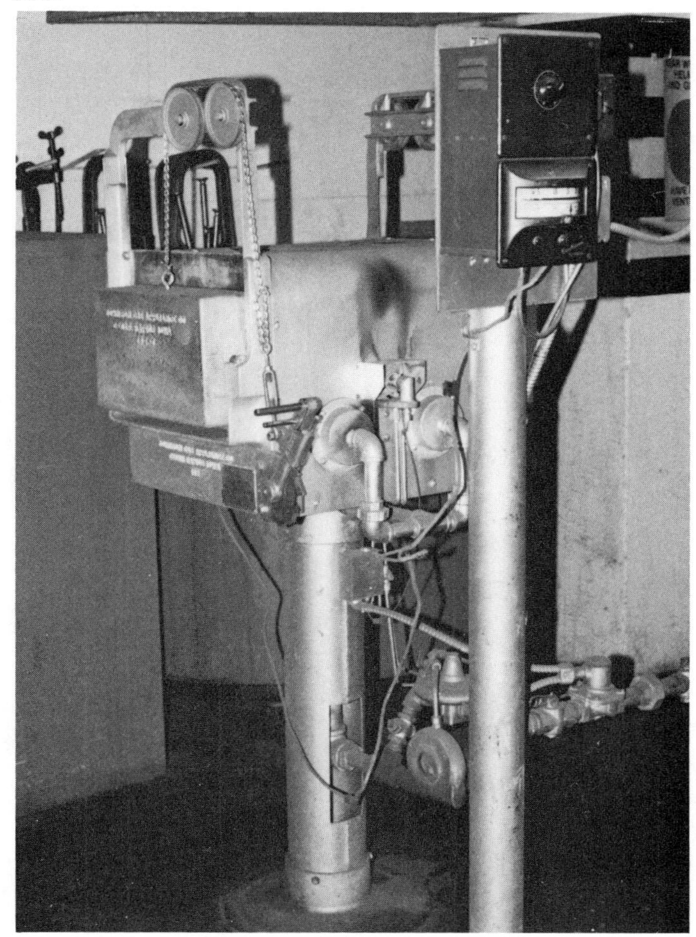

Figure 14. Tempering Furnace.

1. **Stress relieving** is the process of reducing internal stresses by heating the steel to a temperature below the critical range, and holding for an interval sufficient to equalize the temperature throughout the piece. Object of the treatment is to reduce stresses that may have been induced by machining, cold working, or welding.

2. **Normalizing** is a form of treatment in which the steel is heated to a predetermined temperature above the critical range, after which it is cooled to below that range in still air. The purpose of normalizing is to promote uniformity of structure and to alter mechanical properties.

3. **Annealing** consists of heating the steel to a point at or near the critical range, then cooling it slowly at a predetermined rate. Annealing may be used (a) to soften the steel; (b) to develop a particular structure, such as lamellar pearlite or spheroidized carbide; (c) to improve machinability, or to facilitate cold shaping; (d) to prepare the steel for

TEMPERATURE DATA

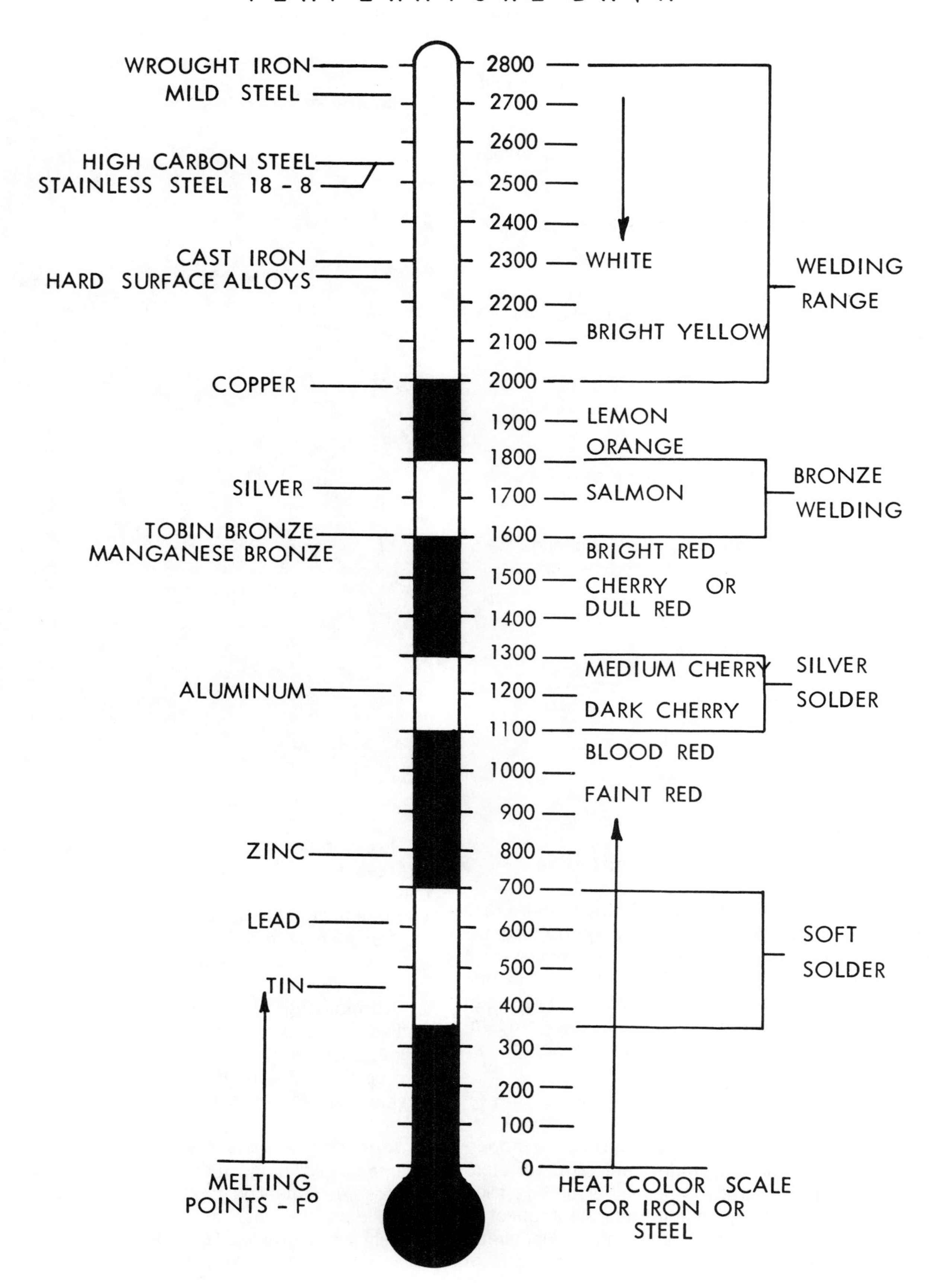

subsequent heat treatment; (e) to reduce stresses; (f) to improve or restore ductility; or (g) to modify other properties.

4. **Hardening and tempering** usually consist of three successive operations: (a) heating the steel above the critical range so that it approaches a uniform solid solution; (b) hardening the steel by quenching it in oil, water, brine, or fused salt bath; and (c) tempering the steel by reheating it to a point below the critical range to get the proper strength and ductility combination.

Two types of steel do not lend themselves to this kind of heat treatment: most soft steels where the amount of carbon content is insufficient to form any appreciable amount of martensite (the hardest micro constituent of steel) on quenching and austenitic stainless or heat resisting steels where the critical range is depressed below room temperatures due to the presence of large amounts of certain alloying constituents.

FUNDAMENTAL METALLURGY

1. Definition of terms

Cementite- Iron carbide Fe_3Ca chemical compound of iron and carbon
Ferrite- Pure iron
Pearlite- Grain structure resulting from a mechanical combination of ferrite and cementite in layer formation
Austenite- When steel is heated to transformation temperature, the grains of ferrite and pearlite become fine grains of austenite. The resulting crystals have a face-centered cubic structure. Change in .83% C steel takes place at 1330° F. Carbon moves freely between face-centered cubic arrangements of iron atom until uniform-mass is formed.
Martensite- Structure that is formed when carbon steel is rapidly cooled by quenching.

2. Most metals have atoms arranged in **body-centered** cubic (9 atoms)or **face-centered cubic**(14 atoms) systems.

3. Carbon steels have body-centered cubic arrangement at room temperature. Complete transformation results in arrangement changes to a face centered cubic system.

4. Transformation begins to take place at 1330° F as illustrated in Figure 15. This is called transformation temperature. Transformation is reversed as steel is cooled.

5. Steel with less than 0.8% carbon does not have enough carbon to form 100% **pearlite** when **austenite** forms. The resulting structure is a mixture of **pearlite** grains and ferrite. Steel with more than 0.83% has too much carbon to be used in formation of **pearlite.** The structure of the austenite is then made up of **pearlite** grains and **cementite** (iron carbide).

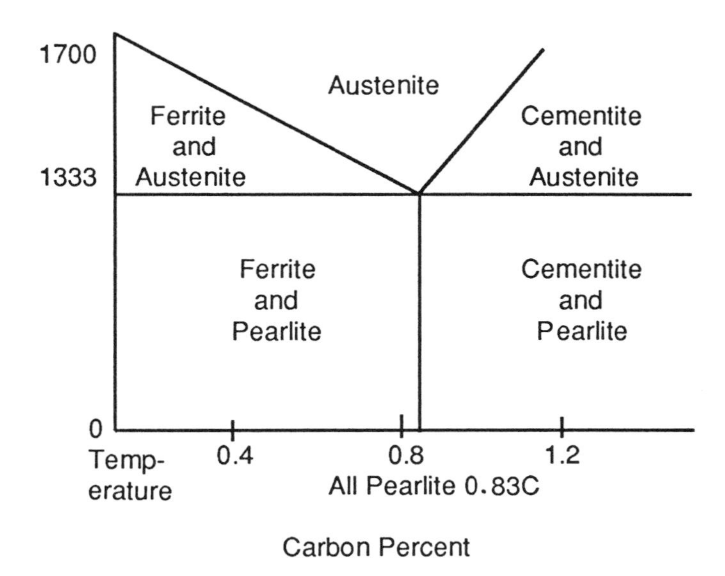

Figure 15. Basic Iron-Carbon Equilibrium Diagram.

THE IRON-CARBON EQUILIBRIUM DIAGRAM

The ICED shown in Figure 16 is complete for the purpose of understanding "layman's metallurgy". As mentioned before, carbon is the one most important alloying element with iron because it determines hardenability of carbon steel. As shown in Figure 16, ferrous metals below the A_1 line (lower transformation temperature) are indicated as ferrite and carbide. This mechanical mixture is known as pearlite and has a space lattice arrangement of a body centered cube with 9 atoms as shown in Figure 17. Iron being an allotropic element is able to rearrange its atomic crystalline structure. This rearrangement begins at the A_1 line and is completed at the A_3 or the A_{cm} line as indicated. On complete transformation, Austenite is formed and the crystalline arrangement is now a face center cube with 14 atoms as shown in Figure 18. Carbon is in combined form below the lower transformation A_1 line. On complete transformation, carbon goes into solid solution and moves freely.

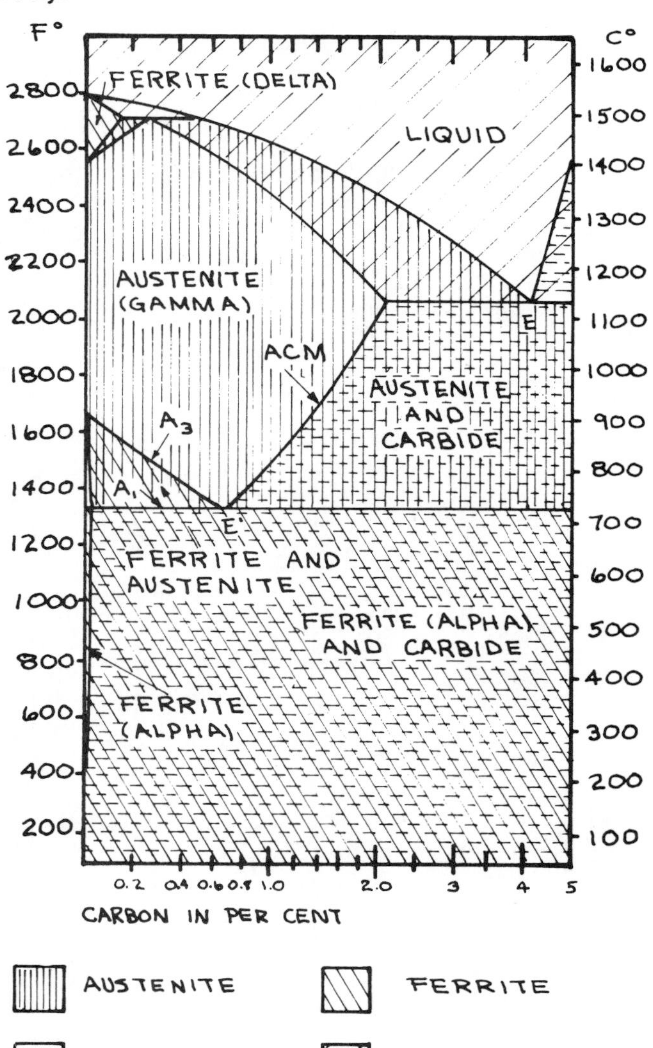

Figure 16. The Iron-Carbon Equilibrium Diagram.

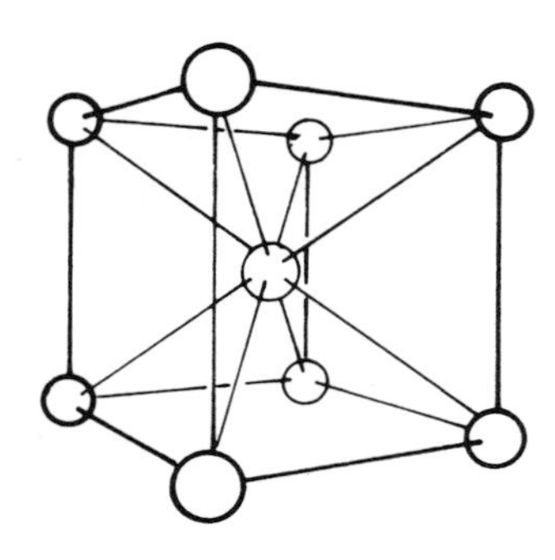

Figure 17. The Body-Centered Lattice of Alpha and Delta Iron.

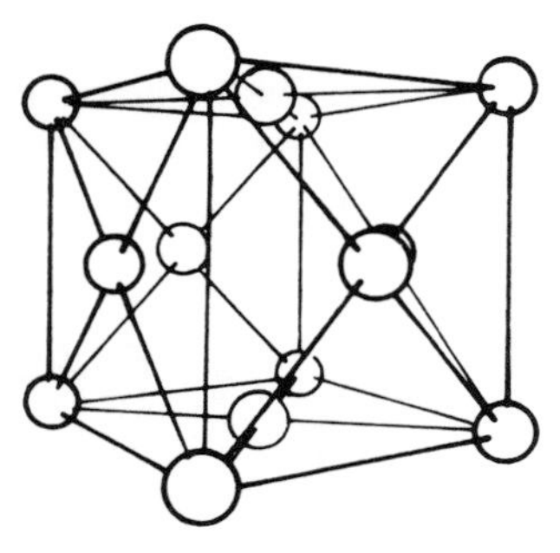

Figure 18. The Face-Centered Lattice of Gamma Iron.

The ICED is useful to the extent that time is stopped thereby removing one of the important variables that affects ferrous metallurgy. Anyone working with ferrous metals involving heat above 1000° F. should be interested in the ICED. This is especially true when the time factor is combined with ICED. Hardenability and weldability are influenced by four factors:

1. Carbon content - Weldable ◄——.35% C——► Hardenable
2. Heating cycle - Maximum temperature
3. Cooling cycle - Minimum temperature
4. Speed of cooling

The addition of certain alloying elements tends to slow down the need for fast cooling in quenching for hardening.

Quenching media are ranked as follows:

1. Brine - severe; fast cooling
2. Cold water - medium rate cooling
3. Warm oil - slow rate of cooling

Rapid cooling also causes shrinkage which may be a factor in crack formation in intricately shaped pieces.

Tempil°
Basic Guide to Ferrous Metallurgy

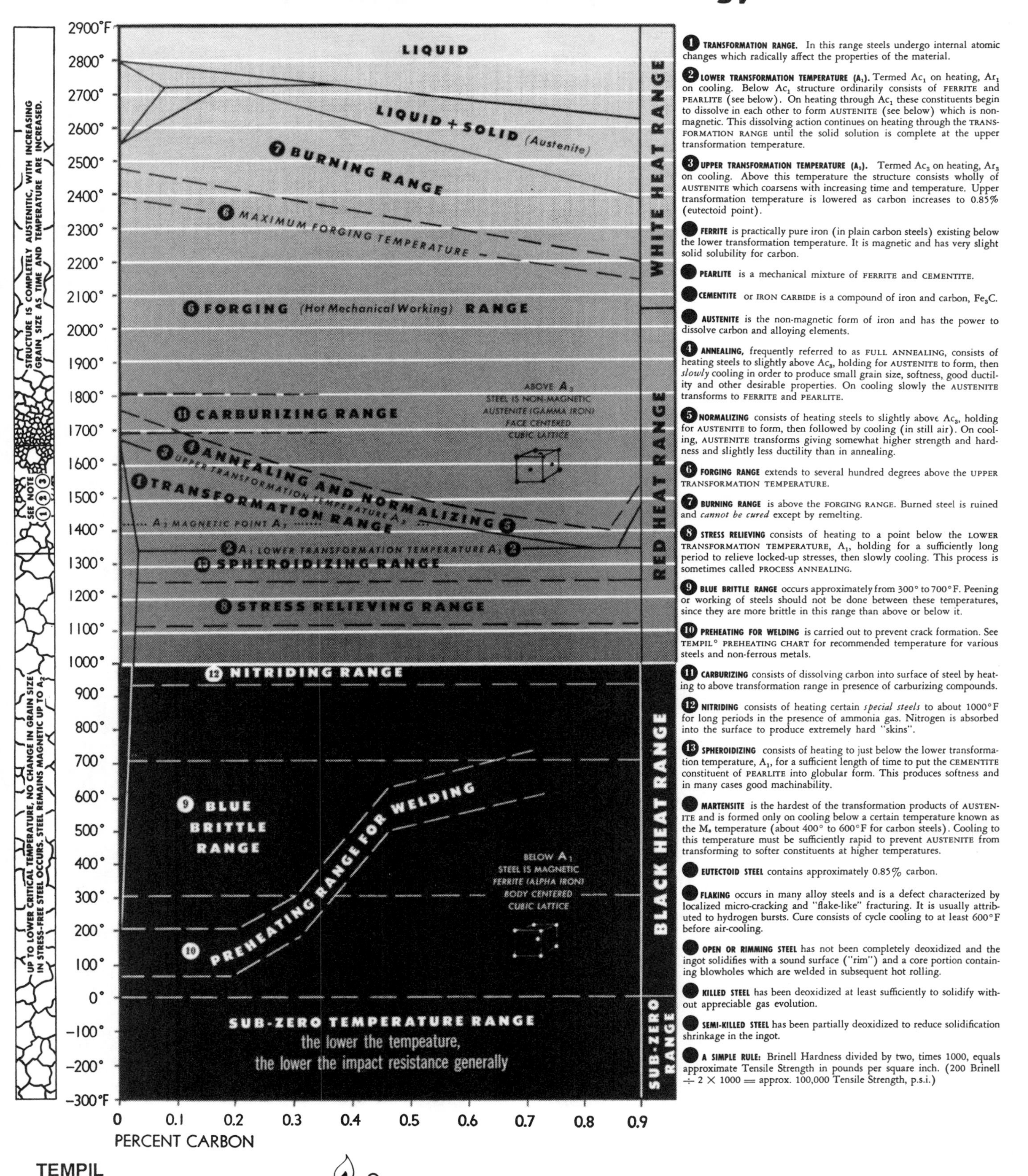

1 TRANSFORMATION RANGE. In this range steels undergo internal atomic changes which radically affect the properties of the material.

2 LOWER TRANSFORMATION TEMPERATURE (A_1). Termed Ac_1 on heating, Ar_1 on cooling. Below Ac_1 structure ordinarily consists of FERRITE and PEARLITE (see below). On heating through Ac_1 these constituents begin to dissolve in each other to form AUSTENITE (see below) which is non-magnetic. This dissolving action continues on heating through the TRANSFORMATION RANGE until the solid solution is complete at the upper transformation temperature.

3 UPPER TRANSFORMATION TEMPERATURE (A_3). Termed Ac_3 on heating, Ar_3 on cooling. Above this temperature the structure consists wholly of AUSTENITE which coarsens with increasing time and temperature. Upper transformation temperature is lowered as carbon increases to 0.85% (eutectoid point).

FERRITE is practically pure iron (in plain carbon steels) existing below the lower transformation temperature. It is magnetic and has very slight solid solubility for carbon.

PEARLITE is a mechanical mixture of FERRITE and CEMENTITE.

CEMENTITE or IRON CARBIDE is a compound of iron and carbon, Fe_3C.

AUSTENITE is the non-magnetic form of iron and has the power to dissolve carbon and alloying elements.

4 ANNEALING, frequently referred to as FULL ANNEALING, consists of heating steels to slightly above Ac_3, holding for AUSTENITE to form, then *slowly* cooling in order to produce small grain size, softness, good ductility and other desirable properties. On cooling slowly the AUSTENITE transforms to FERRITE and PEARLITE.

5 NORMALIZING consists of heating steels to slightly above Ac_3, holding for AUSTENITE to form, then followed by cooling (in still air). On cooling, AUSTENITE transforms giving somewhat higher strength and hardness and slightly less ductility than in annealing.

6 FORGING RANGE extends to several hundred degrees above the UPPER TRANSFORMATION TEMPERATURE.

7 BURNING RANGE is above the FORGING RANGE. Burned steel is ruined and *cannot be cured* except by remelting.

8 STRESS RELIEVING consists of heating to a point below the LOWER TRANSFORMATION TEMPERATURE, A_1, holding for a sufficiently long period to relieve locked-up stresses, then slowly cooling. This process is sometimes called PROCESS ANNEALING.

9 BLUE BRITTLE RANGE occurs approximately from 300° to 700°F. Peening or working of steels should not be done between these temperatures, since they are more brittle in this range than above or below it.

10 PREHEATING FOR WELDING is carried out to prevent crack formation. See TEMPIL° PREHEATING CHART for recommended temperature for various steels and non-ferrous metals.

11 CARBURIZING consists of dissolving carbon into surface of steel by heating to above transformation range in presence of carburizing compounds.

12 NITRIDING consists of heating certain *special steels* to about 1000°F for long periods in the presence of ammonia gas. Nitrogen is absorbed into the surface to produce extremely hard "skins".

13 SPHEROIDIZING consists of heating to just below the lower transformation temperature, A_1, for a sufficient length of time to put the CEMENTITE constituent of PEARLITE into globular form. This produces softness and in many cases good machinability.

MARTENSITE is the hardest of the transformation products of AUSTENITE and is formed only on cooling below a certain temperature known as the M_s temperature (about 400° to 600°F for carbon steels). Cooling to this temperature must be sufficiently rapid to prevent AUSTENITE from transforming to softer constituents at higher temperatures.

EUTECTOID STEEL contains approximately 0.85% carbon.

FLAKING occurs in many alloy steels and is a defect characterized by localized micro-cracking and "flake-like" fracturing. It is usually attributed to hydrogen bursts. Cure consists of cycle cooling to at least 600°F before air-cooling.

OPEN OR RIMMING STEEL has not been completely deoxidized and the ingot solidifies with a sound surface ("rim") and a core portion containing blowholes which are welded in subsequent hot rolling.

KILLED STEEL has been deoxidized at least sufficiently to solidify without appreciable gas evolution.

SEMI-KILLED STEEL has been partially deoxidized to reduce solidification shrinkage in the ingot.

A SIMPLE RULE: Brinell Hardness divided by two, times 1000, equals approximate Tensile Strength in pounds per square inch. (200 Brinell ÷ 2 × 1000 = approx. 100,000 Tensile Strength, p.s.i.)

Tempil° DIVISION, BIG THREE INDUSTRIES, INC.

Tempil° Temperature Indicators, Protective Coatings and Markers.

For 40 years Tempil° has specialized in the manufacture of materials that precisely indicate temperatures, protect against excessive heat buildup or provide effective marking to survive severe service conditions. The accuracy and reliability of Tempil° products is attested to by their inclusion in hundreds of manuals and data sheets of customer companies and government agencies.

Industrial and research applications range from Annealing to Arc-welding, from Bonding plywood to Brazing, from Centrifugal Casting to Ceramics — you get the idea, so to save space we'll end the list with Zinc pots to Zoning in a furnace. We will gladly put our 40 years of experience and know how at your disposal.

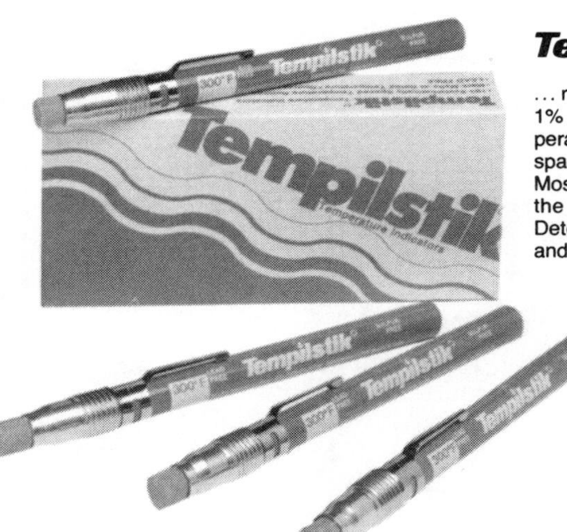

Tempilstik° TEMPERATURE INDICATING CRAYONS

... make marks that are guaranteed to melt within 1% of their rated temperatures. There are 105 temperature ratings, with melting points systematically spaced between 100°F/38°C and 2500°F/1371°C. Most can be certified lead and sulfur free. A few of the hundreds of uses for Tempilstiks° include: Determining surface temperatures during welding and metal fabrication including preheat, postheat, annealing and stress relieving. Determining operating temperatures of bearings, transformers, steam traps, molds, PC board preheaters, motors, electronic components, hydraulic systems, commercial irons, hot plates & heat exchangers.

Tempilaq°

TEMPERATURE INDICATING LIQUID

... is Tempilstik° material suspended in a quick-drying inert vehicle. It is available in the same temperature ratings as Tempilstiks° and with the same ±1% accuracy. Tempilaq° should be used on surfaces which cannot be easily marked with a Tempilskik°, such as polished metal, glass, plastics, rubber, fabrics or electronic components.

Tempilstik°, Tempilaq° and Tempil° Pellets are available in the following temperatures:

°F	°C	°F	°C	°F	°C	°F	°C	°F	°C	°F	°C
100	38	206	97	325	163	575	302	* 1200	649	2050	1121
103	39	213	101	331	166	600	316	* 1250	677	2100	1149
106	41	219	104	338	170	625	329	* 1300	704	* 2150	1177
109	43	225	107	344	173	* 650	343	* 1350	732	* 2200	1204
113	45	231	111	350	177	700	371	* 1400	760	* 2250	1232
119	48	238	114	363	184	* 750	399	1425	774	* 2300	1260
125	52	244	118	375	191	* 800	427	* 1450	788	* 2350	1288
131	55	250	121	388	198	* 850	454	1480	804	2400	1316
138	59	256	124	400	204	900	482	* 1500	816	2450	1343
144	62	263	128	413	212	932	500	1550	843	* 2500	1371
150	66	269	132	425	218	950	510	1600	871		
156	69	275	135	438	226	977	525	1650	899		
163	73	282	139	450	232	1000	538	1700	927		
169	76	288	142	463	239	1022	550	* 1750	954		
175	79	294	146	475	246	* 1050	566	1800	982		
182	83	300	149	488	253	1100	593	1850	1010		
188	87	306	152	500	260	1150	621	* 1900	1038		
194	90	313	156	525	274			1950	1066		
200	93	319	159	550	288			2000	1093		

*Series "R" Tempil° Pellets available in these temperatures. 550°F and below: Regular Pellets can be used for reducing atmosphere. Aerosol Tempilaq° not available below 150°F/66°C.

Tempil Pellet°

TEMPERATURE INDICATING PELLETS

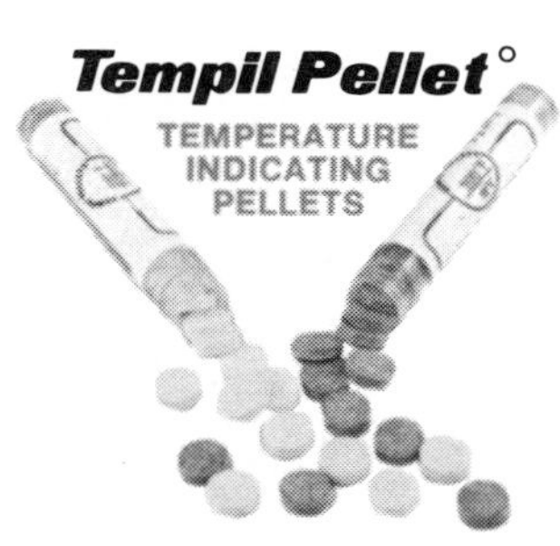

... are Tempilstik° material in tablet form (7/16″ diam.) and available in the same temperature ratings. Major applications include use in industrial ovens and furnaces, and on large heavy objects requiring prolonged heating, where Tempilstik° or Tempilaq° marks might vaporize or be absorbed before melting.

Bloxide°

A DE-OXIDIZING WELDABLE PRIMER

A weldable coating that ensures X-ray quality welds by reducing porosity and pinholing, and prevents rust on workpieces stored outdoors for up to six months ... an excellent primer, Bloxide° is safe for nuclear fabrication.

Anti-heat°

HEAT-SINK COMPOUND

A protective heat-sink compound that confines heat to the welding, brazing, or soldering zone, thus minimizing risk of heat damage, and preventing discoloration, warping, buckling, or distortion of light-gauge metals.

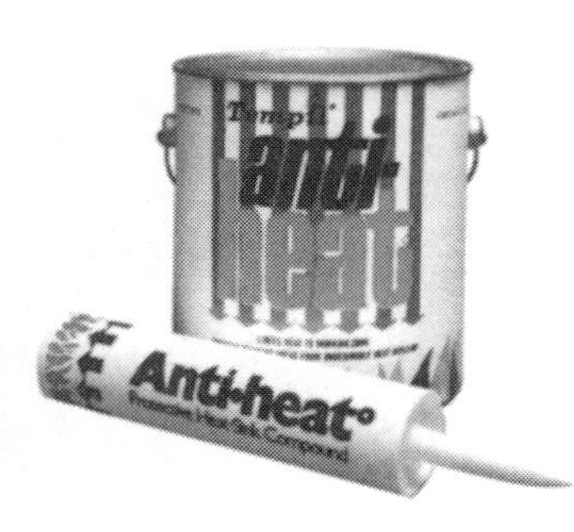

Pyromarker°

HEAT RESISTANT METAL MARKER

A high temperature nuclear grade paint in a ball-point tube. Remains legible on red-hot surfaces (approx. 1100°F/593°C) right through forming and annealing, and is free of lead, sulfur, zinc, mercury, and halogens. 11 colors and two point sizes.

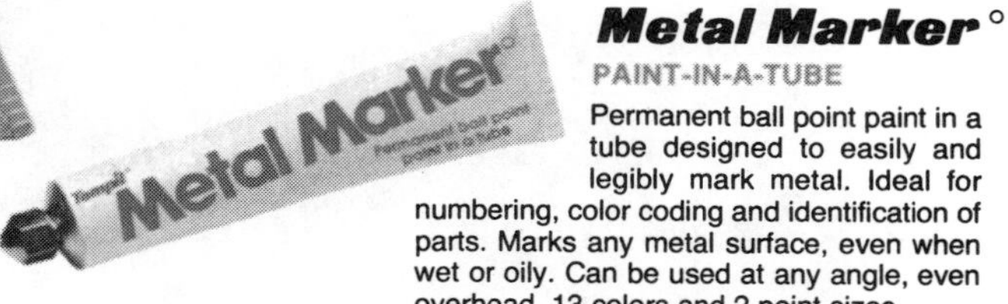

Metal Marker°

PAINT-IN-A-TUBE

Permanent ball point paint in a tube designed to easily and legibly mark metal. Ideal for numbering, color coding and identification of parts. Marks any metal surface, even when wet or oily. Can be used at any angle, even overhead. 13 colors and 2 point sizes.

For detailed literature on each product or our general catalog, contact your local distributor or write Tempil.°

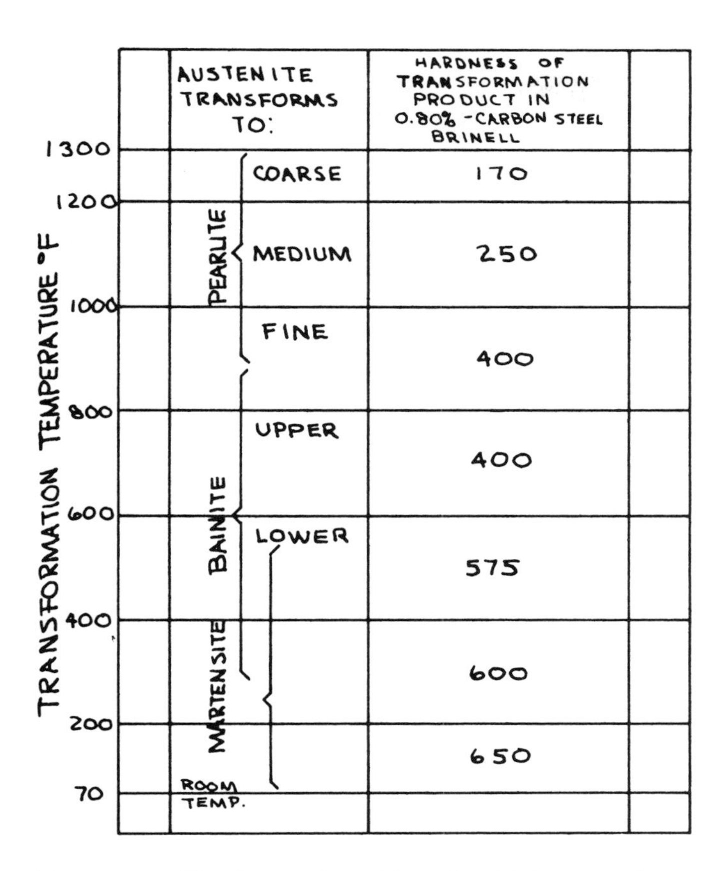

Figure 19. The Formation of Pearlite, Bainete and Martensite is the Result of Cooling to Temperatures as Indicated.

The changing of austenite in the cooling cycle to desired hardness is illustrated in Figure 19. The hardness is an indication of the tensile strength of the steel with the higher hardness index revealing higher tensile strength.

Figure 20 shows that in order to escape the formation of pearlite with its low Brinnell Hardness figure, one must quench quickly (less than 1 second) below about 825° F. In order to obtain complete Martensite, quenching should then continue to below 200° F. Martensite is extremely hard and brittle. Bainite is the intermediate product formed by quenching to slightly above 450° F.

COMPARISON OF QUENCHING AND TEMPERING, MARTEMPERING AND AUSTEMPERING

The salient features of these three hardening methods, quenching and tempering, martempering, and austempering, are depicted graphically in the time-temperature diagrams of Figure 21. In these diagrams, the heating and cooling operations involved in each process are shown superimposed upon a typical isothermal transformation diagram on which the temperature range has been indicated at which the transformation to the hard product bainite or martensite occurs. These diagrams, thus, present a comparative picture of the operations involved in

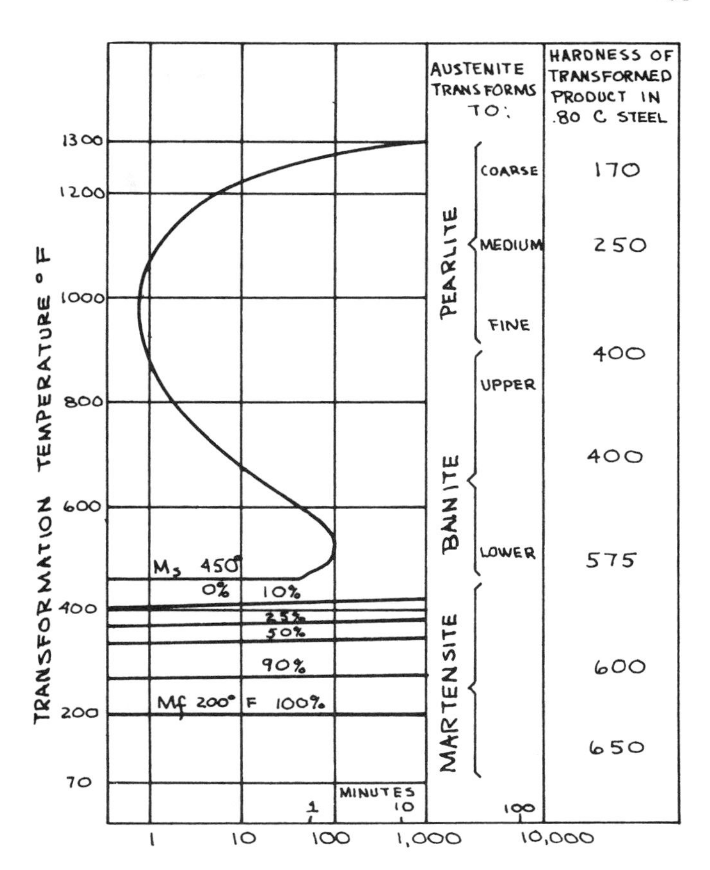

TRANSFORMATION TIME - SECONDS

Figure 20. Time-Temperature-Transformation as Related to Brinnell Hardness of Carbon Steel.

these processes, which shows at a glance both the similarities and the significant differences among the three methods of achieving the full hardening which is the first requisite of heat treatment for optimum strength and toughness.

PROPERTIES OF PEARLITE AND BAINITE

The relationships among strength, ductility, and transformation temperaure are illustrated in Figure 22 which shows the manner in which strength and ductility changes when a eutectoid steel is isothermally transformed at temperatures within the pearlite and bainite temperature ranges. It is apparent that, in general, the steels with bainite microstructures are both stronger and more ductile than those with pearlitic microstructures. This increase in strength and ductility with decreasing transformation temperature is in accord with the trends described above, although at the lowest transformation temperatures in the bainite range the strength increase with decreasing transformation temperature has resulted in a moderate decrease in ductility.

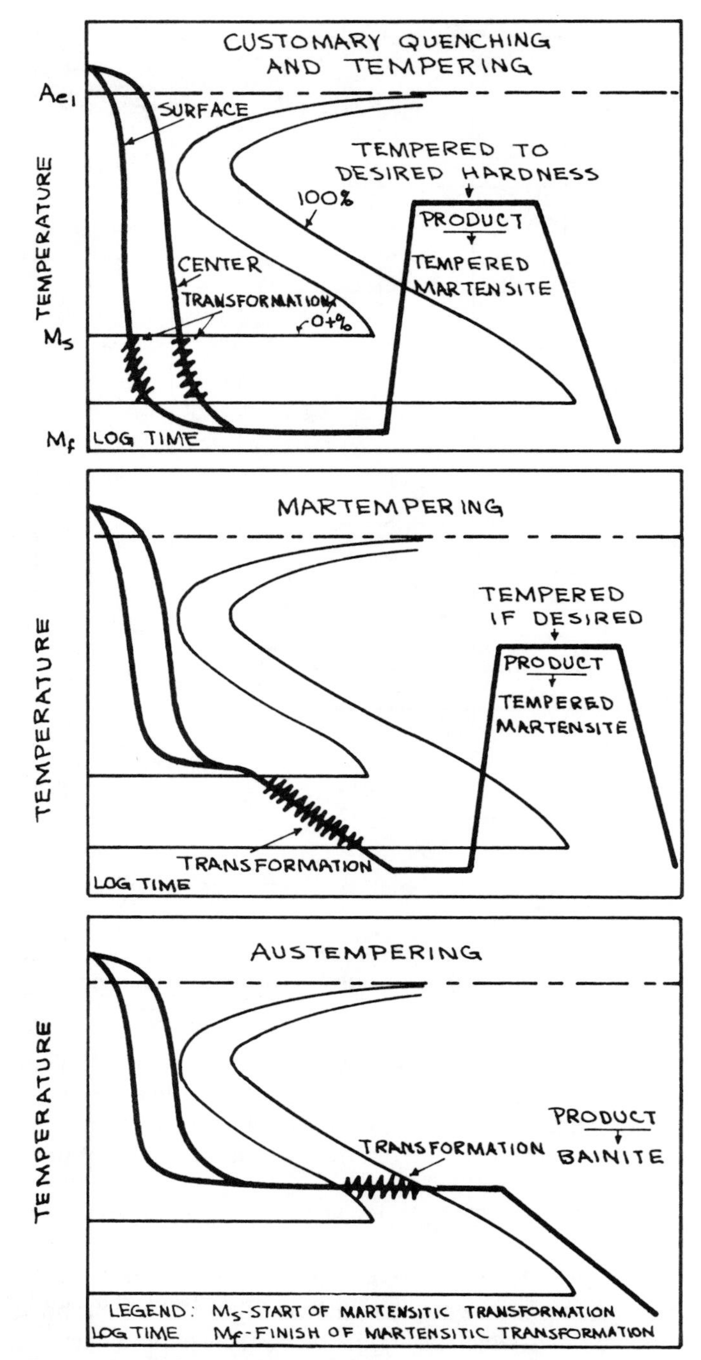

Figure 21. Graphic Comparison of Heat-Treating Processes.

PROPERTIES OF MARTENSITE

Martensite (untempered) is the hardest, and likewise the most brittle of the microstructures obtainable in a given steel. The hardness of martensite as a function of carbon content is shown in Figure 23. The principal importance of this microstructure is as the starting point for tempered martensite structures, which have definitely superior properties. For some applications, however, such as those involving wear resistance, the high hardness of martensite is desirable in spite of the accompanying brittleness.

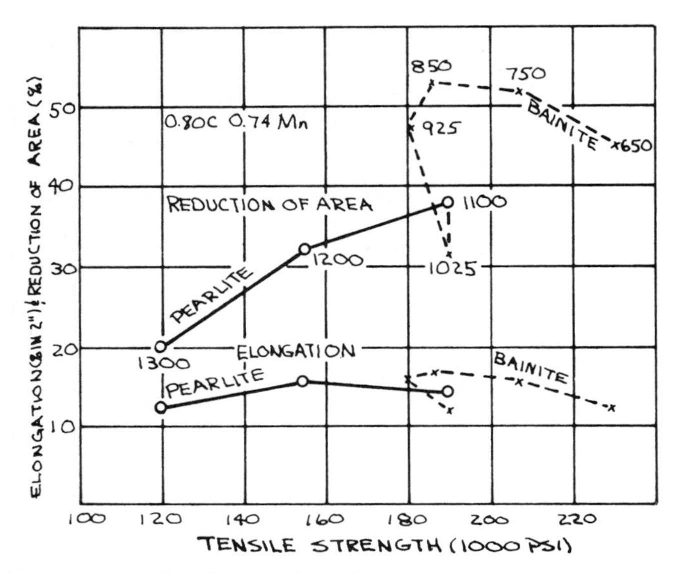

Figure 22. The Properties of Pearlite and Bainite in a Eutectoid Steel. (Numbers at points are temperatures of transformation from austenite.)

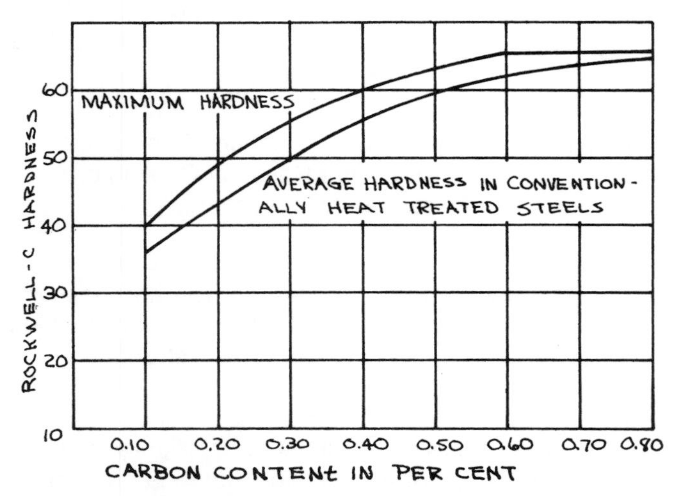

Figure 23. The Hardness of Martensite as a Function of Carbon Content.

PROPERTIES OF TEMPERED MARTENSITE

Tempered martensite microstructures consist of small carbon particles, which tend to be spheriodal in shape, in a fine grained ferrite matrix. This microstructure is a very favorable one in respect to ductility and toughness. Tempered martensite is, therefore, the usual objective of heat treatments aimed at obtaining optimum properties in respect to strength and toughness.

Shown in Figure 24 is the graphic explanation of what takes place during the hardening and tempering of austenite to martensite. This is a similar process to the hardening and tempering of a cold chisel.

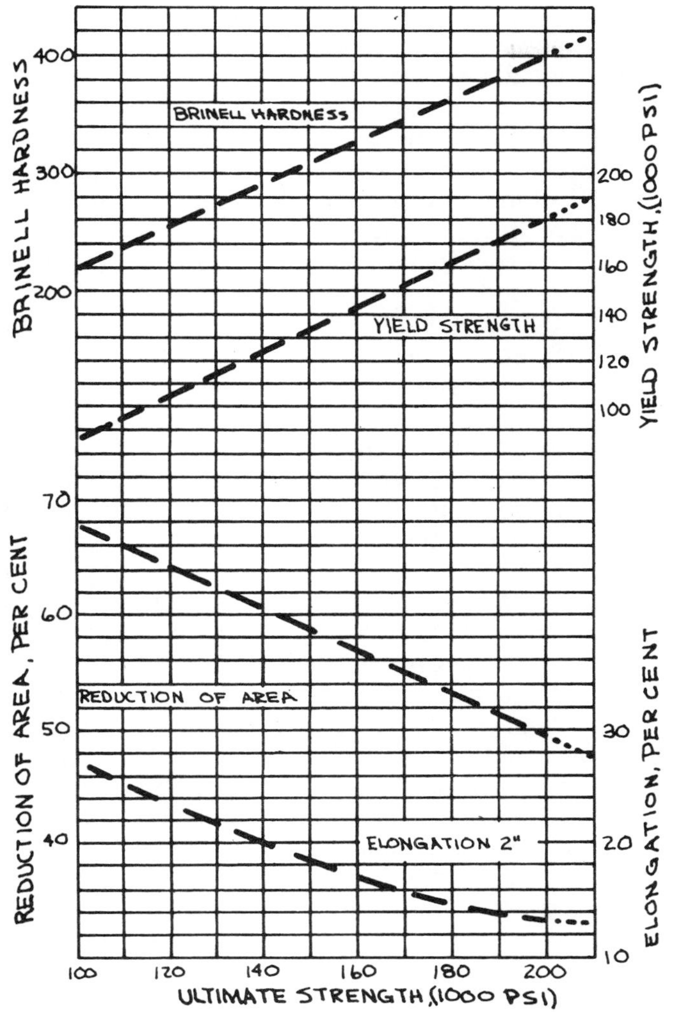

Figure 24. Tempered Martensite as Related to Strength and Ductility.

SURFACE HARDENING

Surface or case hardening is a process of heat treatment whereby the surface hardness of a part is increased. It is applicable to a wide variety of hardenable ferrous alloys and gray and pearlitic malleable cast irons.

1. **Reasons for surface hardening** - surface hardening is done principally to increase the surface hardness of a part to a more useful level without sacrificing other essential and desirable mechanical properties. If the entire mass of a part were of the structure and hardness needed to give the surface the wear resistance and load-carrying capacity desired, it might be so brittle that its usefulness would be reduced or destroyed. Surface hardening then implies that there are at least two distinct zones within a part so processed. Each zone thus has a different hardness and other mechanical properties, for example, yield strength; tensile strength; reduction of area; and elongation, the latter two properties being criteria of ducility.

Increasing the hardness of a given material increases its resistance to wear and deformation. The hardness attainable in ferrous alloys is governed by their carbon content. For example, a steel containing 0.10 percent carbon can be hardened to Rockwell "C"40, while steels containing 0.20 percent, 0.40 percent and 0.60 percent carbon can be hardened to values of Rockwell "C"49, 60, and 65, respectively. Increasing the carbon content of a steel beyond 0.60 percent does not result in a commensurate increase in hardness. However, wear resistance is not solely determined by hardness and carbon content. The addition of alloying elements such as chromium, molybdenum, tungsten, and vanadium which combine readily with carbon to form extremely hard carbides also increases resistance to wear.

2. **Surface hardening by selective hardening methods** -by using steel with sufficient carbon content, a high surface hardness can be produced merely by selectively heating the outer surface layer to the desired hardening temperature to a given depth, and then quenching at a rate that will produce complete hardening. The area beneath the selectively hardened zone will be unaffected by the heating and quenching operations. Flame hardening and induction hardening are surface hardening processes which utilize this principle.

THE FLAME-HARDENING PROCESS

Essentially, flame hardening is a process used to harden steel by heating it above the transformation range (austenitizing) by means of a high temperature flame and then quenching it at a rate that will produce complete hardening. The flame impinges directly on the workpiece.

The essentials for flame hardening comprise the fuel gases, such as oxygen and acetylene, oxygen and propane, or air and natural gas, and a torch or flame head in which the fuel gases are burned. Different fuel gases provide varying temperatures but oxygen and acetylene are the most widely used because they produce the highest flame temperature and equipment to burn them is readily available.

Equipment used for flame hardening ranges from a manually operated hand flame head or welder's torch to custom-built automatic types for the spinning and progressive spinning methods.

The depth of the hardened zone produced by flame hardening will vary from about 1/32" to 1/4" or more, depending upon the fuel used, the flame-head design and time of heating.

Any ferrous alloy capable of being hardened by conventional heating and quenching can be flame-hardened, although AISI-SAE 1035 to 1053 carbon steels are the most widely used. Alloy steels such as AISI-SAE 4140H, 864OH and 434OH are also used when special properties not obtainable with the carbon steels are required. Gray and pearlite malleable cast iron are often flame hardened. Carburized parts are sometimes flame hardened in selective areas. Typical applications of flame hardening include gears, rolls and machine ways, valve stems, push rods, cams and levers.

THE INDUCTION HARDENING PROCESS

Induction hardening is a process of quench hardening in which the heat is generated in the workpiece by electrical induction. In this process, a high frequency alternating current is passed through the inductor or work coil which induces the heating current in the workpiece. Upon attainment of the proper temperature and depth of heating, the workpiece is quench-hardened. From the standpoint of surface hardening, the useful frequency range is from 1,000 to 3,000,000 cycles per second (cps). The actual depth of heating by induction is a function of time, power density, and frequency.

Materials used for induction hardening are similar to those described in the section on flame hardening. Since the maximum hardness attainable with a given steel is governed by its carbon content, the material used must contain the minimum carbon content which will produce the desired hardness. For example, if 55 Rockwell "C" hardness is desired after induction hardening, the carbon content must be 0.40 percent minimum. Hardnesses of 58 and 60 Rockwell "C" minimum require carbon contents of 0.45 and 0.50 percent minimum respectively, and an effective quench.

Literally countless types of workpieces can be surface hardened by induction. Gears (teeth), shafts, cams, crankshafts, cylinders, and levers are but a few examples.

The principal advantages of induction hardening are very similar to those of flame hardening. However, since the cost of induction hardening equipment is substantially greater that that used for flame hardening, its use is more or less limited to items which are produced in large quantities.

3. **Surface Hardening by Diffusion** - the methods of surface hardening described below involve a change in surface chemistry brought about by the addition and diffusion of certain elements at elevated temperatures.

Carburizing- Carburizing is the most widely used method of surface hardening employing the diffusion principle. Since hardness increases with carbon content, increasing the carbon content of the surface of a low carbon steel results in high hardness at the surface and toughness (low hardness) in the core.

Carburizing involves heating the workpiece in the presence of a carbonaceous material, either solid, gas or liquid, holding for a time commensurate with the case depth desired, and then quench hardening. The depth of case acquired is governed by: (1) temperature; (2) time; (3) activity or potential of the carburizing medium; and (4) the analysis of the ferrous alloy uses. Carburizing (and its related processes) is applied generally to low carbon steels of the plain carbon or alloy variety having a mximum of 0.25 percent carbon. Some typical examples are AISI-SAE C1010, C1018, C1024, C1117, B1112, E3310, 4620, 4817, and 8620. Steels of higher carbon content may be carburized to satisfy special requirements.

Pack or box carburizing- The oldest method of carburizing and still in rather common use, employs a solid carburizing compound. The compound generally consists of charcoal and/or coke of high purity to which are added chemicals such as calcium, barium, or sodium carbonate, which act as energizers. The work is packed in a box, pot, or other container with the compound, then sealed and subjected to temperatures ranging between 1550 and 1750° F Carbon monoxide, liberated by the compound, reacts with the steel and releases carbon which dissolves in the austenite and is diffused inward. After the desired case depth is acquired, the work may be cooled in the pot or quench hardened directly therefrom. If the work is cooled in the pot, it is subsequently reheated and quench hardened. Pot cooling affords an opportunity to perform further machining operations or to remove part of the case, if necessary.

While the pack carburizing method is widely used, it has many disadvantages. For example, the compound and the container must be heated; compounds are usually dirty and must be continually replenished with new compound to maintain proper carburizing activity; also, it is difficult to control the carbon content of the case and the uniformity of depth within narrow limits. There are however, some advantages inherent in pack carburizing: Elaborate furnace equipment is not required, and distortion is minimized since the solid compound supports the workpiece. It is particularly advantageous where the quantity of production is low.

Liquid Carburizing is yet another process of surface hardening, in which carbon is provided by molten salts containing cyanides. Nitrogen, in small amounts, may also be absorbed by the steel. The bath is comprised of varying mixtures of sodium cyanide, sodium carbonate, and the chlorides of barium, calcium, potassium, sodium and strontium and is operated at temperatures of 1600° to 1750°F. with a sodium cyanide content of approximately 10 percent. Liquid quenching in brine, oil or water is required to harden the case.

Case depths and hardness attainable by this process are comparable to those of other carburizing processes such as the gas and pack methods. For example, in four hours at temperatures of 1600°, 1650, and 1700°, case depths of approximately 0.033", 0.038" and 0.047" respectively, are obtained. Liquid carburizing baths are generally used for the range of case depths from about 0.032" to 0.250".

Liquid Carburizing With Gas Fired Furnace Before Oil Quenching - a gas fired furnace as shown in Figure 25 is used in heating special carburizing salts. The salts contain finely divided carbon, cyanide and salts that are completely soluble in water. It contains no barium calcium or other types of salts that require special cleaners. Salts are added to the bath to maintain the carbon and the cyanide content.

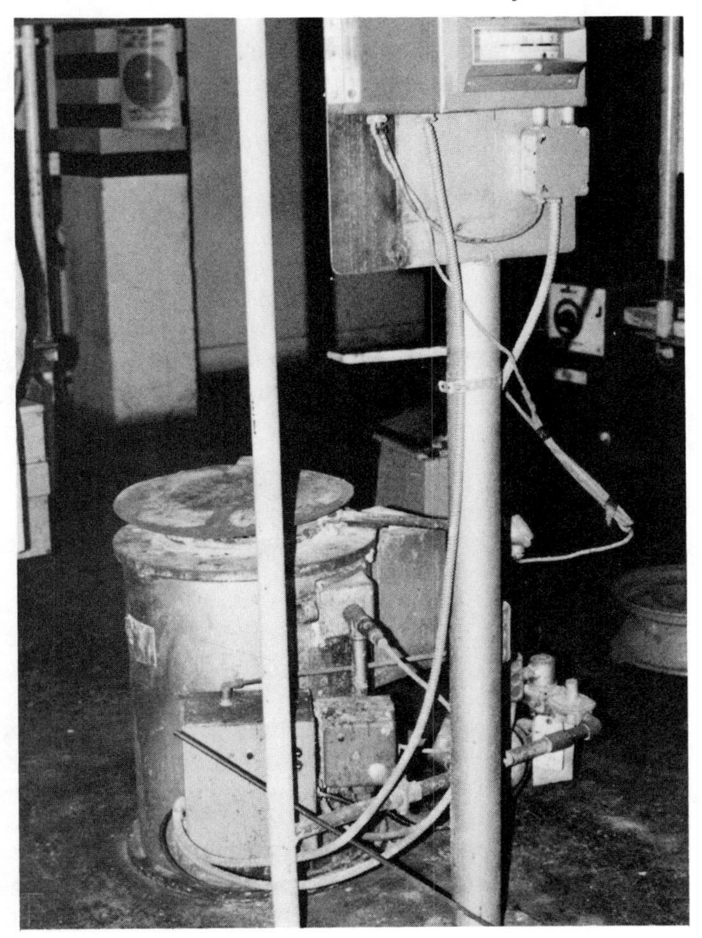

Figure 25. Gas Fired Case Hardening Furnace.

The liquid salts with their components provide a case with high surface hardness and uniformity of carbon penetration is rapid at low carburizing temperatures and relatively short immersion periods. Higher temperatures and longer immersion periods provide deeper case depth.

Typical physical properties of salts:

Melting point	1130° F
Working range	1300-1750° F
Weight per cubic foot	109 lbs. at 1500° F
Appearance	Gray, granulated salts

Information in Table 7 shows penetration rate based on average carburizing values, with 3/8" round SAE 1020 steel pieces. Case depths are a guide only. Cross section, grain size and alloy affect total carbon penetration.

Minutes	1500° F	1550° F	1600° F	1650° F	1700° F
15	.004"	.005"	.007"	.009"	.010"
30	.005"	.008"	.009"	.012"	.013"
60	.009"	.013"	.015"	.018"	.022"
120	.011"	.018"	.020"	.026"	.033"

Table 7. Penetration Rate for Liquid Carburizing.

When many small parts are being treated with short cycles the "drag-out" may be quite rapid. This will require the addition of carburizing salts. These salts must be kept completely dry during storage.

As a general rule, select the highest temperature that gives satisfactory freedom from distortion and reasonable pot life. NOTE - Gas fired furnaces with pots made of boiler plate: Pot life may deteriorate rapidly at temperatures over 1650° F because of the high exterior heat level. This problem does not occur with immersed electrode furnaces because heat is generated in the salt.

The parts after completion of the heat treating period are immediately immersed in oil. Covering the vessel containing the oil will reduce smoke and flame. Caution: Follow special precautions in the handling of products containing cyanide. Avoid any possible addition of water to the carburizing solution. Wear eye and face shield when working with heated furnace. Protect hands with gloves and use tongs when handling parts being carburized. The body must be completely covered while working around the case hardening furnace.

Dry Powder Carburizing - Dry powder products as illustrated in Figure 26 may be used for case carburizing. The material to be carburized is prepared to the desired shape and then heated to a bright cherry red color (1650° F.) either in a furnace or using an oxy-acetylene torch. The heated part is then rolled in the carburizing powder until the powder cakes to the stock. The carburizing process then begins. Reheat the stock with the fused shell of carburizing compound to 1650° F. Quench the stock in water using a scrubbing motion to help remove the shell. Remove the scale from part and it is case hardened. Spark test the surface of the stock to check the carbon content or use a file to check the surface for hardness.

Figure 26. Using a Carburizing Compound for Case Hardening.

HEAT TREATING DEFINITIONS

Austenite- A solid solution formed when carbon and certain alloying elements dissolve in gamma iron. Gamma iron is formed when carbon or constructional alloy steel is heated above the so-called critical range, and the ferrite (alpha iron, with a body-centered crystal structure) is transformed to a face-centered crystal structure. Austenite does not exist in most ordinary steels at room temperature.

Austenizing- The process of forming austenite by heating a ferrous alloy to temperatures in the transformation range (partial austenizing) or above the transformation range (complete austenizing).

Carbonitriding- Introducing carbon and nitrogen into a solid ferrous alloy by holding above Ac_1 in an atmosphere that contains suitable gases such as hydrocarbons, carbon monoxide and ammonia. The carbonitrided alloy is usually quench hardened.

Carburizing- Introducing carbon into a solid ferrous alloy by holding above Ac_1 in contact with a suitable carbonaceous material.

Case Hardening- Hardening a ferrous alloy so that the outer portion or case is made substantially harder that the inner portion or core. Typical processes used for case hardening are carburizing, cyaniding, carbonitriding, nitriding, induction hardening and flame hardening.

Core- In a ferrous alloy, the inner portion that is softer than the outer portion or case.

Cyaniding- Introducing carbon and nitrogen into a solid ferrous alloy by holding above Ac_1 in contact with molten cyanide of suitable composition. The cyanided alloy is usually quench hardened.

Decarburization- The loss of carbon from the surface of a ferrous alloy as a result of heating in a medium that reacts with the carbon.

Differential Heating- Heating that produces a temperature distribution within an object in such a way that, after cooling, various parts have different properties as desired.

Direct Quenching- Quenching carburized parts directly from the carburizing operation.

Flame Annealing- Annealing in which the heat is applied directly by a flame.

Flame Hardening- Quench hardening in which the heat is applied directly by a flame.

Fog Quenching- Quenching in a fine vapor or mist.

Gas Cyaniding- A misnomer for carbonitriding.

Hardening- Increasing the hardness by suitable treatment, usually involving heating and cooling. When applicable, the following more specific terms should be used: Age hardening, case hardening, flame hardening, induction hardening, precipitation hardening, and quench hardening.

Heat Treatment- Heating and cooling a solid metal or alloy in such a way as to obtain desired conditions or properties. Heating for the sole purpose of hot working is excluded from the meaning of this definition.

Hot Quenching- Quenching in a medium at an elevated temperature.

Induction hardening- Quench hardening in which the heat is generated by electrical induction.

Induction Heating- Heating by electrical induction.

Interrupted Quenching- Quenching in which the metal object being quenched is removed from the quenching medium while the object is at a temperature substantially higher than that of the quenching medium. See also "time quenching".

Martempering- Quenching an austenitized ferrous alloy in a medium at a temperature in the upper part of the martensite range, or slightly above that range, and holding in the medium until the temperature throughout the alloy is substantially uniform. The alloy is then allowed to cool in air through the martensite range.

Nitriding- Introducing nitrogen into a solid ferrous alloy by holding at a suitable temperature (below Ac_1 for ferritic steels) in contact with a nitrogenous material, usually ammonia or molten cyanide or appropriate composition. Quenching is not required to produce a hard case.

Preheating - Heating before some further thermal or mechanical treatment. For tool steel, heating to an intermediate temperature immediately before final austenitizing. For some nonferrous alloys, heating in order to homogenize the structure before working.

Quench Hardening - Hardening a ferrous alloy by austenitizing and then cooling rapidly enough so that some or all of the austenite transforms to martensite.

Quenching - Rapid cooling. When applicable, the following more specific terms should be used: Direct quenching, fog quenching, hot quenching, interrupted quenching, selective quenching, spray quenching, and time quenching.

Selective Heating - Heating only certain portions of an object so that they have the desired properties after cooling.

Selective Quenching - Quenching only certain portions of an object.

Spray Quenching - Quenching in a spray of liquid.

Tempering - Reheating a quench-hardened or normalized ferrous alloy to a temperature below the transformation range and then cooling at any desired rate.

Time Quenching - Interrupted quenching in which the duration of holding in the quenching medium is controlled.

METAL IDENTIFICATION

The ability to identify materials is constantly a problem for manufacturers, warehouse workers, and maintenance personnel. Often in the higher echelons of the flow of materials, identification is made by color coding, labeling securely or possibly by chemical analysis.

When repair people are faced with machinery and welding operations during maintenance, they are usually "on their own".

The following are useful ways of identifying metals:

1. **How was part used** - mechanical and physical requirements of job.
2. **How was part made** - casting, rolling, stamping, forging pressing (sintering).
3. **Surface appearance** - rough as from molds in foundry. Smooth with "squeeze out" evidence and legible part numbers as from drop forging.
4. **Fracture appearance** - texture of grain structure, color of new and/or old break, uniformity of grain structure. Degree of bending before breaking.
5. **Chip test** - carefully remove with chisel and observe how chip separates from piece.
6. **File test** - the file test is called "a pauper's" hardness tester. It is an inexpensive and effective way to determine hardness of a metal therefore an indication of weldability and machineability. If the file cuts the metal easily it's low carbon steel. High Carbon steel can not easily be cut by the file.
7. **Magnetic test** - some alloys are nonmagnetic. Degree of magnetic attraction.
8. Oxy-acetylene neutral **welding flame** shows degree of heat conductivity, speed of melting, color change during heating.
9. **Spark test** - careful observation of sparks at grinding wheel under subdued light.
   a. Surface feet per minute - 5000

   $$\text{Sfm} = \frac{\text{Circumference} \times \text{RPM}}{12}$$

   b. Grinding wheel should be clean, medium grain, medium hard.
   c. Pressure of metal on wheel medium and uniform.
   d. Keep samples of known metal types available to be able to match known with the unknown.

Metal identification is a reliable skill if practiced in an orderly manner. There is no substitute for experience by a careful observer.

GRAY CAST IRON

I. Formation

A. Approximate composition
   1. Iron
   2. Carbon 1.7 - 4.5%
   3. Silicon
      a. 1.9 - 2.6%
      b. Encourage graphite formation
   4. Sulfur
      a. .06 - .12%
      b. Undesirable element
   5. Phosphorus
      a. .06 - .25%
      b. Undesirable element
   6. Manganese
      a. .70 - .90%
      b. Neutralizes sulfur
   7. Other elements which are added for special properties - copper, nickel, aluminum, titanium, vanadium, molybdenum, and chromium.

B. Procedure
   1. Metal is heated to 2750° F.

COMPARING CAST IRON AND STEEL

Physical Properties	Cast Iron	Steel
1. Surface appearance	Rough from sand in mold One or more "Gates" Ridge where two halves of mold met	Cast steel is rough scaley, part numbers more clear and ridges more often ground smooth
2. Sound	Low dull tone when hit Dull, chalky tone tone	Higher pitched tone Clearer pitched
3. Sparks	Red color close to wheel. Small volume and short length of yellow sparks	White color close to wheel. As carbon increases there are shorter shafts but volume of exploding sprigs increase.
4. Fracture Appearance	From dark gray to bright silver. Brittle clean break. Graphite may rub off.	Definite grain pattern. Silver gray color. Crystalline surface
5. Melting Points	2050 to 2300° F. About 600° F. below steel	About 2750° F. for mild steel
6. Carbon Content	1.7 to 5.0% Free particles of carbon	0.05 to 1.7% Carbon chemically combined with iron
7. Most Common Alloy	Iron, Carbon, Silicon	Iron, Carbon, Manganese

Mechanical Properties	Cast Iron	Steel
1. Tensile strength	Low 20,000 - 53,000 psi	High 46,000 - 200,000 psi
2. Compressive Strength	Several times the tensile strength	High
3. Ductility	Very little will not stretch	Usually high
4. Brittleness	High	Usually low
5. Hardness	Varies considerably 140 - 510 Brinell	100 - 130 Brinell
6. Elasticity Modulus	10,000,000 to 25,000,000 psi	30,000,000 psi
7. Yield Point	None	Varies as to carbon content
8. Fatigue strength	Low	High
9. Notch Toughness	Varies	Varies
10. Elastic Limit Yeild Strength	Doesn't stretch	High
11. Elongation % in 2 inches	Hardly any; Nodular exception	12 - 35%
12. Toughness	Low	High
13. Impact Strength	Low	High
14. Corrosion Resistance	Low	Low
15. Electrical Resistance	Low	Low
16. Thermal Conductivity	Low 11.0% compared to silver at 100%	Low - 10.9%
17. Thermal Expansion	Low	Low

2. Poured into casting
3. Cooled moderately slow to slow to encourage formation of
   a. Pearlite and graphite
   b. Pearlite and graphite and free ferrite
   c. Ferrite and graphite
4. Carbon is distributed between
   a. Combined carbon - In pearlite
   b. Uncombined or free carbon - In graphite

C. Chip Test - no curl - free break

D. Rough surface - gate and riser markings (from mold)

II. Identification

A. Spark test - short steamers which are brick red and follow straight lines, with numerous fine, repeated yellow sparklers as shown in Figure 27.

B. Appearance of fracture - dark gary porous structure

Average stream length with power grinder to 25 inches varies with pressure

Volume - small

Many sprigs followed by appendage

Figure 27. Spark Appearance for Gray Cast Iron.

WHITE CAST IRON

I. The formation of white cast iron.

A. White cast has no free graphite.

B. All of its carbon is combined as cementite or as pearlite.

C. A low silicon content and rapid cooling rate are responsible for the structure of white cast iron.

II. The identification of white cast iron.

A. The fracture of white cast iron will disclose a fine, silvery white, crystalline formation.

B. The spark test will show short streamers that are red in color, see Figure 28.

C. There are fewer sparklers than in gray cast iron and these are small and repeating.

D. The surface has the appearance of a sand mold and the surface is dark gray.

E. Outer part of spark stream shows whiter spark from combined carbon.

Average stream length with power grinder to 20 inches varies with pressure

Volume - very small

Sprigs- finer than gray iron, small and repeating

Figure 28. Spark Appearance for White Cast Iron.

CHILLED CAST IRON

I. Formation, identification and properties

A. Its chemical composition is similar to gray, but has a lower silicon content.

B. Chilled cast has thick surface section which is extremely hard and is resistent to wear - its interior structure is typical of gray cast.

C. The casting of chilled cast
1. Cast against chilled iron surface
2. The chilled surface quickly freezes the outer surface
3. Creates a thick surface zone of extremely hard white iron free from graphite carbon.
4. Chilled zone may be changed
   a. To increase chilled zone, decrease silicon content.
   b. To decrease chilled zone, increase silicon content.

D. Identification similar to white cast iron.

MALLEABLE CAST IRON

I. Formation

A. Made by annealing white cast iron
1. Temperature of 150 - 1650° F.
2. Annealing period of about 72 hours.
3. Cementite decomposes leaving ferrite and graphite nodules.
4. Most carbon is eliminated by diffusion and resultant product is ferritic.
5. Carbon content varies from 2-5 percent.
6. Existing carbon in the rounded or nodular form.

II. Identification and Properties

A. Physical properties
1. Dark Gray color except for lighter outer layer (skin of steel).
2. Evidences of sand mold on surface.
3. Spark tests (note Figure 29)
   a. Straw yellow color
   b. 30" stream length with power grinder
   c. Moderate volume of sprigs
   d. Longer shafts than gray iron ending in numerous small sprigs.
4. Tensile strength of 30,000 to 53,000 psi.
5. Melting point of approximately 2300°.
6. Malleable is stronger, tougher, more ductile than gray cast.
7. Chip test - chip curls some before breaking.

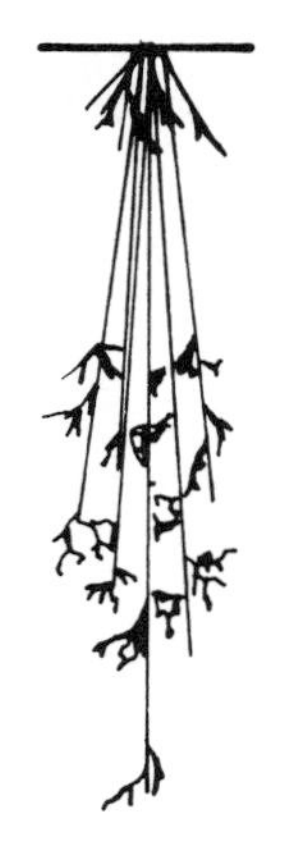

Color - Straw Yellow

Average stream length with power grinder to 30 inches, varies with pressure.

Volume - Moderate

Longer shafts than gray iron ending in numerous small, repeating sprigs.

Figure 29. Spark Appearance for Malleable Iron.

NODULAR CAST IRON

I. Formation

A. By addition of magnesium and/or cerium as desulfurizer
1. Added after the sulfur level was reduced to about 0.02%.
2. Added in the form of alloys
   a. Containing between 8 and 15% magnesium
   b. The rest of it is either nickel, nickel-silicon, copper-silicon or ferrosilicon.
3. Amount of magnesium added is influenced by the nature of the molten iron at the time of addition.
4. Above is followed by innoculation

B. Carbon precipitates as balls or spheres of graphite
1. In a matrix of basically pearlite
2. Some of the ferrite and free cementite is present

C. Three manufacturing processes
1. By reladling into another ladle containing the magnesium alloy, then reladled back into original ladle as ferrosilicon innoculent is added to the stream.
2. By pouring into ladle, which contains the alloy, from the furnace:
   a. Innoculant is added as it is poured into mold.
   b. Magnesium is volatile and precautions must be taken to prevent its loss.
3. Two to three pounds of metallic magnesium can be added under pressure of 10-15 atmospheres to the molds.
4. Basic iron is formed in cupola:
   a. Can produce sulfur contents of less than 0.02%

b. Can utilize scrap metal
c. Optimum carbon content range of 3.5% -3.8%.

II. Identification

A. Microstructures appear as spheres of graphite in a pearlite matrio.

B. To the eye, it appears similar to medium carbon steel.

C. Tensile strength, 1. 60,000 to 150,000 psi which is higher than most cast irons.

D. Impact resistance as great as medium carbon steel.

E. More ductile than cast iron.

F. Properties closely resemble medium carbon steel.

G. As much as 10% elongation before fracture.

H. Spark appearance like gray cast iron except whiter color.

I. Chip test - chip curls some before breaking.

CARBON STEEL (HOT ROLLED)

I. Formation

A. Approximate composition
1. Iron - approximately 90% - depends on carbon content
2. Carbon - low to high .10% to 1.7%
3. Manganese - approximately .60%
4. Phosphorus - approximately .02%
5. Sulfur - approximately .02%
6. Silicon - approximately .08%

B. Procedure
1. Purification of pig iron - remove P, S and Si.
2. Decarburization to desired amount
3. Oxidation to speed process - BOF
4. Pour in ingot or continuous cast
5. Roll to desired shape.

II. Identification

1. Chip Test - Curl
2. Fine Grain - fracture - light gray
3. Spark Test - white sparks - short lines with many sparks for high carbon - longer lines with fewer sparks for low carbon, (Figure 30.)
4. Smooth surface from rolling.

LOW CARBON STEEL

Color - white

Average length of stream with power grinder varies with pessure

Volume - moderately large

Shafts shorter than wrought iron and in forks and appendages

Forks become more numerous and exploding sparks increase as carbon content increases

*These data apply also to cast steel

HIGH-CARBON STEEL

Color - White

Average stream length with power grinder to 55 inches varies with pressure.

Volume - large

Numerous small and repeating exploding sparks.

Figure 30. Spark Appearance for Low and High Carbon Steels.

WROUGHT IRON

I. Formation

A. Approximate composition

1. Iron 97.0%
2. Carbon .02%
3. Manganese .03%
4. Phosphorus .12%
5. Sulphur .02%
6. Silicon .15%
7. Slag 3.0%

B. Procedure

1. Careful purification of pig iron
2. Decarburization
3. Molten metal poured into slag.
4. Spongy mass squeezed in press
5. Iron silicate evenly distributed in fibrous form (wood-like)
6. Rolled to shape.

II. Identification

A. Very ductile (like a nail)
B. Highly corrosion resistant
C. Chip test - definite curl
D. Sharp test - long lines - few sparks as illustrated in Figure 31.

WROUGHT IRON

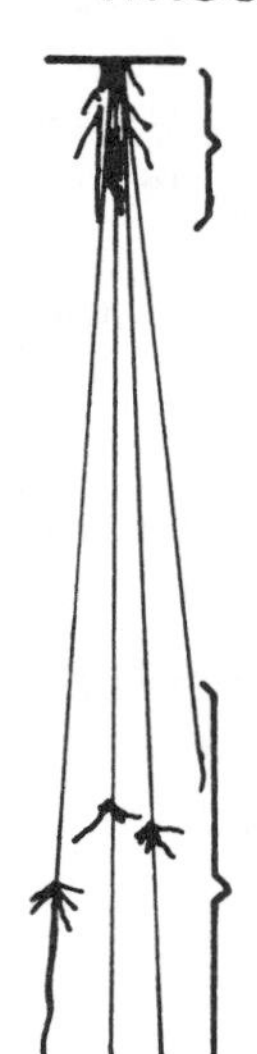

Color - Straw yellow

Average stream length with power grinder varies with pressure

Volume - large

Long shafts ending in forks and arrow like appendages

Color - white

Figure 31. Spark Appearance for Wrought Iron.

LOW AND HIGH ALLOY STEELS

I. Formation

A. Approximate composition

1. AISI-SAE classification chart shows many analyses. Iron and carbon content varies considerably.
2. Alloying elements added to carefully made carbon steel for specific chemical, physical or mechanical characteristics.

II. Identification

A. Corrosion resistance (no rust)
B. Nonmagnetic (18-8 stainless steel)
C. Work harden - manganese steel
D. Spark test - seldom spark explosions as in carbon steel. Note Figure 32.
E. Sparks follow grinding wheel
F. Volume of sparks usually low

ALLOY STEEL

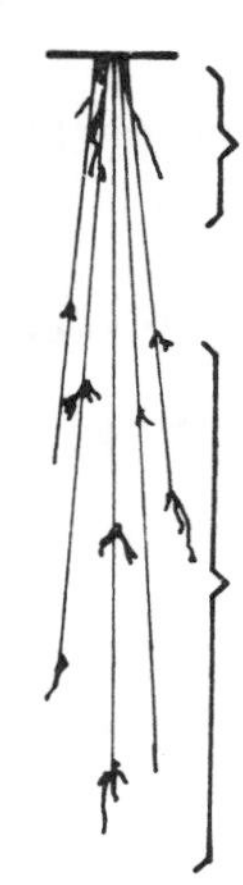

Color - varies: yellow, red, white

Stream length varies with type and amount of alloy content and pressure

Shafts may end in forks, buds, or arrows frequently with break between shaft and arrow. Few if any sprigs.

Color - varies: white, red, yellow

**Spark shown is for stainless steel.

NICKEL AND NICKEL ALLOYS

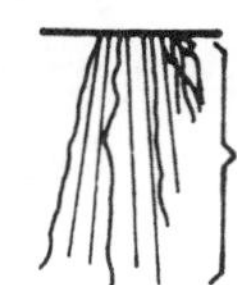

Color - Orange

Average stream length with power grinder to 10 inches varies with pressure

Short shafts with no forks

Figure 32. Spark Appearance for Alloy Steels.

HARDNESS TESTING

Hardness is a property of a material which denotes resistance to permanent deformation. Thus, hardness can be related to tensile strength in a non-destructive manner by heat treaters. Hardness can be indicated by the following methods:

1. Penetration hardness
   a. Rockwell (most common in production)
   b. Brinell
   c. Vickers
2. Rebound (Vickers)
3. Scratch (Moh's)

The Rockwell hardness test measures the depth of penetration of a given penetrator under a specificied load. The Rockwell B test uses a 1/16" steel ball with loading applied in a manner that measures the amount of indentation. The Rockwell C Test uses a diamond penetrator in a similar manner. Obviously, the C test is used on harder materials than the B test. The C test measures to C76 which indicates beyond 387,000 psi TS. The B test measures to B111 which indicates 177,000 psi TS.

BRINELL HARDNESS - NUMBER CONVERSION. The Rockwell hardness numbers on the two standard hardness scales, B and C, can be converted into Brinell hardness numbers by the following equations, developed by Petranko at the U.S. Bureau of Standards. These equations are semiempirical and are accurate to within + 10 percent.

For Rockwell numbers B 35 to B 100:

$$\text{Brinell hardness number} = \frac{7300}{130 - \text{Rockwell B number}}$$

For Rockwell numbers C 20 to C 40:

$$\text{Brinell hardness number} = \frac{1{,}420{,}000}{100 - \text{Rockwell C number}}$$

For Rockwell numbers greater than C 40:

$$\text{Brinell hardness number} = \frac{25{,}000}{100 - \text{Rockwell C number}}$$

Tensile Stength can be approximated as follows:

$$\frac{\text{Brinell test number}}{2} \times 1000 = \text{TS psi in thousands}$$

METAL WORKING TOOLS AND PROCEDURES

Metal cutting tools are great labor savers for separating and/or shaping metal. Power metal cutting tools and some hand metal cutting tools can be ruined or badly damaged by improper use. The ability to identify metals and knowledge of some of their physical properties is essential before using any cutting tool in order to avoid damage to tools and materials, or injury to the operator. Power tools either electric or pneumatic must be operated properly for efficiency and safety.

Hand Metal working tools as illustrated in Figure 33 are a must for efficient and quality metal work while working in the metal shop. The tools shown are identified as: hammer, file, chisel, center punch, pliers, combination square, vise-grip pliers, screwdriver, awl, adjustable wrench and bench brush. These tools will be discussed in the following pages along with types and procedures for using.
Figure 33. Hand Metal Working Tools.

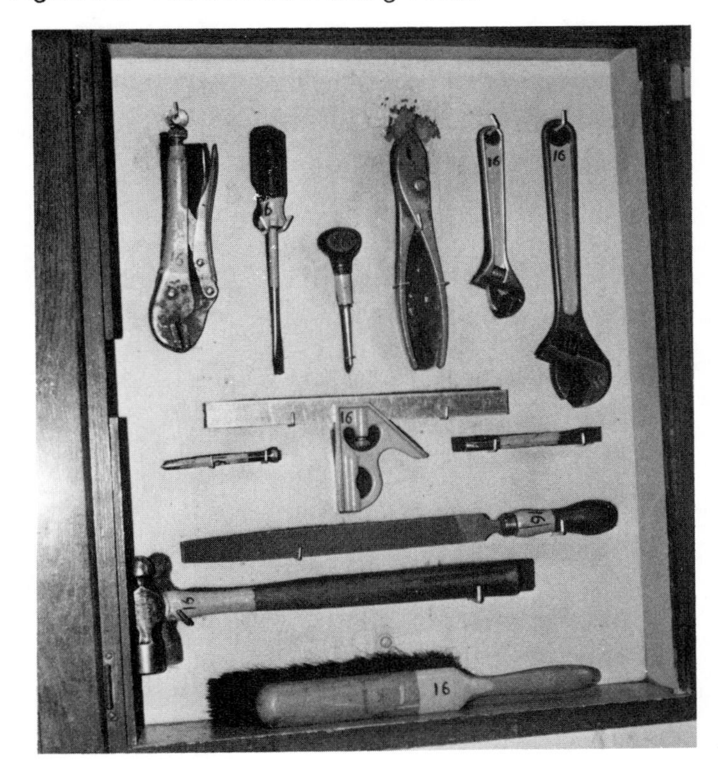

HAMMERS

HAMMERS

The marking cutting and aligning of metals involves chisels, punches and hammers. Hammers (note Figure 34) of varying weights and types of peens (See Figure 35) are used in these processes. Hammers are made for many trades and uses. Bricklayers, drywallers, tinners, and upholsterers are examples of tradesmen who use special hammers. The mechanics ball peen hammer is the most popular for general metal work. They normally are in weight from 8 to 32 ounces with "rim tempered" heads. The "rim tempering" minimizes chipping of the face. Handles are from 13 to 16 inches in length depending on the weight of the hammer head. See Figure 34 for the parts of the ball peen hammer.

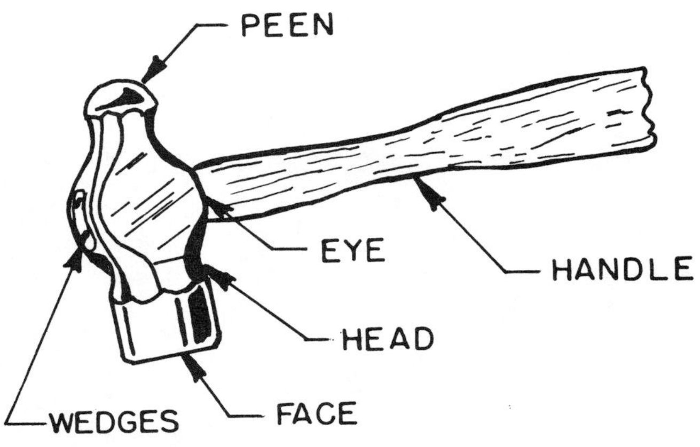

Figure 34. The Ball Peen Hammer

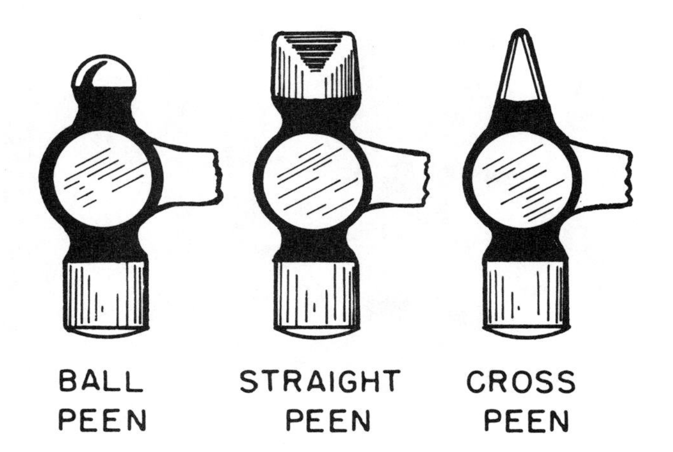

Figure 35. Types of Peens.

HACKSAWS AND HACKSAWING

1. **Power hacksaws** - Two types are available. The band hacksaw which adjusts for horizontal or vertical cutting (withtable) is preferred. The economy reciprocating straight blade will suffice for lighter duty work. Power hacksaw blades are specified by length, width, tooth pattern and number of teeth. Look for the following features in a power hacksaw: speed variation, feed (pressure) device, angle adjustment on secure clamp and automatic shut-off. See Figure 36.

Floor Model Metal Cutting Saw - Vertical or horizontal cutoff capacity 3 1/2 to 6 inches, vise and automatic shut-off, 1/2 inch blade width, 1/4 hp, 115 volt A.C. motor with 3 wire cord. Wet-cut metal cutting band saws are usually larger capacity. The coolant permits faster cutting and increases blade life. See Figure 36.

Portable Band Saw - Variable speed 80-240 fpm cutoff capacity 3 3/8 inch x 4 1/8 inch rectangular stock, weight 17 pounds. Saw blades available for all types of metal cutting. There is a saw stand for job applications where work support is required for accurate cutting. See Figure 37.

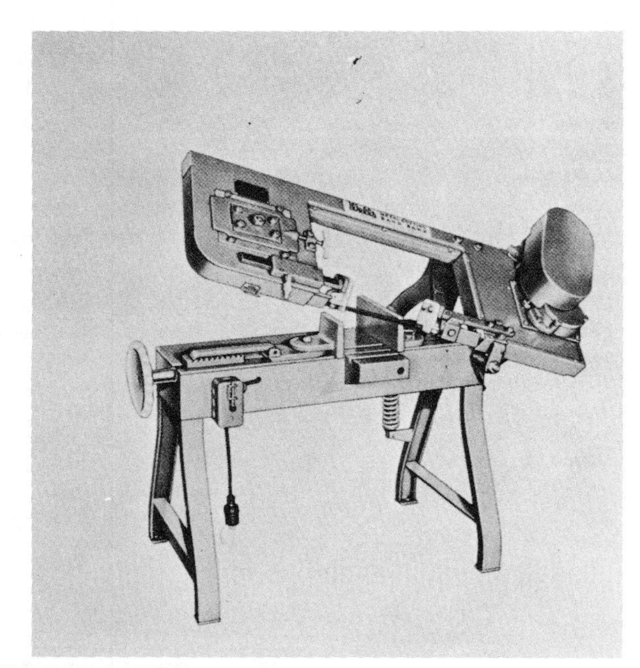

Figure 36. The Floor Model Band Saw.

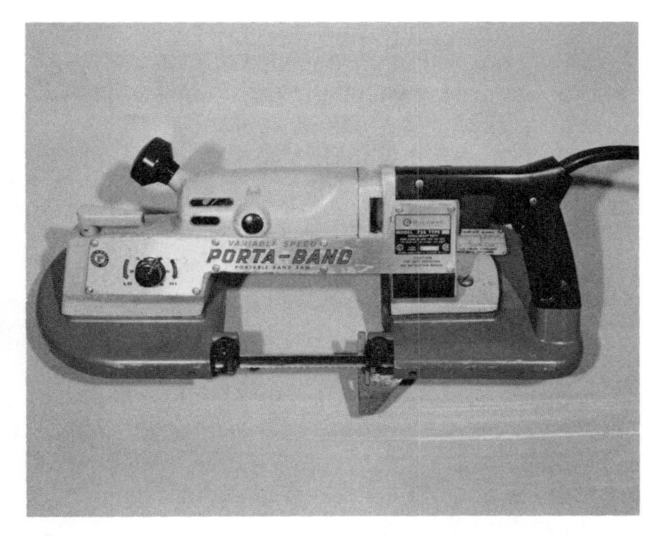

Figure 37. The Variable Speed Portable Band Saw.

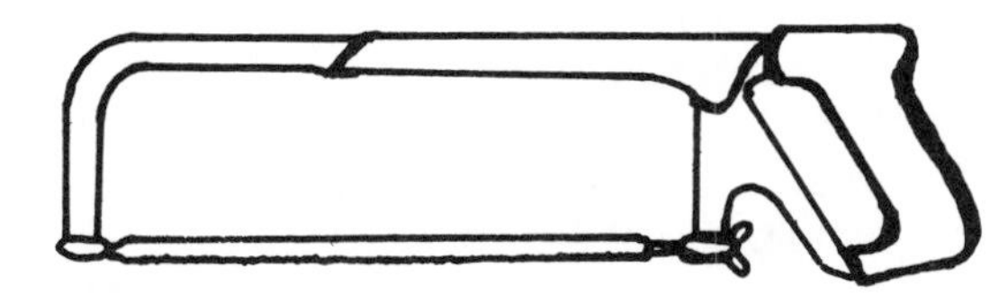

Figure 38. Adjustable Frame Hand Hacksaw.

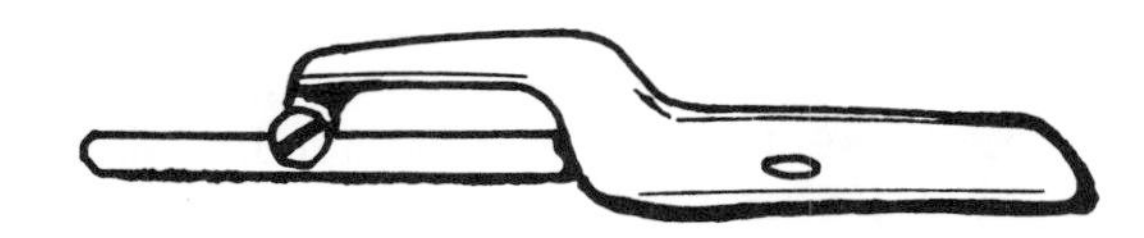

Figure 39. Mini Hacksaw.

2. **Hand hacksaws**
   a. Frames
      (1) **Adjustable** - This is the most popular frame for general use. They vary in depth from 2-1/2" to 3-1/2" and will take blades from 8" to 12" in length. By loosening the wing-nut, the mounting posts may be turned so that the blade will cut at right angles to the frame, see Figure 38.
      (2) **Solid** - Frames may be had that vary in depth and length. They remain very rigid and for specialized work are very desirable.
      (3) **Mini** - The mini hacksaw as shown in Figure 39 is used in difficult sawing situations. It uses regular blades.
   b. Blades
      (1) A blade should be selected that will be best suited to the work being done. In most cases it is desirable to select 10" or 12" blades for use in adjustable frames.
      (2) The blade should have enough teeth per inch so that two or more are in contact with the metal at all times. Standard blades are 7/16" to 9/16" wide with 14 to 32 teeth per inch and 8" to 12" long. Blades with 18, 24 or 32 teeth per inch are most common.
      (3) Blades are made with three kinds of set:
         (a) **Alternate** - alternate teeth are bent slightly in opposite directions.
         (b) **Raker set** - every third tooth remains straight, the other two are alternately set.
         (c) **Undulated** - alternate short sections are bent in opposite directions - blades with 24 or 32 teeth only.
      (4) Quality of the blade
         (a) **Carbon type** - straight carbon steel
         (b) **Tungsten steel** (common type) - used for general sawing of mild steel and soft non-ferrous metals, not satisfactory for sawing tool steel, drill rod etc.
         (c) **High speed steel** - for cutting at high speed, hard to cut metals. Molybdenum steel is most economical to buy for all types of sawing if used properly. Desirable for cutting hard steel alloys.
         (d) Select a blade that has been **heat treated** to provide it with the toughness and hardness required to withstand various uses:
            a'. Flexible or soft blades with only the teeth hardened do not break as readily as hard blades when sawing thin cross-sections of tubing, angle, channel, babbit, brass and similar relatively soft materials.
            b'. All-hard blades do not have the tendency to buckle or run out of line when under pressure.

3. **Cutting metal with a hacksaw**
   a. Mount the blade in the frame so that the teeth point forward.
   b. Tighten the tension nut so that when struck with a piece of metal, the blade gives a slight ring.
   c. Clamp the work securely so that two or more teeth are in contact with the metal at all times.
   d. Saw close to the vise to prevent vibration.
   e. Thin material may be placed betwen two blocks of wood before sawing.
   f. Grip saw firmly; one hand on handle and other holding front of frame. Make each stroke go full length of blade.
   g. **Apply pressure only on the forward stroke then lift slightly on the return stroke.**
   h. Always start a new cut when replacing a blade.
   i. Use approximately 35 to 50 strokes per minute - 60 strokes should be the maximum. Hard materials overheat and dull blade when sawed too fast.

4. Most common causes of blade breakage or failure:
   a. Using a saw with large teeth to saw thin metal.
   b. Using too much pressure.
   c. Sawing too fast.
   d. Using short strokes, causing excess wear and loss of set.
   e. Insufficient tension on blade.
   f. Failure to give work proper support.
   g. Failure to hold saw firmly and push straight forward on the cutting stroke and return straight back.
   h. Applying pressure on return stroke.

COLD CHISELS

1. **Types of chisels** - A cold chisel as illustrated in Figure 40, is made of high carbon tool or alloy steel and is used for chipping or cutting metals. If properly tempered, it will cut any metal that is softer than itself. If a trial cut with a sharp file slides across the metal without cutting, it is too hard to cut with a chisel or hacksaw. The following are several of the commonly used metal cutting or cold chisels: Note Figure 40.
   a. **Flat chisel** - It is made in several widths with two beveled faces forming a 60 to 70° degree cutting edge. It may be ground straight for light work or slightly convex for heavy chipping. It is the most common type and has a wide variety of uses.
   b. **Side chisel** - It is similar to the flat chisel, differing only in that it is beveled only from one side. It is used in chipping side of slots, smoothing keyways, and cutting off rivet heads.

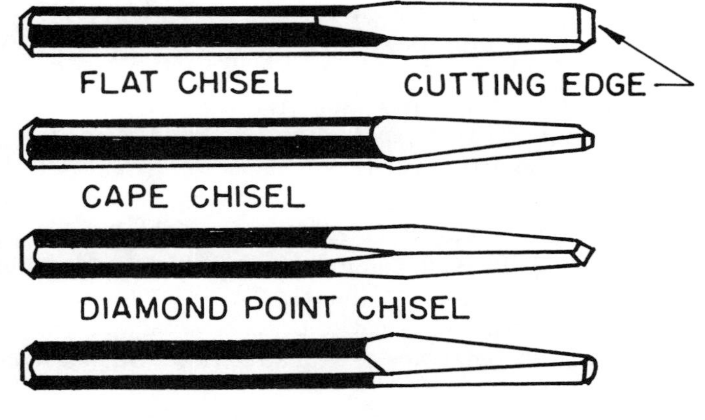

Figure 40. Cold Chisels Commonly Used in Metal Work.

   c. **Cape chisel**- The cape chisel has a narrower cutting point than the flat chisel. Slots and grooves are cut with this type of chisel. It is ground similar to the flat chisel.
   d. **Diamond-point chisel**- A special chisel with four-sided bevel and a diamond shaped point. The cutting point, ground at 60°, is used for cutting oil grooves and also as an emergency "easy-out" or bolt extractor.
   e. **Splitting chisel**- Similar in shape to the flat chisel except that it is not ground to a sharp cutting edge. The cutting edge is flat and slightly concave. The area just back of the edge is slightly thinner so that it will not bind in cutting. This chisel is used in cutting heavy sheet metal, as oil drums, range boilers and similar material.
   f. **Round-nose chisel**- Has a round cutting edge, ground at a 60° angle. It is used in cutting grooves that have a round contour.
2. **Cutting metal with a chisel**
   a. Heavy wire or small round stock
      (1) Mark the point where the stock is to be cut.
      (2) Place the stock on a soft metal block or the chipping block of the anvil.
      (3) Holding the chisel in a vertical position, strike a light blow with a hammer, then check to see if a cut is being made at the right point.
      (4) Complete the cutting. Use light blows at the last to avoid marring the chipping block. It is best to break the last remaining portion so that other workers are not endangered by flying chips or pieces. Heavy pieces of metal are cut in the same manner only the cut is made from two or more directions toward the center.
   b. Cutting heavy sheet metal or light plate 3/8" thickness or less.
      (1) Mark the stock where the cut is to be made.
      (2) Clamp the stock securely in a vise in an upright position with the line even with or slightly below the top of the jaws. Waste material should extend above the jaws.
      (3) Hold the chisel at a 20° to 45° angle in relation to the piece to be cut. The lower beveled face of the flat chisel should rest smoothly on the jaws of the vise. Held in this manner the chisel and the jaws of the vise produce a shearing cut.
      (4) Holding the chisel firmly, strike heavy blows to start cut. Lighter blows may be used as the cut advances. The lower beveled edge of the chisel must be kept flat on the jaws of the vise at all times.

PUNCHES

There are many types and sizes of punches that have been developed for specific purposes. The starting, pin and aligning punches, as shown in Figure 41 are used in machinery and equipment repair. They vary greatly in length and diameter. The prick and center punches, Figure 42 are primarily metal marking tools. The prick punch is often used in accurately centering, followed by center punching the metal before drilling.

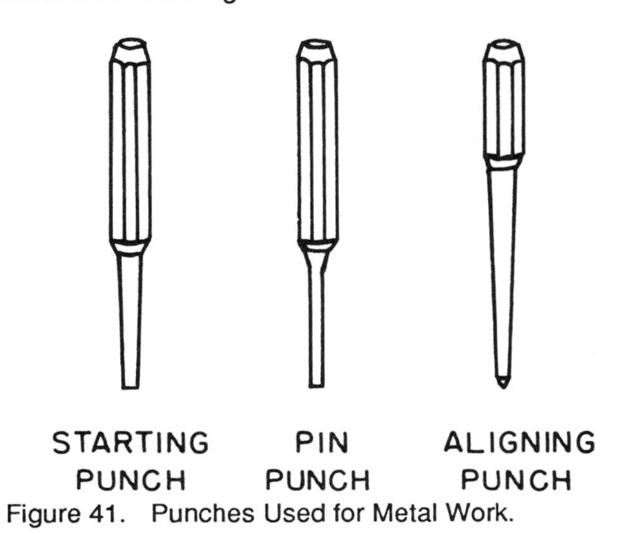

Figure 41. Punches Used for Metal Work.

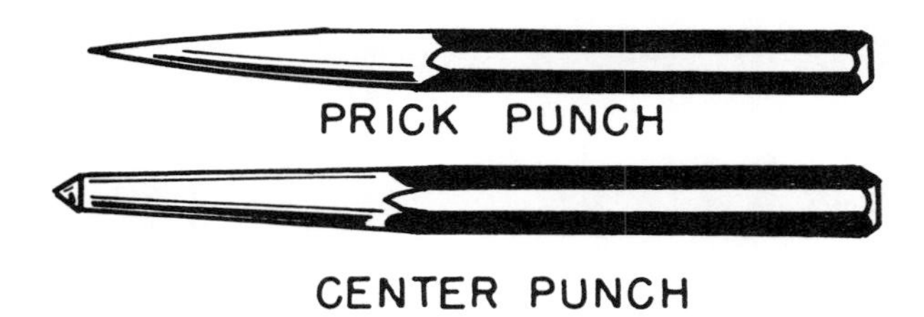

Figure 42. Prick and Center Punch.

BOLT CUTTERS

Bolt cutters are used for cutting bolts or bars of mild steel or iron. They should never be twisted in cutting. All cuts should be made perpendicular to the material. High carbon and alloy steel should never be cut with a bolt cutter. They are made in several sizes:

Size	Length	Maximum Capacity soft rods or bolts
0	18"	5/16"
1	24"	3/8"
2	30"	1/2"
3	36"	5/8"
4	42"	3/4"

SHEARS

Shears are made for mounting on the floor or on benches, depending on their size or capacity. The maximum capacity in thickness of material should be known before using the shear. The metal should always be held perpendicular to the jaws. This prevents wedging or spreading the jaws.

The **foot-powered squaring shear** is illustrated in Figure 43. It is used to cut soft mild steel, 18 gage or thinner.

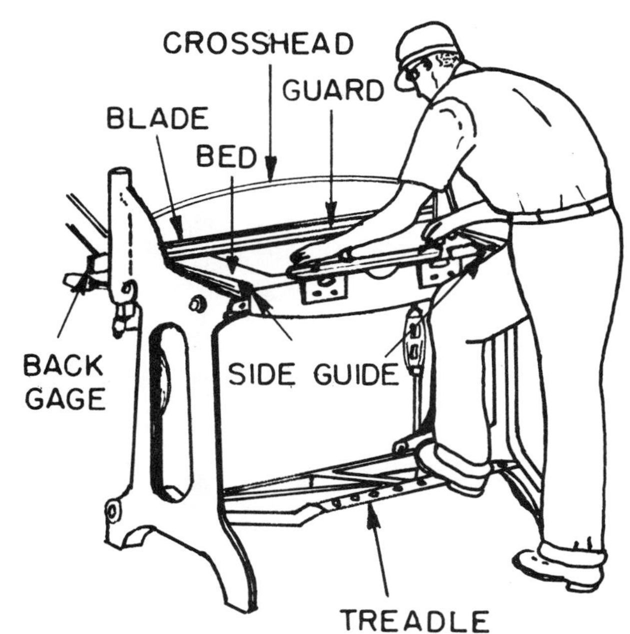

Figure 43. Foot Powered Squaring Shear.

Illustrated in Figure 44 is a **power shear**. This power shear has a capacity of 3/8" thick mild steel up to 8" in width. In addition to square shearing it can be used for notching metals and for parting round stock and angle iron. As with bolt cutters, high carbon or alloy steels should never be cut in the shears.

General precautions on the use of metal cutting tools: Use a file to check for hardness before using any metal cutting tool. Use corner of a flat file. If file cuts easily, proceed with other tools. If the file cuts with difficulty, be careful for damage to other cutting tools. If the file does not cut, soften by annealing or use abrasive cut-off saw.

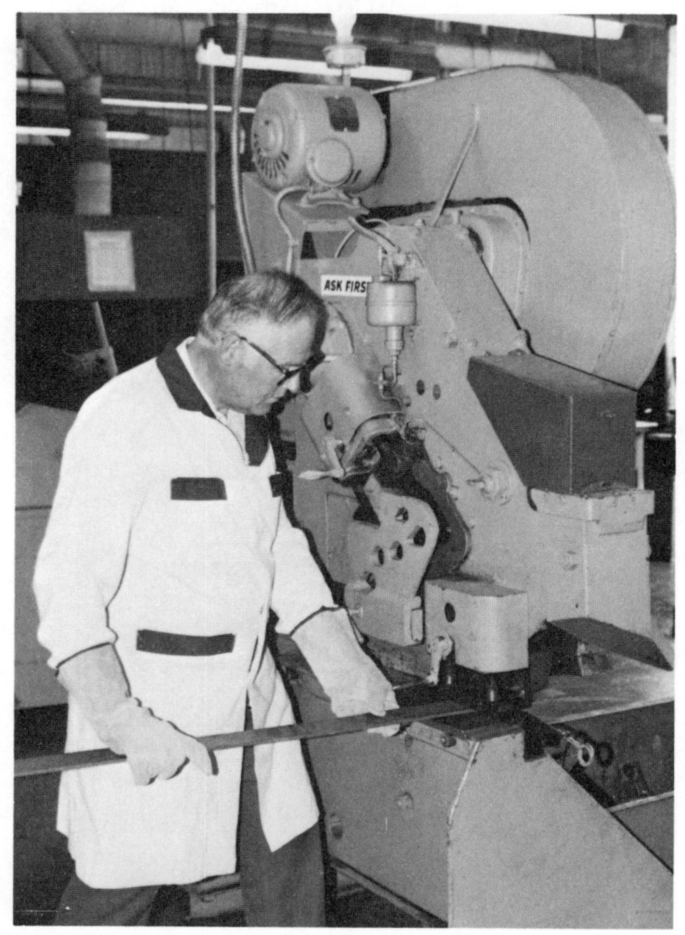

Figure 44. Power Shear.

DRILL BITS - DRILLING AND COUNTERSINKING

The drill is one of the most important metal cutting tools found in the metal shop. It is important that we understand how to select, maintain and use the drill for cutting holes in metal.

There are three common kinds of drills. The flat drill is usually a homemade drill, made of low cost, but a good grade steel, flattened, hardened, and sharpened. The straight-fluted or Farmer drill is used for drilling brass, copper and other soft metals. Twist drills are made with two, three or four cutting lips. The two-lip drill is used for drilling holes in solid metal while three and four-lip drills are used to enlarge already drilled holes.

Twist drills are made of carbon steel or high speed steel. If the drill shank is not stamped "HS" meaning high speed, it is made of carbon steel. High speed drills cost two to three times as much as carbon drills; however, they may be run twice as fast as the carbon type. Faster cutting could be done with the high speed drill.

Parts of the twist drill - The main parts of the twist drill are illustrated in Figure 45. The **body** is the part in which the **grooves** or **flutes** are cut. The drill is held in the drill press by the **shank**. The cone-shaped cutting end is the **point**. The margin is the narrow edge along side the flute. It keeps the drill from binding or rubbing on the hole being drilled. The margin also aids from drawing the temper out of the drill by overheating. The **web** is the metal in the center running lengthwise between the flutes. It gets thicker near the shank. The **tang** is the flattened end of the shank which fits into a drill press **spindle slot** and keeps the shank from slipping.

The **lips** are the cutting edges of the drill. The **dead center** is the end at the point of the drill and should always be in the exact center of the point. The **heel** is the part of the point behind the cutting edges.

The flutes are shaped to: (1) help form the cutting edge at the point, (2) curl the chips into small spaces, (3) form passages for the chips to come out of the hole, and (4) allow the lubricant to travel to the cutting edges of the drill.

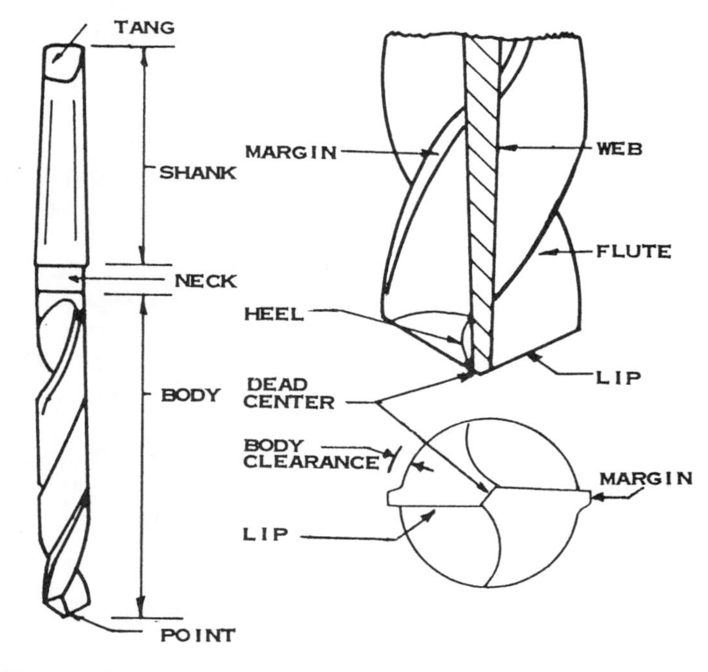

Figure 45. Parts of a Twist Drill.

Kinds of drill shanks- Five types of drill shanks are manufactured as illustrated in Figure 46. The most common types are the straight shank used in the regular drill chuck and the tapered shank which requires a drill press having a tapered spindle. Tapered shanks are most common on larger diameter drills.

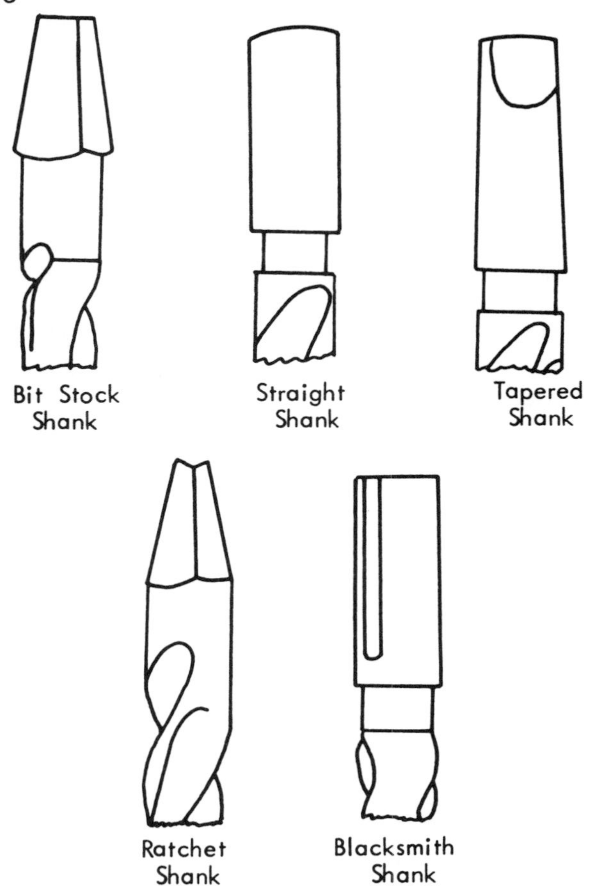

Figure 46. Kinds of Drill Shanks.

Sizes of drills- Small drills are usually purchased in sets. The size of the drill is known by its diameter and may be purchased by: (1) **Gage number** - number drills range in size from No. 80 (0.0135") to No. 1 (0.228"), (2) **Letter drills** - letters A (0.234") to Z (0.413") and (3) **Fractional drills** - range in size from 1/64" to 4" or larger. The sizes increase by 64th of an inch in the smaller sizes and by 32nds and 16th in the larger sizes. Note that the letter drills begin where the number drills leave off. The diameter of the drill is stamped on the shank except on the very small drills which must be measured by a size gage to determine exact size.

Drill sharpening- Nearly all drill problems are caused by improper sharpening. The correct sharpening of a drill requires attention to four very important points: (1) lip clearance, (2) length of lip, (3) angle of lips, and (4) angle and location of dead center. If the first three are correct, the fourth will also be correct.

1. **Lip clearance**- Lip clearance is made by grinding away the metal behind the cutting edge. If there is no lip clearance, it would be impossible to cut into the metal for the bottom of the drill would rub and not cut. Too much lip clearance will cause the drill to heat and burn quickly. Also the corners of the drill may be broken due to excessive forces because the drill tends to cut too deep. Proper lip clearance angle is 12 - 15° as illustrated in Figure 47.

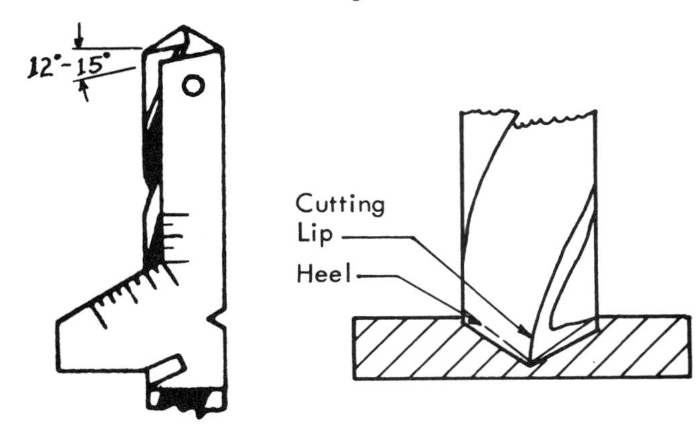

Figure 47. Lip Clearance.

2. **Length of lips**- The lips should be the same length. Unequal lengths will cause the hole to be larger than the drill. This also causes undue strain on the drill press or excessive wobbling of the work piece. Use the drill grinding gage to determine the length of the lips as shown in Figure 48.

3. **Angle of lips**- The two lips must have the same angle. For ordinary work, this is 59°; however, a flatter angle is recommended for drilling hard or treated metal while a steeper angle may be recommended for drilling wood or fiber products. The proper angle and equal length of lips are shown in Figure 48.

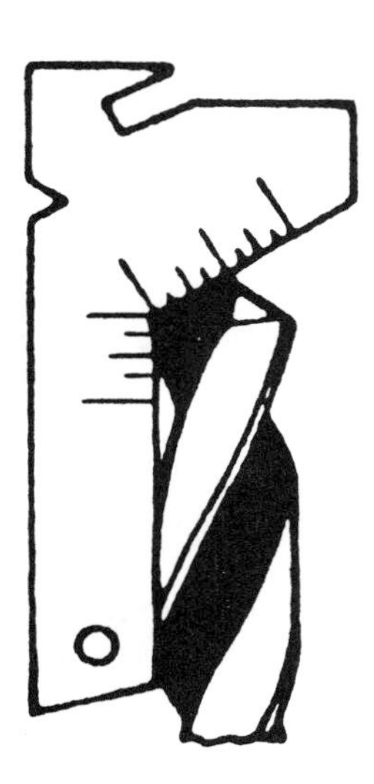

Figure 48. Checking Lip Angle and Length.

4. **Angle of dead center**- As stated, if the first three measurements are correct, the angle of dead center will also be correct. As shown in Figure 49, the angle should be 120° to 135° with the cutting edge. This angle will decrease as the cutting lip clearance angle is decreased.

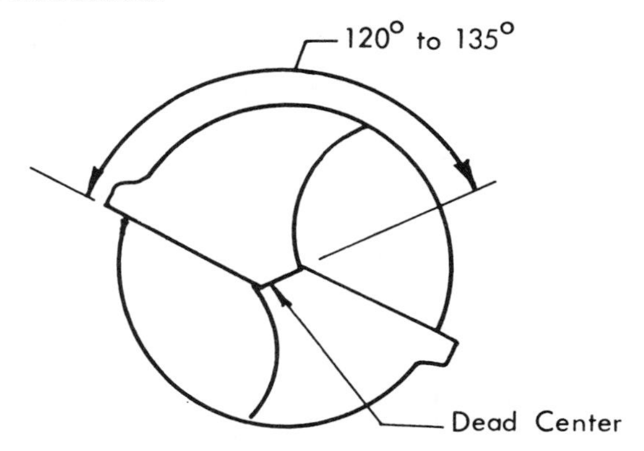

Figure 49. Angle of Dead Center with Cutting Edge.

Drilling- In performing the drilling operation, a number of points must be emphasized.

1. **Drill Speed**- Drill speed is the distance the drill would travel in one minute if laid on its side and rolled. The larger the drill, the greater the distance in feet it would travel. Other things beside the diameter must be considered in speed, these being the type of metal the drill is made from and the hardness of the metal being drilled. Table 8 shows drilling speeds and feeding rates for various drills.

2. **Feed**- The feed of the drill is the distance it cuts into the metal on one turn. It varies with drill size and materials being drilled; however, the drill should be fed into the metal just as fast as it can cut. The feed should be the same for the high speed drills as carbon steel drills. Hand feeding is more difficult to regulate than machine regulated feeding.

3. **Putting the drill in the drill press**- The tapered drill should fit snugly into the drill press spindle. The tang should fit into the slot of the spindle. A straight shank drill may be held in a drill chuck which is held in the drill press spindle. The drill chuck should be tightened using the proper chuck key and the jaws should be tightened in the three places to get a better and more even grip on the drill.

4. **Center punch**- Layout of the hole is important for the quality finished job. Always center punch to help assure that the drill will start in the correct position.

5. **Drilling large holes**- In drilling large holes, a pilot hole should be drilled. The larger the drill, the thicker is the web, and the wider is the dead center. The dead center does not cut; therefore, on large drills, it interferes with the drilling process. The pilot hole drill should be slightly larger than the thickness of the web. The pilot hole also serves as a hole guide if it is located accurately.

6. **Drilling lubricants**- Heat is produced in drilling due to the chips rubbing over the lips of the drill. A lubricant should be used to reduce drawing of the temper from the drill and to wash away the chips. Some metals should not be lubricated such as cast iron and brass. Lubrication here will cause glazing which is making the metal smooth, bright, shiny and hard like glass. Glazing can ruin a drill. Steel should be lubricated. A lard base oil (cutting oil) is most commonly used.

7. **Holding the work piece**- Holding the work piece in place while drilling is important not only for accuracy but also for the safety of the operator. Some of the more common devices or methods for fastening the work in place are strap clamps, C-clamps, parallel clamps, V-blocks and drill press vises.

COUNTERSINKING

Countersinking is required in the use of oval or flat head wood or metal screws and stove bolts. The metal is first center punched and bored the size of the screw shank, interior diameter of the metal screw or the diameter of the stove bolt. Countersinking is best done at slow speeds using a drill press or hand drill. Use Care, countersink only the depth of the head of the screw or stove bolt. Rose countersinks as shown in Figure 50 are the most popular type. They usually have a 1/4 inch shank. They may be purchased in carbon steel for wood or soft metal or high speed for harder metals. Use drilling lubricant in countersinking metal. Comparative angles - drill bits are 118°, countersinks are 82°. Accurate countersinking can only be done with a countersink.

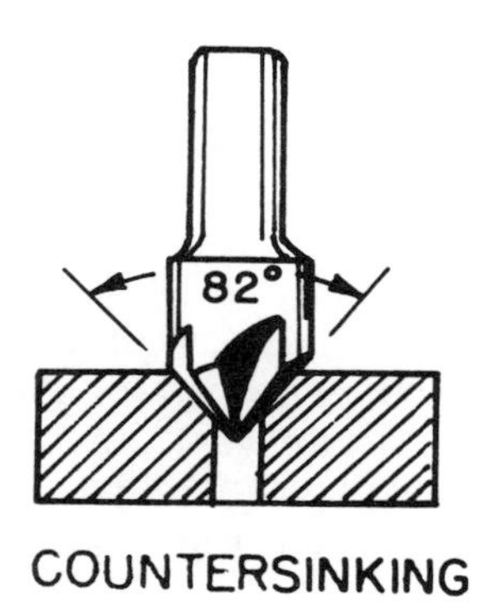

COUNTERSINKING

Figure 50. The Rose Countersink.

POWER DRILLS

Drilling holes is a major operation in construction and repair of equipment. Three drills are usually required to completely meet this need.

Drill Press - Bench or floor models are available in sizes from 14 to 20 inches. This measurement refers to the maximum diameter of a circle in which a hole may be drilled in the center. A 3-jaw key chuck with a capacity from 0 to 1/2 inch straight shank bits is standard equipment. If there is major need for drilling holes larger than 1/2 inch, it is suggested that the drill be equipped with a No. 2 Morse taper spindle. The 3-jaw chuck will be fitted on a Morse taper shank to handle drill size up to 1/2 inch. Larger drills will have Morse taper shanks that fit directly into the Morse taper spindle. One-half inch round shank drills are available in sizes from 1/2 to 1 inch for limited service. Step pulley models usually give spindle speeds from 425 to 5500 RPM with 1725 RPM motors. Spindles driven by variable speed pulleys usually have speed ranges from 350 to 4700 RPM. The slower speeds are desired in most metal drilling operations. The speed of variable speed pulleys should be changed only after starting the motor.

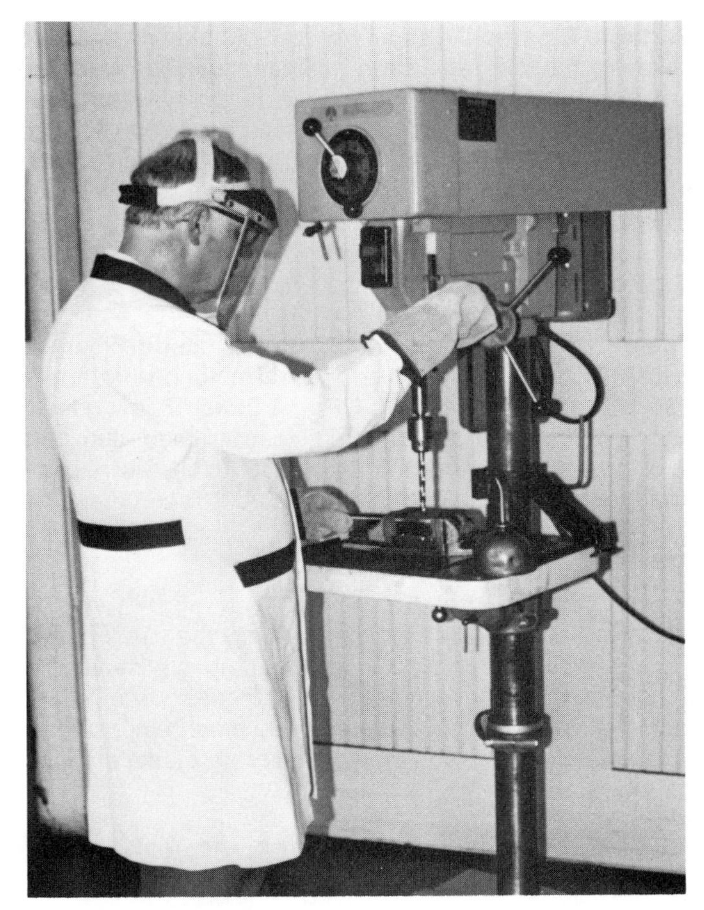

Figure 51. A 15-Inch Drill Press with Variable Speed Pulley and 3-Jaw 1/2" Key Chuck.

TABLE 8. DRILLING SPEEDS AND FEEDS (High-Speed Steel Drills)

SPEED IN REVOLUTIONS PER MINUTE (RPM) FOR HIGH-SPEED STEEL DRILLS
(Reduce RPM One-Half for Carbon-Steel Drills)

Diameter of Drill	Low-Carbon Steel Cast Iron(soft) Malleable Iron	Medium-Carbon Steel Cast Iron(hard)	High-Carbon Steel High-Speed Alloy Steel	Aluminum And Its Alloys Ordinary Brass Ordinary Bronze	Feed Per Revolution Inches
	80-100 Ft. Per Min.	70-80 Ft. Per Min.	50-60 Ft.Per Min.	200-300 Ft. Per Min.	
1/8"	2445-3056	2139-2445	1528-1833	6112-9168	0.002
1/4'	1222-1528	1070-1222	764-917	3056-4584	0.004
3/8"	815-1019	713-815	509-611	2038-3057	0.006
1/2"	611-764	534-611	382-458	1528-2292	0.007
3/4"	407-509	357-407	255-306	1018-1527	0.010
1"	306-382	267-306	191-229	764-846	0.015

Portable Electric Drills - Two portable electric drills are essential tools in repair and maintenance. The following are suggested specifications of standard duty, intermittent service drills: 3/8 inch variable speed drill - 0-1000 RPM by varying finger pressure; 3/8 inch 3-jaw geared chuck; needle and ball thrust bearings; single pole variable speed trigger switch; about 2.4 amps at 115 volts. Double insulated or a grounding system with 10 feet of 3 wire cable and grounding plug.

The 1/2-inch Portable electric drill as illustrated in Figure 53 is a 1/2 inch reversible drill - Universal motor no load speed 600 RPM, load speed 300 RPM; 1/2 inch 3-jaw geared chuck; ball thrust bearing on spindle; polished aluminum housing; about 5.0 amps at 115 volts; 10 feet 3 wire cable and grounded plug or double insulated is recommended.

METAL FILES

Parts of files - All files have essentially the same parts - Figure 54. One exception is that the tang is eliminated in handy fills. Safe edges, types and coarseness of cut, shape and length are common variations that must be considered in selecting a file. Common types of metal files are shown in Figure 55.

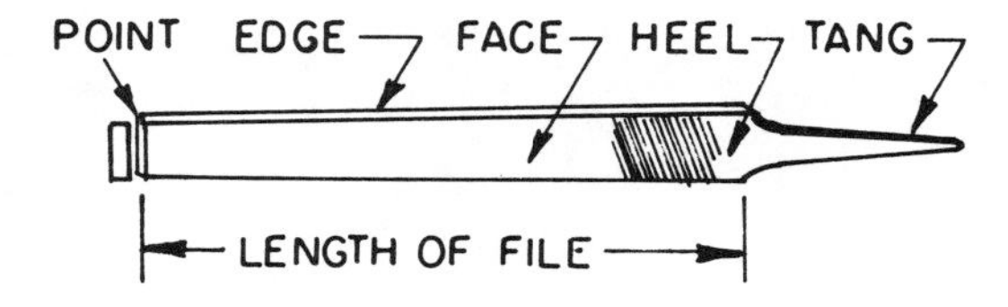

Figure 54. Parts of Hand File.

HALF - ROUND METAL FILE

They vary in length from 6 to 12 inches. Their teeth are finer than wood files or rasps. Bastard is a common degree of coarseness.

Figure 55 A.

HAND METAL FILE

These files taper only in thickness. They are available in lengths from 6 to 14 inches and in bastard, second cut and smooth degrees of smoothness.

Figure 55 B.

Figure 52. A 3/8-Inch Portable Electric Drill is One of the Most Used Shop Tools.

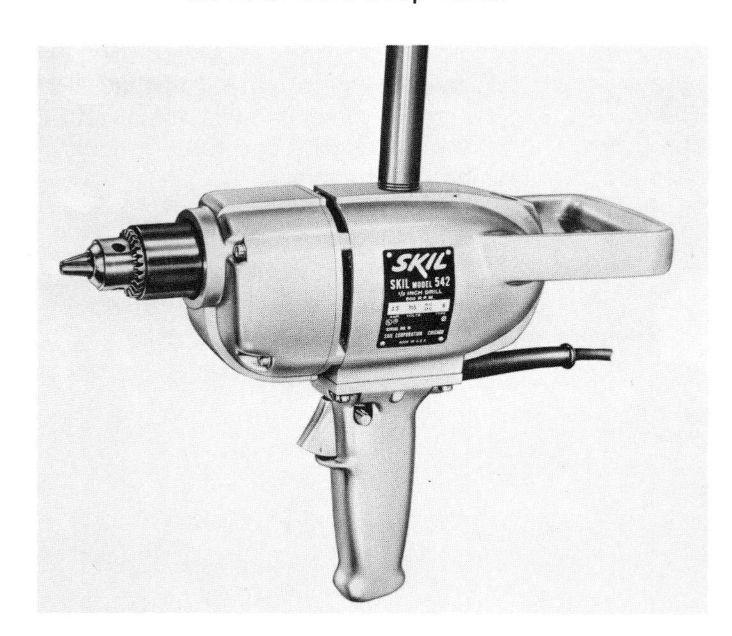

Figure 53. The 1/2 Inch Portable Electric Drill.

MAGICUT FILE

This unique pattern has narrow chip breakers created by steep angled serrations. This gives rapid removal of metal and smooth finish. Excellent for all types of metals and plastics where speed in removal of material is desired.

Figure 55 C.

MILL FILES

These files are used for sharpening circle saws, draw filing and finishing metal. All sizes are tapered slightly in width and are single cut on sides and edges. They are available in bastard and second cut and in lengths from 6 to 12 inches. A 10 inch mill bastard is perhaps the most used metal file.

Figure 55 D.

HANDY FILES

Handy files are single cut on one side for sharpening edge tools and double on other side for rapid removal of metal. One edge is cut, the other is safe or uncut. Eight inch is usual length measured exclusive of handle. This file is often chosen for sharpening garden tools and for use about the home.

Figure 55 E. Common Metal Files.

SELECTION OF FILES

1. **Classification and use of files**

Type of Cut (Figure 56)	Coarseness of Cut# (Figure 57)	Shape ## and Use (Figure 58)
Single	Bastard	Mill - General tool sharpening and draw filing
Double	2nd cut	Flat - General coarse cutting of metal and also for wood
Rasp*	Smooth	Hand - Same use as flat but has one safe edge**
Milled tooth***		Square - or rectangular holes Taper**** Sharpen saws, for angles 60 - 90° Round -Filing curved sufaces Half-round filing curved surfaces Knife - Filing narrow slots, notches and grooves.

\# Coarseness of cut - also varies with the length of the file. Most files are available in 4, 6, 8, 10, 12, 14 inch lengths which gives considerable range in coarseness of teeth. Length does not include the tang.

\## Many files are available in either blunt or taper shapes. In addition, there are numerous special purpose files such as the auger bit file, lead float, brass, aluminum, swiss pattern, rotary and disc.

* Rasp- for very coarse cuts usually for wood.

** Milled tooth (curved tooth) - for cutting soft metals such as lead, solder, and aluminum

*** Safe edge - One edge has no teeth. Thus a hand file is useful where a flat file cannot be used. Other special edges are available such as one round edge or two round edges. Mill-bastard files with rounded edges are desirable for sharpening circle saws.

**** Taper (triangular files) - Available in regular, slim, extra slim, double extra slim and blunt shapes.

2. **Specifications for purchase include:** length, coarseness, type of cut and shape. Example - 12", bastard, single cut, mill.

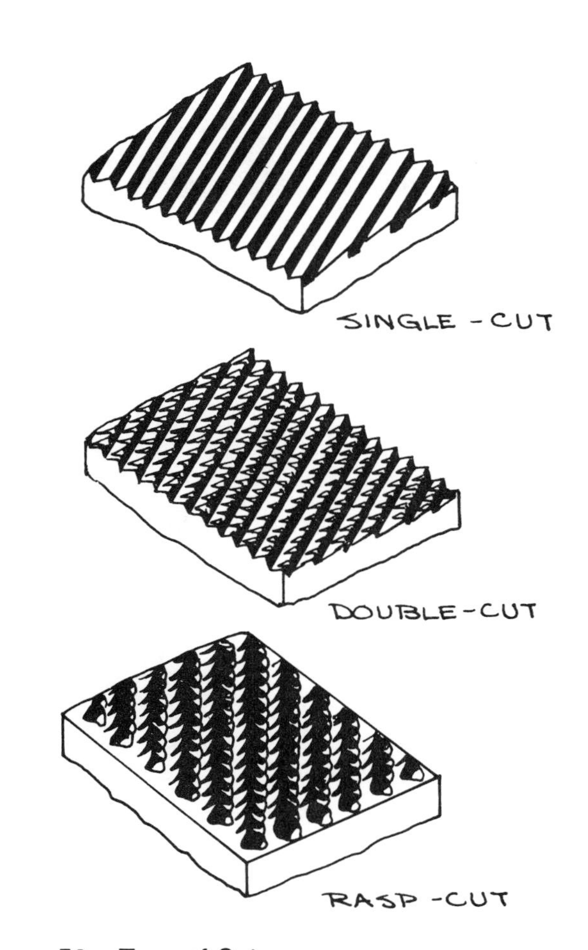

Figure 56. Type of Cut.

3. **Choosing a file**
   a. For rough, fast cutting - use a 10" or 12", bastard, double cut, flat or other shape as needed.
   b. For smooth, fine work - use an 8" or 10" bastard, single cut or other shape as needed.
   c. For smooth, highly polished work - use a 6" or 8", second cut or smooth, single cut, mill or other shape as needed.

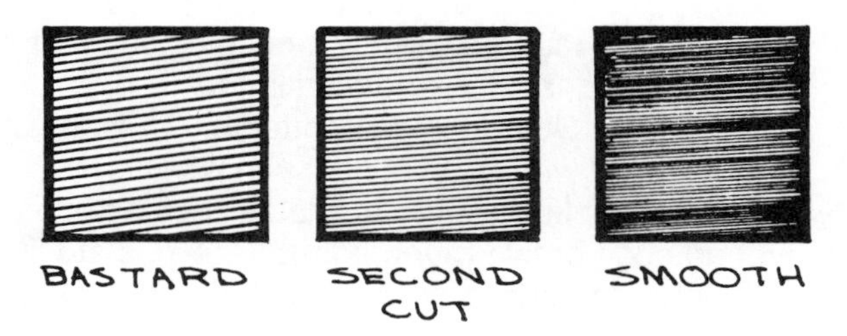

Figure 57. Coarseness of Cut.

PROCEDURE FOR FILING

1. Select correct file for the job, note Figure 58.

2. Fit the file with the proper handle, see Figure 59.

3. Clamp material close to vise jaws to prevent vibration.

4. Apply pressure only on the forward stroke.

5. Use just enough pressure to keep the file cutting.

6. Use a full steady stroke - not too fast.

7. Raise the file slightly on the return stroke to prevent dulling when filing hard material. In the case of soft non-ferrous metals, draw the file back along the metal as an aid to cleaning; a side-sweeping motion will also help in removing cuttings.

8. Draw filing
   a. Single cut file is drawn across the surface in a perpendicular fashion.
   b. Relieve the pressure on the back stroke as in ordinary filing, except in the case of soft metals. The angle at which the file is held with respect to its line of movement must vary with different files.

9. Check for smoothness and squareness with a straight edge or square.

10. When filing soft metal, it may be necessary to rub the cutting edges with chalk to help prevent clogging.

11. Once the file becomes clogged, it should be cleaned with the file card, using a small scribe to pick out lodged cuttings, see Figure 60.

12. Store files so that the cutting edges are not in contact with each other or other tools.

13. Keep files clean, dry and free from oil.

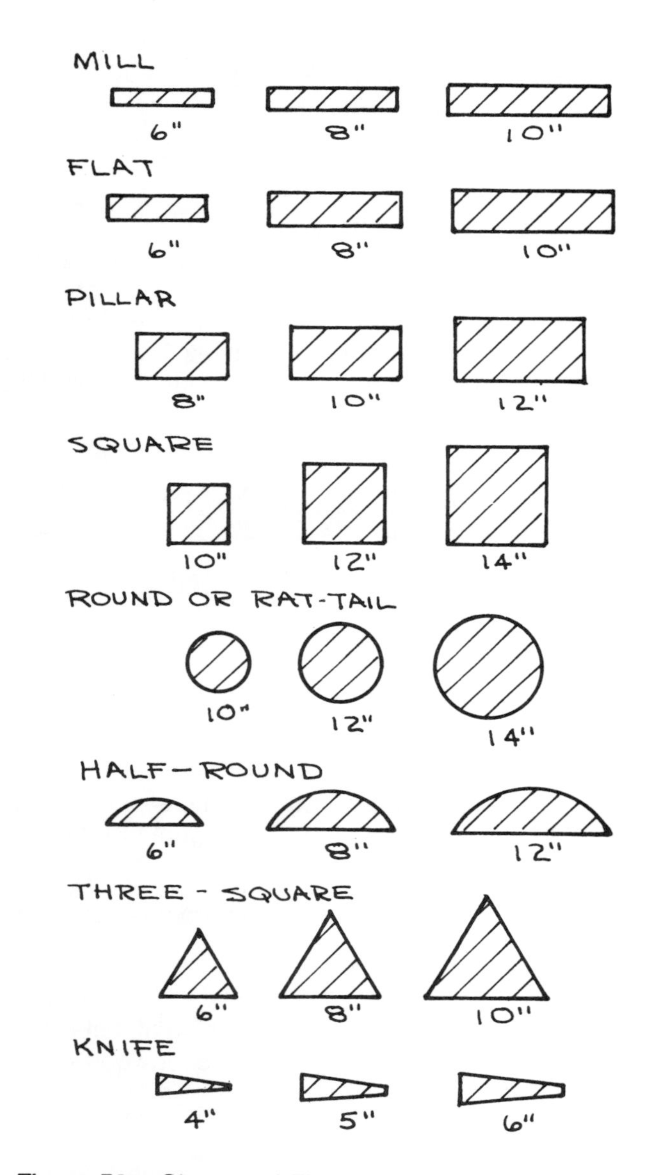

Figure 58. Shapes of Files.

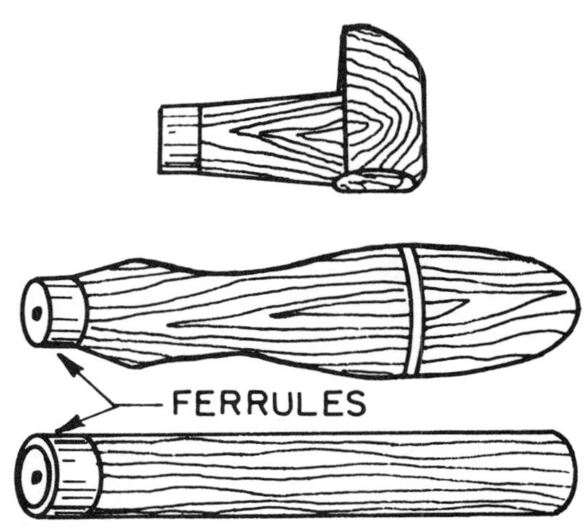

Figure 59. File Handle

File Card and Brush - The teeth of a file must be kept dry, free of oil and clean if the file is to cut satisfactorily. Moisture causes rust and oil results in the teeth filling with metal cuttings and foreign material. The File card and brush shown in Figure 60, if used freely, reduces the dulling of the teeth and results in the faster cutting of the file. The cutting life of the file is also increased.

Figure 60. File Card and Brush.

THREADING AND TAPPING

Joining materials by mating external and internal threads is an applicatiion of one of the basic machines of Physics. There are many types of threads used for specific service. Only the most common types will be covered here. Since taps and dies are hardened and tempered metal cutting tools, the user must remember to identify the metal being worked before breakage occurs.

SELECTION OF THE SCREW PLATE (Taps and Dies)

Figure 61. The Screw Plate Tap and Die Set.

1. Care should be used in the selection of the correct type of thread for the job at hand. The hardness of the steel is of prime consideration in making a choice of thread. Fine threads are cut in the harder metals. Since threads are cut forming a 60 degree angle, the finer threads have more threads per inch (pitch).

2. Standard taps in common use
   a. **National Coarse (N.C.)** - formerly called United States Standard (U.S.S.). Used on soft or mild steel. Common in farm or general repair shops.
   b. **National Fine (N.F.)** - formerly know as Society of Automotive Engineers (S.A.E.). Used on harder and higher carbon, cold drawn or rolled steel.
   c. **Machine Screw** - both fine and coarse for small diameters up to 1/4".
   d. **American National Pipe Thread (N.P.T.)** - differs from the above in that the threads are tapered 1/16" per inch of threads. This causes the threads to bind in tightening, forming a water tight joint.

3. National Coarse and National Fine screw plates, as shown in Figure 61 are sold in sets. The set may be N.F. or N.C. or a more complete set made up of both types. Diameters may range from 1/4" to 1-1/2" depending on completeness of the set. The number of threads per inch of N.F. and N.C. varies with different diameters. Taps and dies are marked with manufacturer's name, size, type of thread and number of threads per inch. A die marked 1/4" - 20 N.C. will cut 20 N.C. threads per inch on 1/4" stock.

USE OF TAPS

Taps are tools used to cut internal threads in nuts, pipe, steel plate, etc.

1. Center punch, then drill a hole in the material slightly larger than the minor diameter. (A hole theoretically equal to the minor diameter of the screw to work in the tapped thread would, however, require that the cut be full depth of thread which is not practical since too much force would be required and breakage would be excessive.)

   a. Major diameter - commonly know as the outside diameter of the thread.
   b. Minor diameter - root diameter.
   c. Three-quarters of the full depth (75% thread) will provide sufficient strength (equal to 95% of a full, 100%, thread) for ordinary commercial purposes, since the bolt will usually break before the thread will strip. Note Figure 62 showing percent of thread.

2. Determine the tap-drill size by one of the following methods:
   a. Subtract the pitch from the major diameter of the screw (or tap) and select the drill of a size

nearest under this value (1/2 - 1/13 gives 27/64 as the nearest fraction, or to figure by decimal equivalent, .500 - .077 equals .423).

b. Drill size = tap size - $\frac{*1.3 \times (\%\ thread)}{Pitch\ (No.\ threads\ per\ inch)}$ (For N.C.)

$$= 1/2" - \frac{1.3 \times .75}{13}$$

$$= 1/2" - \frac{.975}{13}$$

$$= 27/64$$

*For N.F. Use a factor of 1.4

OR

c. Use the table on page 85. It is not always necesary to calculate the size of a tap drill to use with each tap size since this information is often available in table form.

3. Select the correct tap for the job. Taps are of three types:
   a. **Taper**
   b. **Plug**
   c. **Bottom**

4. Insert the tap in the tap-wrench (always start with taper tap) and tighten securely.

5. Clamp the material securely in a vise. See that the tap enters the hole straight.

6. Use thread cutting oil or emulsion to prevent overheating.

7. Using a taper tap for thru-holes, cut the threads in the tap-drill hole.

8. **Do not force**- if the tap becomes clogged with cutting, back up just enough to break them loose, otherwise excess breakage will occur. The tap should be backed off a half turn every turn or two.

9. When tapping blind hole, use a tapered tap followed by a plug tap, then a bottom tap.

USE OF DIES

Dies are used for cutting outer threads on round stock, pipe, etc. Two types of dies are in use; **adjustable round split**, and **two-part dies**, with collet and guide. The latter is preferred.

1. Pick a die of the same size as the stock to be threaded.

2. Check to see that the die is adjusted properly to size as indicated by the gage marks on the face of the die.

3. The screw guide must be sufficiently tight to hold the die-halves securely. Depth adjusting screws are found on the outside diameter of the collet.

4. Place the die in the die stock with the screw guide inserted first. Lock the die with the set screw.

5. Slightly chamfer the end of the material to be threaded so the die will start easily.

6. Pass the stock through the guide and turn the die stock in a clockwise direction for right hand threads. If no guide is provided, the die must be aligned by sight.

7. Use thread cutting oil or emulsion.

8. When it is necessary to cut full threads up to the shoulder of a bolt, the die is backed off and reversed so that the tapered end and guide point away from the work.

CAUSES OF POOR THREADING

1. Feeding too fast.
2. Uneven pressure.
3. Dull dies or taps.
4. Stock too large or dies too small.
5. a. Load too heavy tending to strip threads.
   b. Tendency to make 100% threads.
6. Lack of cutting oil or emulsion.
7. Under size material - flat shallow threads.
8. Stock too hard or out-of-round.
9. Guide not tight against dies.

TAP DRILL SIZES (To PRODUCE APPROX. 75% FULL THREAD)

National Coarse (N.C.) (Formerly U.S.S.)				National Fine (N.F.) (Formerly S.A.E.)			
Tap		Drill Size inches		Tap		Drill Size inches	
Diam.	Threads per inch	Decimal	Nearest Fraction	Diam.	Threads per inch	Decimal	Nearest Fraction
1/4	20	.201	13/64	1/4	28	.213	7/32
5/16	18	.257	1/4	5/16	24	.272	9/32
3/8	16	.3125	5/16	3/8	24	.332	21/64
7/16	14	.368	3/8	7/16	20	.3906	25/64
1/2	13	.4219	27/64	1/2	20	.4531	29/64
5/8	11	.5312	17/32	5/8	18	.5781	37/64
3/4	10	.6562	21/32	3/4	16	.6875	11/16

INFLUENCE OF TAP DRILL SIZE

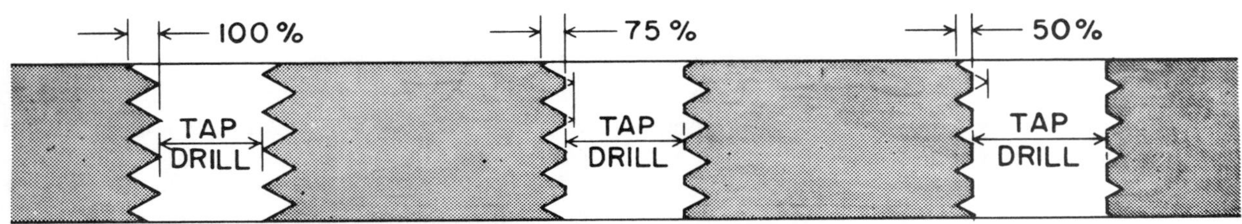

Full threads takes more effort to cut. It is hard on taps and is only 5% stronger than a 75% thread

75% thread is recommended standard for most work. except machine tapping, which is done best with 50-60% thread

50% thread is 80% as strong as 75% thread. It is easy to cut and good for all jobs except thin work less than one tap diameter thick.

Figure 62. Percent of Thread.

GRINDING WHEELS

Abrasives are probably man's earliest tool. Hard stones were used to shape softer stones. Sand was used to polish by rubbing. Biblical references site the use of abrasives by Moses and Solomon.

Natural abrasives in the form of quarried "grindstones" (actually sandstone) are familiar at antique displays. The important difference in use between the old abrasive wheel and the modern wheel is speed of rotation. Accidents have happened when the natural abrasive sandstone wheels became powered by mechanization with the tragic result of an exploding grinding wheel.

Today's grinding wheels are man-made to operate at maximum speeds that make them very efficient tools. Quality control in manufacturing makes modern grinding wheels uniform. However, the buyer must still be able to specify the product he wants for the job he wants to do.

Grinding wheels are made in many combinations of coarseness and hardness to meet the variety of conditions under which they are used. These combinations are known as "Grain" and "Grade". Grain is the size or number of abrasive grain and is used to determine the wheel's degree of coarseness. Grade is the degree of hardness of the bonding agent that holds the abrasive particles together.

GRAIN SIZE

The numbers used to designate the **grain size** refer to the meshes per lineal inch in the screens over which the crushed abrasive is passed. For example, No. 24 grain passes through a screen having 24 meshes per lineal inch or 576 per square inch and its retained on a screen having 30 or more meshes per lineal inch. The following are common grain sizes:

Very Coarse	Coarse	Medium
8-10	12-14-16-20-24	30-36-46-60
Fine	**Very Fine**	**Flour Sizes**
70-80-90-100-120	150-180-200-240	280-320-400-500-600

GRADES

Grade is a term used to indicate the hardness of a wheel. It represents a measure of the strength of the bonding agent holding the abrasive grains together. It does not express in any way the hardness of the abrasive itself. A grinding wheel may be made of a very hard abrasive and still be a very soft wheel. The following indicates common grades of Vitrified process grinding wheels. Wheels with this type of bonding agent are supplied unless otherwise specified.

Very Soft	Soft	Medium
G	H-I-J-K	L-M-N-O

Hard	Very Hard
P-Q-R-S	T-U-W-Z

The harder grades cut fast and do not wear down as rapidly but tend to heat the material much more than the softer grades. It is a common practice to select wheels of medium or soft grades for tool grinding. Grades from I to O work well as tool grinders while grinding wheels of grades from P to T are usually used on the large grinders in welding rooms and in similar places where fast cutting is desirable.

TYPES OF BONDING AGENTS

There are five distinct types and kinds of bonds used in wheel manufacturing.

1. **Vitrified Wheels** - (Ceramic Clay) for general purpose work. The majority of wheels on the market are of this type. This type of wheel is supplied unless specifications call for another type.
2. **Silicate Process Wheels** - (Silicate of Soda) used largely in cutlery field. Wheels run cool.
3. **Rubber Bonded Wheels** - thin wheels for cut-off work, or wide wheels for snagging. Run at higher speeds.
4. **Shellac Bonded Wheels** - used for stone cutting and finishing of cast and hardened steel. Produces a fine finish.
5. **Resinoid Bonded Wheels** - (Synthetic Resin) ideal for cutoff and saw gumming. May be used in stone cutting, and metal finishing.

ORDERING GRINDING WHEELS

In ordering , the diameter of the wheel, the width of the face and the diameter of arbor must be given. Wheels with straight sides and faces are supplied unless specified otherwise. A wide variety of sizes and shapes are available, note Figure 63.

Grinding wheels should be handled carefully so that the edges are not chipped or damaged, as this throws the wheels out of balance and thus causes them to vibrate when put into use. The ginding wheel dresser should be used to maintain a desirable shape. The maximum operating speed, which is indicated on all wheels, should be carefully checked. Some serious accidents have occurred due to operating the wheels at over the maximum recommended speed.

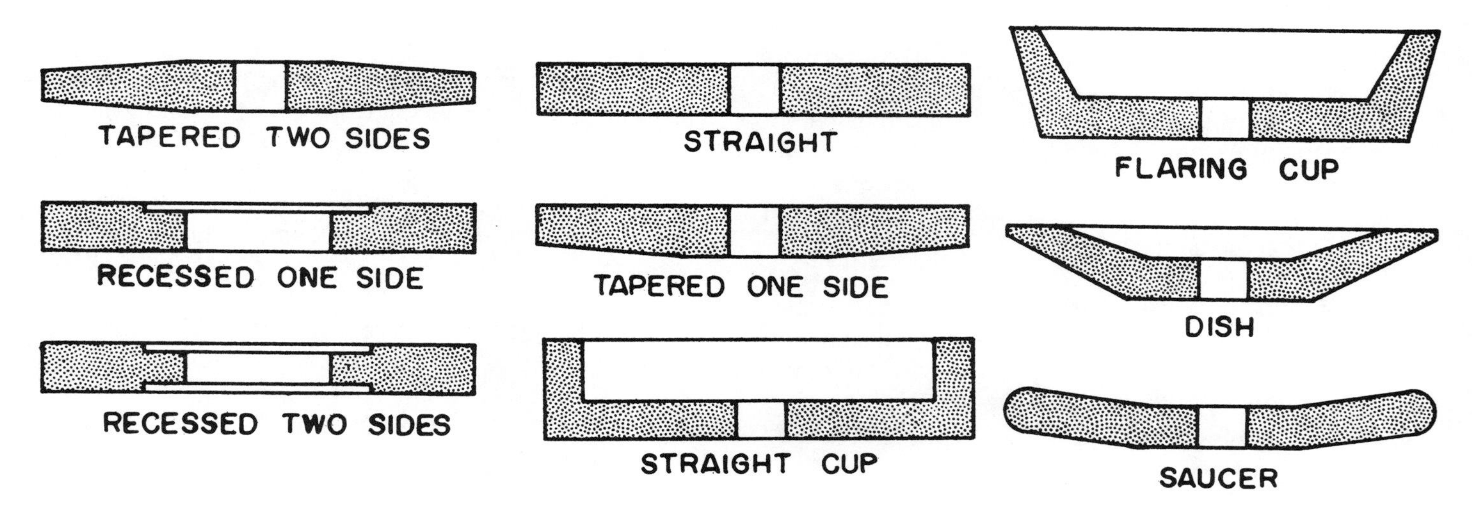

Figure 63. Grinding Wheel Shapes.

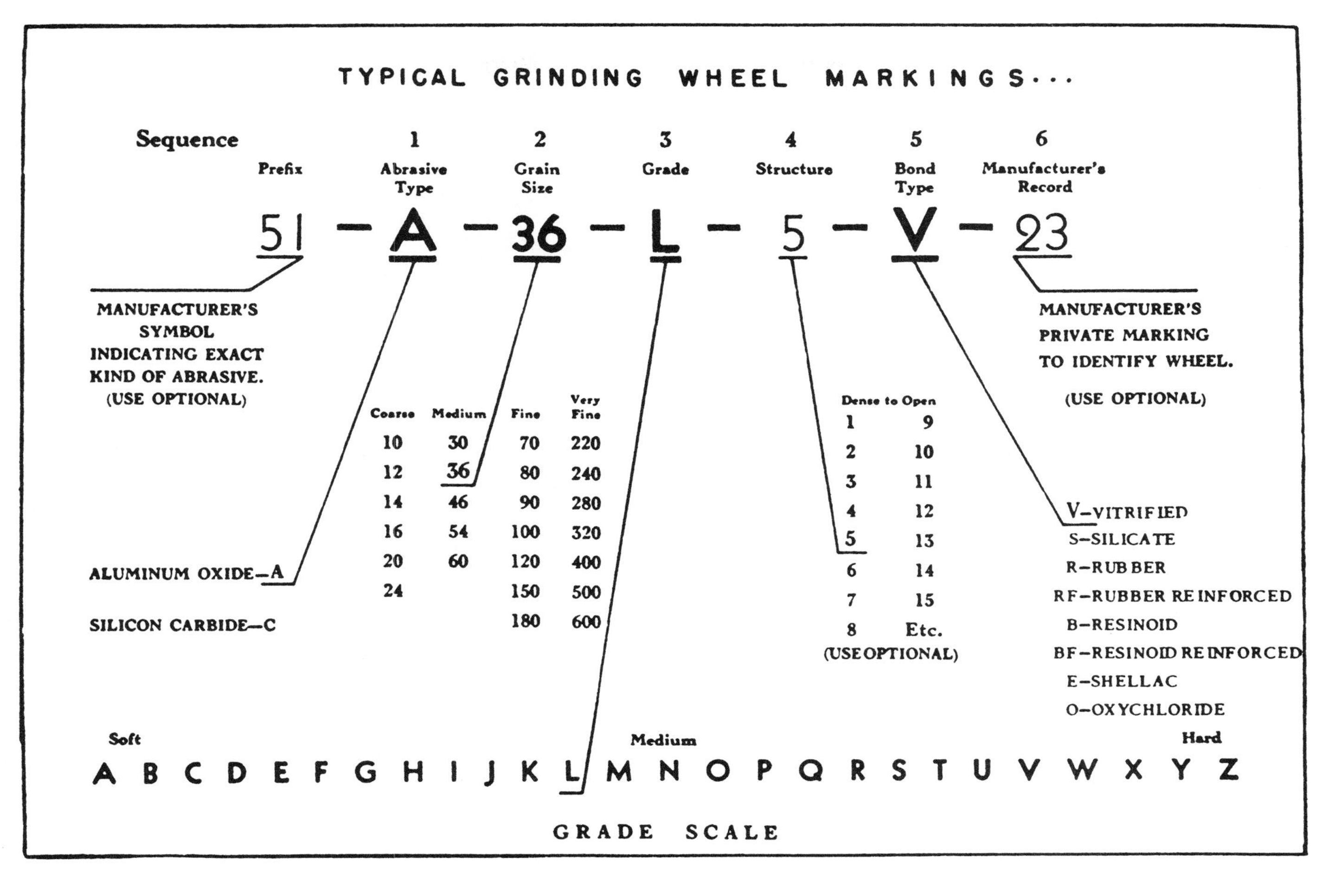

Figure 64. Grinding Wheel Markings.

GRINDING WHEELS ARE SAFE...USE BUT DON'T ABUSE

DO AND DON'T

1. Do always handle and store wheels in a careful manner.
2. Do visually inspect all wheels before mounting for possible damage in transit.
3. Do check maximum operating speed established for wheel against machine speed.
4. Do check mounting flanges for equal and correct diameter. (Should be at least 1/3 diameter of the wheel and relieved around hole.)
5. Do use mounting blotters supplied with wheels.
6. Do be sure work rest is properly adjusted. (Center of wheel or above; no more than 1/8" away from wheel.)
7. Do always use a guard covering at least one-half of the grinding wheel.
8. Do allow newly mounted wheels to run at operating speed, with guard in place for one minute before grinding.
9. Do always wear safety glasses or some type of eye protection when grinding.
10. Do turn off coolant before stopping wheel to avoid creating an out-of-balance condition.

1. Don't use a wheel that has been dropped.
2. Don't force a wheel onto the machine or alter the size of the mounting.
3. Don't ever exceed maximum operating speed established for the wheel.
4. Don't use mounting flanges on which the bearing surfaces are not clean and flat.
5. Don't tighten the mounting nut excessively.
6. Don't grind on the side of the wheel unless the wheel is specifically designed for that purpose.
7. Don't start the machine until the wheel guard is in place.
8. Don't jam work into the wheel.
9. Don't stand directly in front of a grinding wheel whenever a grinder is started.
10. Don't grind materials for which the wheel is not designed.

EXAMPLE OF GRINDING WHEEL SPECIFICATION AND INTERPRETATION See Figure 64.

Carborundum Aloxite - Operating Speed 3600 R.P.M.
A-60-M-5-V 7 x 1 x 5/8

3600 R.P.M. - Maximum safe speed.

Position

1	Abrasive Type	A - Aluminum Oxide
2	Grain Size	60 - about medium grit
3	Grade	M - about medium grade
4	Structure	5 - fairly dense
5	Bond Type	V - Vitrified (Ceramic Clay Bond)

7 x 1 x 5/8 - 7" diameter of wheel, 1" width of face, 5/8" diameter of arbor.

POWER GRINDERS

A bench or pedestal mounted grinder should be one of the first pieces of power equipment purchased. The following are suggested minimum sizes and specifications:

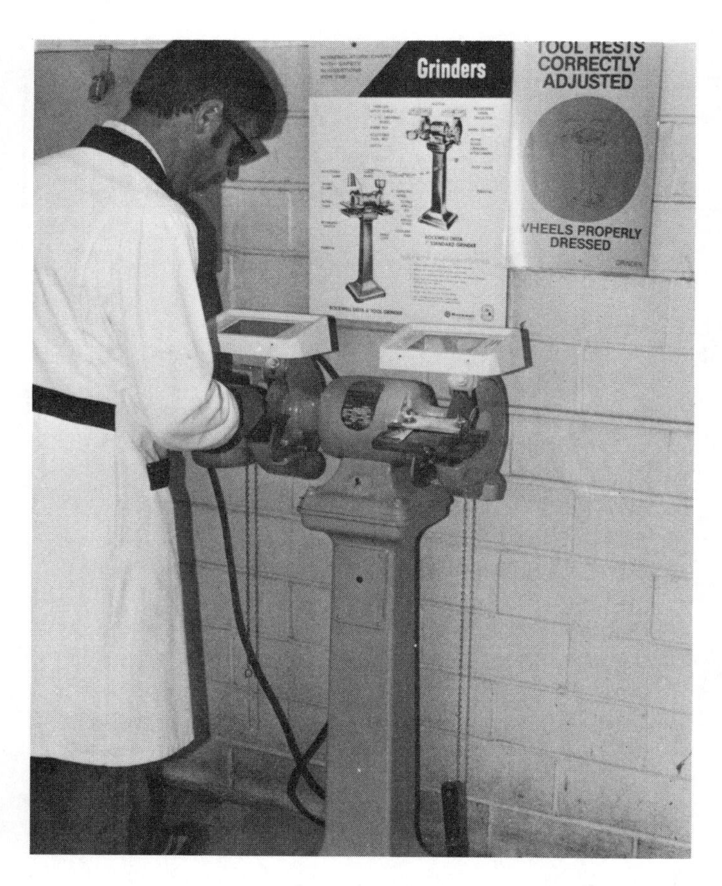

Figure 65. This Pedestal Tool Grinder with Twin Lamps and Shatter Proof Eye Shields Will Sharpen Tools and do a Limited Amount of Maintenance Grinding.

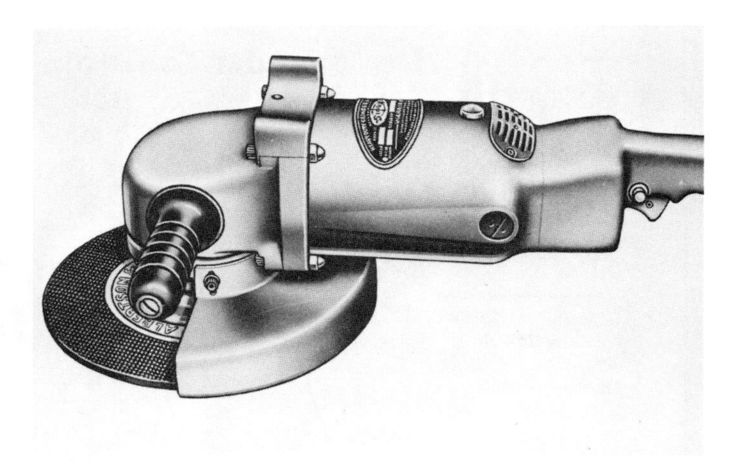

Figure 66. The Angle Grinder With its Depressed Wheel is used Extensively for Veeing Metal in Preparation for Welding and for Smoothing Welded Surfaces.

Bench or Pedestal Grinder - 1/2 h.p. - single phase, 115/230 volt, 1725 r.p.m., capacitor start, sealed ball bearings—7" x 1" x 5/8", wheels with guards, tool rests and safety, illuminated eye shields. The illuminated shatterproof shields throw light directly on grinding wheel and work and protect the operator's eyes. The above grinder is primarily a tool grinder. If a larger grinder with 3/4 h.p., 8" x 1" wheels or 1 h.p., 10" x 1" wheels is purchased, equipped with coarse and medium-fine wheels, the grinders uses will be greatly extended. See Figure 65.

Portable Electric Grinders - These grinders are very desirable since they can be taken to the work. They may be used for wire brushing and buffing in cleaning metal surfaces. They are also used in removing rust scale paint and grinding off rivets and bolts. They are used extensively for veeing metal in preparation for welding and for smoothing welded surfaces. The angle grinder is preferred for the latter operation. Suggested specifications:
Heavy-duty Angle Grinder - depressed center wheel 7" x 1/4", 8000 RPM, guard assembly, spindle lock, clamp washer and spanner wrench. See Figure 66.

Portable Grinder - Wheel size 5" x 3/4" x 1/2" - no load speed 4500 RPM, sealed ball bearings, enclosed switch, shaped grips, adjustable wheel guard. Weight about 13 pounds. See Figure 67.

COATED ABRASIVES

A coated abrasive consists of a flexible-type backing upon which a film of adhesive holds and supports a coating of abrasive grains. Present day coated abrasives include **flint, emery, crocus, garnet aluminum oxide** and **silicon carbide** held by various types of resins and glues and varnishes on backings such as paper, cloth, vulcanized

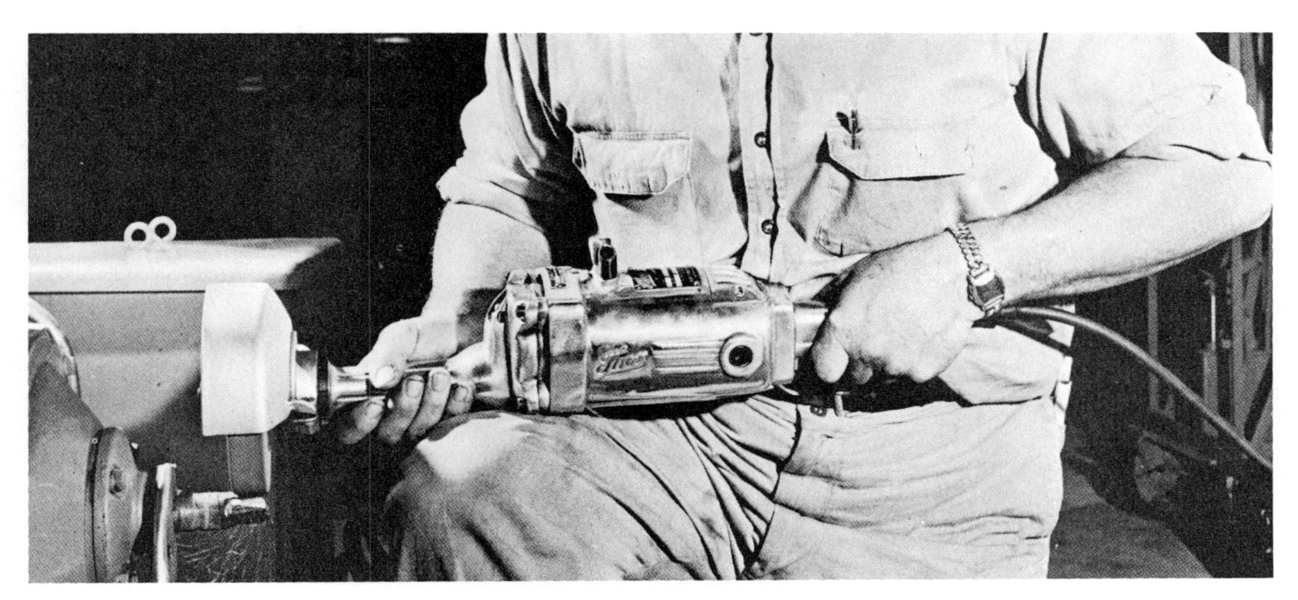

Figure 67. This Portable Grinder is used for Shaping a Drive on a Power Unit.

fiber or combinations of these. The earliest record of coated abrasives goes back to the 13th century in China. The earliest coated abrasives were made of natural abrasive particles. It was not until the man-made abrasives aluminum oxide and silicon carbide were discovered that speed and efficiency became the advantage for using coated abrasives.

The cutting action of a coated abrasive and a grinding wheel are quite similar. The great difference is in the heat build up of the two abrasive devices. Coated abrasives run cooler than grinding wheels. However, coolants are used on grinding where the heat build-up is severe.

Coated abrasives shapes include belts of all sizes, drums, and rigid and flexible discs. Abrasive grinders are specified by size, shape, grain size, type of abrasive, type of backing. Belts are available in almost any length and width. For example, the range of width is said to be from 1/8" to about 88". The belts are arranged over a seemingly endless arrangement of pulleys for industrial operations. The simple two-pulley type found in repair shops provides low cost, safe, versatile grinding and polishing facilities. Figure 68 illustrates the use of a belt grinder.

WIRE BRUSHES AND BRUSHING

Wire brushes for manual and power use are essential tools for shop uses such as weld cleaning, paint removal, rust removal and general cleaning. Hand brushes are available with wood or metal handles that are curved or straight. The wire can be carbon steel, brass or stainless steel. Brass is used where sparks are not allowed. Stainless steel is used to clean aluminum or stainless steel. Brushes vary in length of bristles, rows of bristles, length and width of rows. Power brushes are selected with greater care due to greater variety of types and applications.

KINDS OF WIRE- The standard type of wire (fill) is untempered carbon steel, brass and stainless steel.

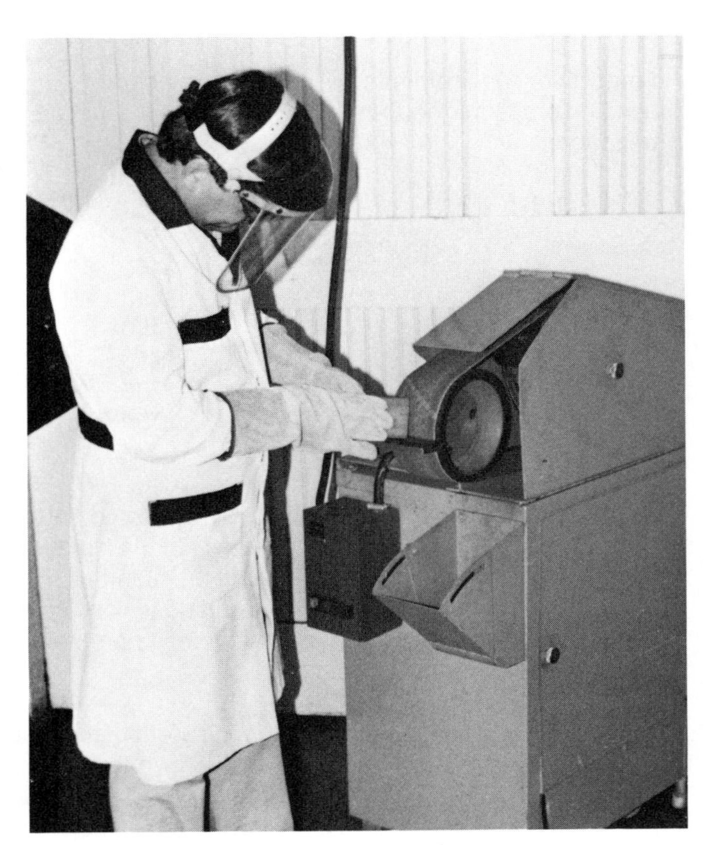

Figure 68. Coated Abrsaive Belt Grinder.

TYPES OF POWER BRUSHES

Radial brushes are circular in shape, in a variety of diameters and face widths. The three styles are crimped wire, straight wire and twisted tuft, the most aggressive.

Cup brushes are available with twisted tuft for heavy-duty work or crimped wire for lighter work.

End brushes are for light or heavy use in confined areas often used with portable tools.

BRUSH DIMENSIONS

1. Outside diameter is the overall diameter. Always use the largest diameter possible. However, portable tools are limited to 8" diameter.
2. Trim (fill) length is the length of the material entending beyond the brush bask or flange. A short trim is more aggressive than the more flexible long trim.
3. Arbor hole is the actual size of the mounting shaft. Arbor hole adapters are available for slower speed wheels.
4. Operating face width will vary with the accessibility. The wider the brush, the greater the efficiency.

POWER BRUSH ENGINEERING FACTORS

Horsepower requirements will vary with the speed, resistance and pressure requirements. The following is for a 1" brush face:

Brush Diameter	Motor Size	RPM
4"	1/4 hp	3450
6"	1/2 hp	3450
8"	3/4 hp	3450
10"	1 hp	1750
12"	1 hp	1750
15"	1 1/2 hp	1750

Higher speed gives greater efficiency.

Pressure Control- Brushes should be operated at the highest possible pressure. The sharp ends of the filaments do the work. Excessive pressure bends the filament. This results in a wiping action rather than a cutting action. It also causes excess flexing of the filament leading to premature breakage and shortened wheel life.

Brush Face Contact- Always use the full width of the face of the wheel to avoid grooving and uneven wear.

Wire or Fill Arrangements- The following are some of the more common fill arrangements for power brushes:

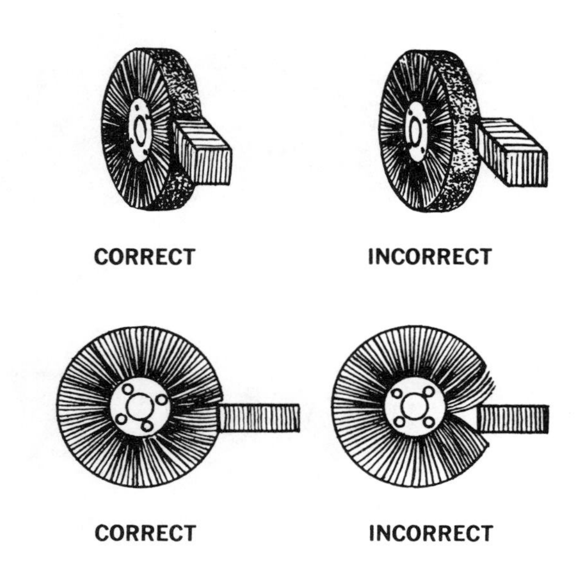

Figure 69. Correct Procedure for Using a Power Brush.

Knot wire wheels consist of groups of wire tightly twisted and firmly anchored around a circular center. These may be arranged in a single row or in specified multiple rows. These wheels provide maximum impact for severe applications.

Straight or crimped wire wheels provide a more gentle cleansing and polishing action.

POWER BRUSH OPERATION

1. Operate the brush with **lightest possible pressure** and at the **highest practical speed**. See Table 9.
2. **Periodic reversing** of the direction of the brush rotation will maintain efficiency by self-sharpening action. The sharp tips of the brush fill do the work.
3. Use the **finest gage wire** that does the work. This increases brush life by reducing wire fatigue.
4. Have the brush contact the work so that the **full brush face** is utilized as shown in Figure 69.

TABLE 9. TABLE OF SURFACE SPEEDS.

(Recommended speeds are given in RPM. The following table gives approximate linear speeds of the brush face in feet per minute for standard diameters at various RPM)

	DIAMETERS							
RPM	2	3	4	6	8	10	12	15
1,000	525	775	1,050	1,575	2,100	2,625	3,150	3,925
1,500	775	1,175	1,575	2,350	3,150	3,925	4,725	5,900
1,750	925	1,375	1,800	2,750	3,650	4,550	5,500	6,800
2.000	1,050	1,575	2,100	3,150	4.200	5,250	6,275	7,850
2,500	1,300	1,950	2,625	3,925	5,250	6,550	7,850	9,825
3,000	1,575	2,350	3,125	4,725	6,275	7,850	9,425	11,775
3,450	1,700	2,600	3,600	5,400	7,200	9,000	11,000	13,500
3,750	1,950	2,950	3,900	5,900	7,800	9,800	11,800	—
4,000	2,100	3,150	4,175	6,275	8,375	10,475	—	—
4,500	2,350	3,525	4,700	7,075	9,425	—	—	—
5,000	2,625	3,925	5,225	7,875	—	—	—	—
6.000	3,125	4,700	6,275	9,425	—	—	—	—
8,000	4,200	6,275	8,375	—	—	—	—	—
10,000	5,250	7,850	—	—	—	—	—	—
15.000	7,850	11,775	—	—	—	—	—	—
20.000	10,450	15.700	—	—	—	—	—	—

CLASSROOM EXERCISE METALS

Name ____

Date ______________________ Grade ______________________

1. Metallurgy is defined as ____

2. Tensile strength and compression strength are considered ______________ properties of metal.

3. Tensile strength is the resistance of metal to being:
   a. pulled apart
   b. pushed together.

4. Circle the below properties of metal that would be considered physical properties.
   a. Toughness
   b. Tensile Strength
   c. Cost
   d. Hardness
   e. Electrical Resistance
   f. Compression Strength
   g. Ultimate Strength
   h. Corrosion Resistance
   i. Brittleness
   j. Tungsten

5. Medium carbon steels have a carbon range of ________ to ________ percent and are commonly used for ____.

6. List 5 metallic elements used for producing alloy steels.

   a. ______________________
   b. ______________________
   c. ______________________
   d. ______________________
   e. ______________________

7. The main alloying elements in stainless steel are ____________ and ____________ which allows it to resist corrosion.

8. Brass, copper, tin, aluminum, silver are considered ____________ metals because they do not contain ____.

9. Metals containing a carbon content of ______________ percent or higher are considered cast iron.

10. Select a type of steel or iron that would be best used for the following applications:

    a. Drill bits ______________________________
    b. Bolts or rivets ______________________________
    c. Engine block ______________________________
    d. Taps and dies ______________________________
    e. Bearings ______________________________
    f. Angle iron ______________________________

11. The ingredients to make Pig Iron include: ____________, ____________, ____________ and ____________.

12. The majority of the steel produced in the U.S. today is produced through the ______________________ ______________________ process.

13. Alloy steels such as stainless steel and tool steel are produced through the ______________________________ ______________________________ process.

14. ______________________________ iron is produced in a blast furnace.

15. Steels such as angle iron or I-beams are produced by
(a) casting
(b) rolling.

16. The classification system for identifying steel was established by ______________ and ______________.

17. A C1035 would have a carbon content of ______________ percent carbon.

18. List tools and equipment that would be used to forge and shape a cold chisel.

a. ______________________________ d. ______________________________
b. ______________________________ e. ______________________________
c. ______________________________ f. ______________________________

19. A hardy fits into the ______________ ______________ of the anvil and is used for ______________ metal.

20. When forging metal it would best be held with a:
a. blacksmith pliers.
b. vise grips.
c. tongs.
d. bare hand.

21. Identify the main parts of the anvil.

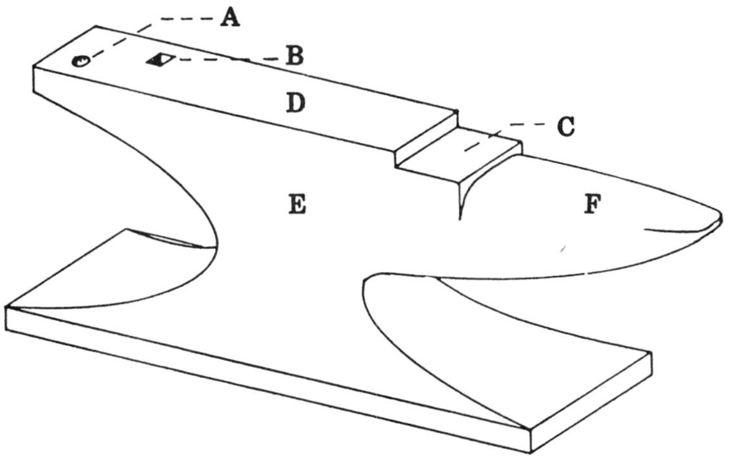

A
B
C
D
E
F

a. ______________________________
b. ______________________________
c. ______________________________
d. ______________________________
e. ______________________________
f. ______________________________

22. The blacksmith's hammer for most forging jobs should weigh approximately ______________ ounces.

23. Metal being forged at a cherry red color would have a temperature of ______________ degrees F or ______________________________ degrees C.

24. Using the temperature data chart, list the degrees that the below conditions take place:

a. Mild steel melts ______________degrees F.
b. Silver soldering range ______________degrees C.
c. Welding mild steel ______________degrees F.
d. Cast iron melts ______________degrees F.
e. Bronze welding range ______________degrees F.
f. Tin melts ______________degrees C.

25. List 5 purposes or reasons for annealing steel during the heat treating process.
    a. ____________________
    b. ____________________
    c. ____________________
    d. ____________________
    e. ____________________

26. The hardening or tempering process consists of three successive operations, list in order.
    a. ____________________
    b. ____________________
    c. ____________________

27. Pure iron is referred to as ____________________.

28. ____________________ is the structure that is formed when carbon steel is rapidly cooled.

29. When steel is heated to its transformation temperature, the grains of ferrite and pearlite become fine grains of ____________________.

30. Rank the following quenching media in reference to their rate of cooling during quenching. No. 1, fast and No. 3 slowest cooling.
    __________ a. Cold water
    __________ b. Brine
    __________ c. Warm oil

31. Surface or ____________________ hardening is a process of heat treatment whereby the surface hardness of a part or tool is increased.

32. List 6 methods for surface hardening steel.
    a. ____________________
    b. ____________________
    c. ____________________
    d. ____________________
    e. ____________________
    f. ____________________

33. Using the liquid carburizing method of case hardening, cooking a piece of round stock for 60 minutes at 1650° F should yield a case depth of ____________ inches.

34. One of the most accurate and most commonly used methods for metal identification is the
    a. file test.
    b. magnetic test.
    c. how used test.
    d. spark test.
    e. fracture appearance test.

35. List the three characteristics or conditions to look for during the spark test.
    a. ____________________
    b. ____________________
    c. ____________________

36. In spark testing, metal producing white colored streamers with a large volume and an average streamer length would be ______________ ______________ steel.

37. Identify the following characteristics for gray cast iron.

a. Color ____________________

b. Streamer length ____________

c. Volume ____________________

d. Number of sprigs ____________

38. The most common penetration hardness test is the ______________________ hardness test.

39. Identify the hand metal working tools illustrated below:

a. ______________________________

b. ______________________________

c. ______________________________

d. ______________________________

e. ______________________________

f. ______________________________

g. ______________________________

h. ______________________________

i. ______________________________

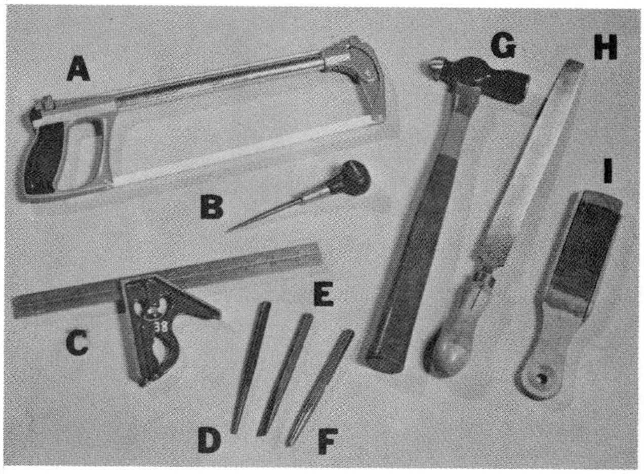

40. Identify the parts of the hammer by matching the names of the parts with the letters on the hammer.

__________ Handle

__________ Face

__________ Wedges

__________ Peen

__________ Eye

__________ Head

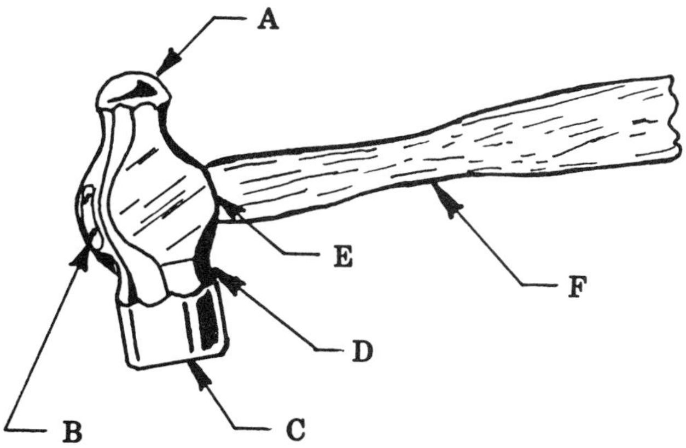

41. List the three common types of hacksaw blade sets.

a. ____________________ b. ____________________ c. ____________________

42. For most sawing jobs in metal work, select a blade having ____________________ teeth per inch and a ____________________ set.

43. List two advantages of the wet-cut band hacksaw over the dry-cut band hacksaw.
a. ____________________
b. ____________________

44. List one use for each of the metal cutting chisels illustrated.

a. ____________________

b. ____________________

c. ____________________

d. ____________________

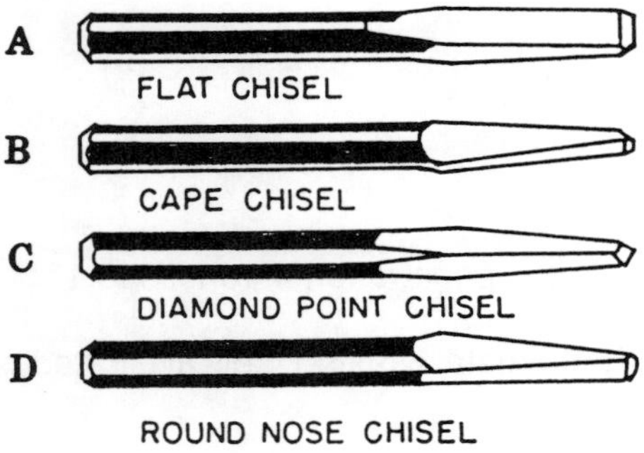

A FLAT CHISEL

B CAPE CHISEL

C DIAMOND POINT CHISEL

D ROUND NOSE CHISEL

45. The cutting edge of the flat cold chisel is sharpened at a ____________________ degree angle.

46. The ____________________ punch is used to establish a point to start a drill bit.

47. Identify the parts of the twist drill bit.

a. ____________________

b. ____________________

c. ____________________

d. ____________________

e. ____________________

f. ____________________

g. ____________________

h. ____________________

i. ____________________

j. ____________________

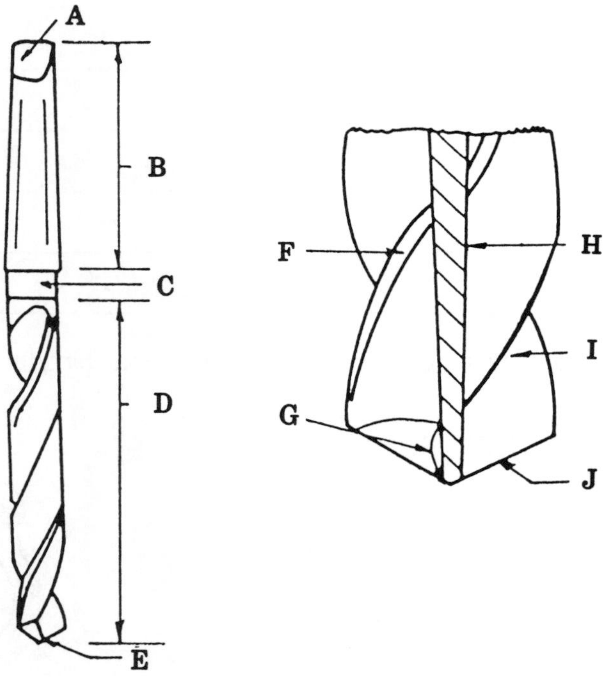

A B C D E F G H I J

48. Drill bits should be sharpened with a lip clearance angle of __________ degrees, lip angles of __________ degrees and angle of dead center of __________ degrees.

49. Twist drills are sized in three classifications, they are:
(a) ____________________, (b) ____________________, & (c) ____________________.

50. The speed for drilling in medium carbon steel with a 1/2" twist drill should be ____________________ to ____________________ RPM.

51. List 4 specifications or characteristics that should be considered when selecting a 3/8" portable electric drill.
    a. ______________________ c. ______________________
    b. ______________________ d. ______________________

52. Identify the 6 parts of the file.

    a. ______________________
    b. ______________________
    c. ______________________
    d. ______________________
    e. ______________________
    f. ______________________

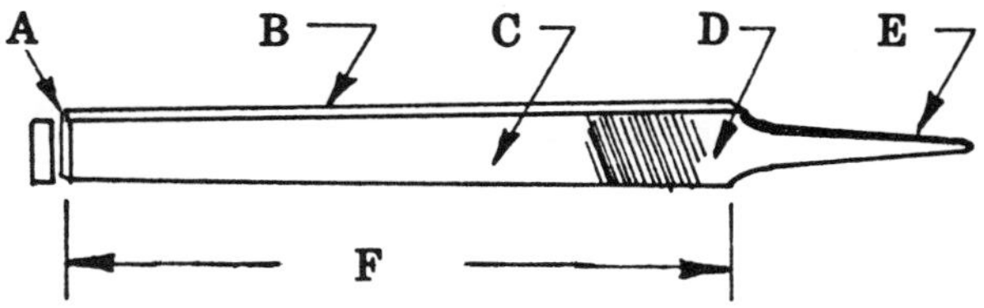

A B C D E F

53. Illustrate the three common coarseness of cuts used on files by filling in the boxes.

SECOND CUT BASTARD SMOOTH

54. Match the shapes of files with names of shapes and letters on the illustrations.

Square ______________________

Round ______________________

Three-square ______________________

Mill ______________________

Knife ______________________

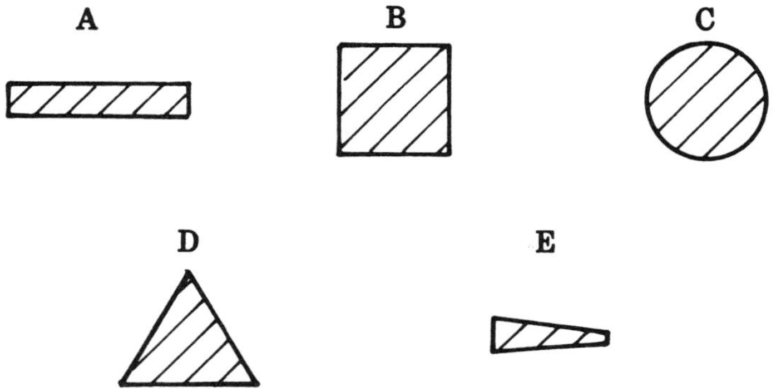

A B C D E

55. When files become clogged, they should be cleaned with a ______________________ ______________________.

56. Standard taps and dies have two classes of treads, these being N.C ______________________ ______________________ and N.F. ______________________ ______________________.

57. Another name for a tap and die set is a ______________________ plate.

58. The percent of thread when tapping threads in a hole is determined by:
    a. the size of the tap.
    b. the size of the drill bit.
    c. the type of tap.

59. The recommended percent of thread for most tap and die work is ______________________ percent thread.

60. Identify the three types of taps illustrated.

A

B

C

a. ______________________________

b. ______________________________

c. ______________________________

61. A die for cutting threads on a piece of round stock has the following number and letters. Identify or define each term.
3/8 ______________________________
NC ______________________________
16 ______________________________

62. Match the following grain size numbers found on a grinding wheel with the name of the grain size.

a. Very Coarse	____________	60
b. Coarse	____________	320
c. Medium	____________	10
d. Fine	____________	150
e. Very Fine	____________	24
f. Flour Size	____________	8

63. Grinding wheels are graded by letters. A letter M would be a _______________grade while a very hard grade could have a letter _______________.

64. The bonding agent for ceramic clay grinding wheels used for general purpose grinding is a ___________________________ bond.

65. Identify or define the grinding wheel markings found on a typical grinding wheel.
A - ______________________________
80 -______________________________
J - ______________________________
4 - ______________________________
V - ______________________________
7 - ______________________________
x
1 - ______________________________
x
3/4 -______________________________
3600 - ______________________________

66. Identify the shapes of grinding wheels illustrated.

a. _______________

b. _______________

c. _______________

d. _______________

e. _______________

A B C D E

67. Discuss where coated abrasives would be used in a metal working laboratory. _______________________

____

68. At 3,750 RPM an 8 inch diameter wire brush will have a face speed in feet per minute of ____________ fpm.

69. List 4 specific safety practices that a person working in a metals lab should follow when using the metal working tools listed.

a. Grinder

____

____

____

____

b. Drill Press

____

____

____

____

c. Gas Forge

____

____

____

____

d. Metal Shear

____

____

____

____

e. Hand Tools

____

____

____

____

70. List 6 items of personal protective equipment that a person working in a metals laboratory must have and use while completing metal projects or activities.

a. ______________________________ d. ______________________________

b. ______________________________ e. ______________________________

c. ______________________________ f. ______________________________

LABORATORY EXERCISES METALS

LAB EX-1 - AX OR HAMMER HANDLE WEDGES OR HASP PINS

1. Material: - Wedges - 1/4" flat mild steel, width and length optional.
   Pins - 5/8" round mild steel, length optional.

2. Construction Procedure:

   a. Wedges
      (1) Heat to a uniform cherry-red to orange-yellow color.
      (2) Work to shape rapidly, starting at the end and working back to approximately 1" taper with a 1/16" to 1/8" tip. Use drawing blows.
      (3) Cut deeply on the hardy, cool and break over the edge of the anvil.
      (4) Finish ends on grinding wheel to 1" length. **Do not finish edges or faces**.
      (5) Grind cutting edge to 60 degree angle.

   b. Hasp Pin
      (1) Heat outer end to an orange-yellow color.
      (2) Starting at outer end draw to about 3/4" square taper. Outer end should be approximately 3/8" square. Use drawing blows.
      (3) Draw to an octagon.
      (4) Finish round by using light blows.
      (5) Finished taper should be 3-1/2" long with 3/16" to 1/4" tip.
      (6) Cut to 4-1/4". Use hardy.
      (7) Using rounded edge of anvil face flatten upper end to 1/4" thickness, 5/8" width and 1" length. Only the upper 3/4" of stock should contact the anvil in forging the flat end.

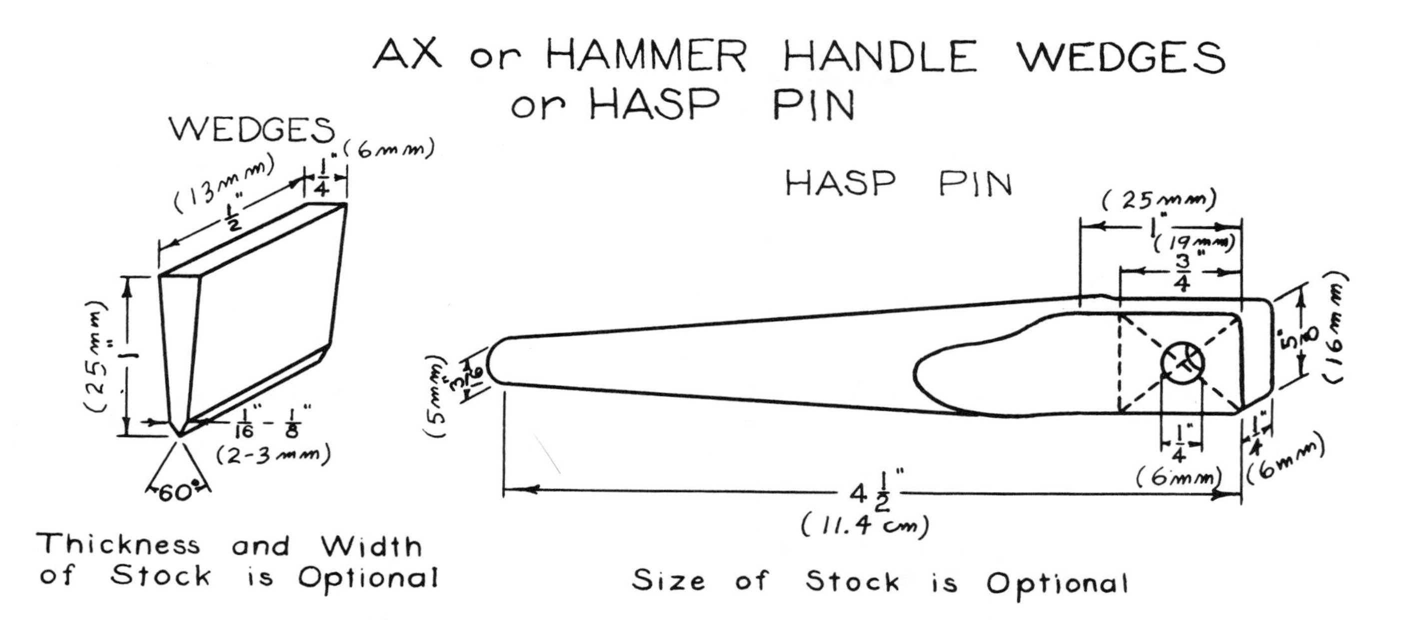

LAB EX-2 - EYEBOLT AND NUT

1. Material - 1 - 3/8" x 9-3/8" (24.0 cm) round mild steel.
   1 - 3/8" x 1"x1" (25 mm x 25 mm) flat mild steel.

2. Construction Procedure:
   a. Estimate the amount of stock needed for turning a 1" eye by the use of the following formula: Length of stock = (inside diameter) + (diameter of stock) x 3-1/7.
   b. Mark length of eye with center punch.
   c. Heat area to be bent to a uniform light cherry-red to orange-yellow color.
   d. Using the rounded edge of the face of the anvil bend to 90 degrees angle at the mark.
   e. Start bending the tip end around the small part of the horn. Strike bending blows. Gradually work back from the end to the square bend. Turn eye over and close.
   f. Eye should be centered with the bolt stock.
   g. Fabricate nut from 3/8" x 1" flat stock.
   h. Square stock carefully to measurements with a file.
   i. Determine tap-drill size from table found on page 85.
   j. Establish center by the use of diagonals and mark with center punch.
   k. Drill hole for the tap.
   l. Cut threads according to procedure outlined on pages 83 to 84.
   m. Cut 1-1/2" of N.C. threads on bolt; - check depth of threads.

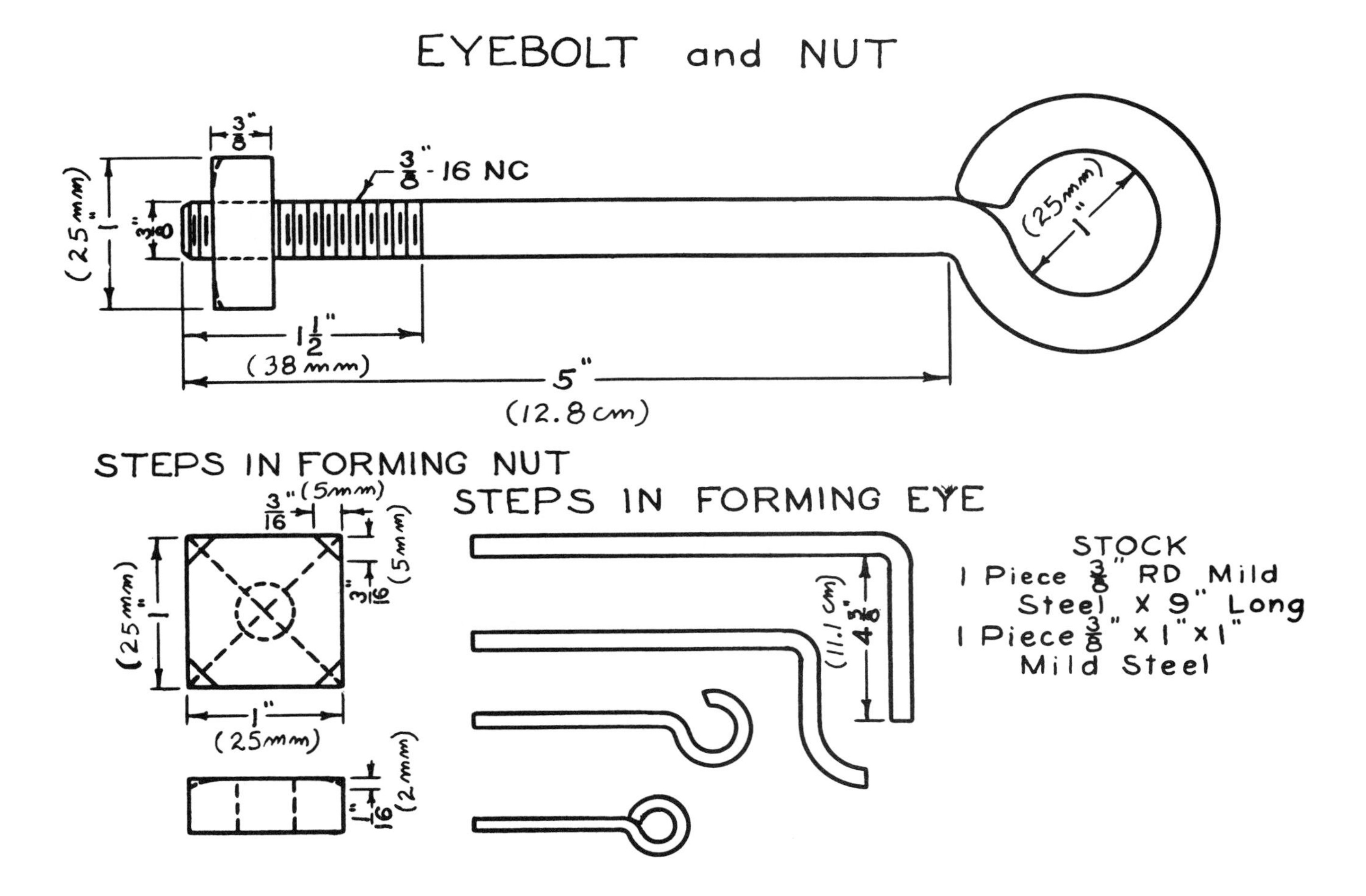

LAB EX-3 - COLD CHISEL

Construction Procedure:

1. Heat outer 2-1/4" of stock to a uniform cherry-red color.
2. Place one side against anvil face. Using drawing blows, work to shape rapidly starting at end and work back to 2-1/4" taper.
3. Finish to 3/16" at tip and 5/8" width. Do not work below a dull-red color.
4. Anneal - heat to cherry-red and cool slowly (12-24 hours) in lime or sand.
5. File and polish forged faces - Do not grind.
6. Temper with water: (Practice on old cold chisel).
   a. heat 2" to 3" of tip to uniform cherry-red color
   b. cool 3/4" to 1" until drops cling to tip when removed from water
   c. move tip to avoid cracks at water line
   d. quickly remove scale with steel brush or file
   e. observe color changes - quench lower 1/4" on purple color. Color order is light straw, dark straw, brown, purple, dark blue and light blue
   f. final cooling should be done gradually to avoid cracking
7. Grind cutting edge to 60 degrees angle. Use tool gage to check angle.
8. Chamfer end opposite point approximately 1/2" by 7/16" to prevent mushrooming.

Questions:

1. What is the carbon content of tool alloy steel?____________
2. What steel making process is used in making tool steel?
   ____
3. The color in the color coding system for tool steel is
   ____
4. Define annealing ____________
   ____
5. At a dull cherry red, the temperature in the metal is approximately ____________ degrees F.
6. If the color is stopped on light straw the cutting edge will be:
   a. brittle or b. soft.
7. Define tempering ____________
   ____
8. What is the purpose of the chamfer? ____________
   ____

Materials:

1 - 1/2" x 5" octagon tool steel, .7 - .8 percent carbon content. A proven successful material is manufactured by Carpenter Steel Co., Reading, PA. The product is Solar (Water Tough) AISI, Type S2.

Evaluation Score Sheet:

Item	Points Possible	Points Earned
1. Correct dimensions		
a. Chisel is 5-1/2" long	10	____
b. True taper and correct length	10	____
c. Chamfer dimensions and squareness	10	____
d. Cutting edge is 5/8" x 3/16"	10	____
e. Cutting edge ground to 60 degrees, edges even and straight	15	____
2. Tempering - tip correct hardness	25	____
3. Overall appearance	10	____
4. Attitude and work habits	10	____
Total Points	100	____

Name____________

Date____________

Grade____________

LAB EX-4 - TRACTOR DRAWBAR HITCH PIN

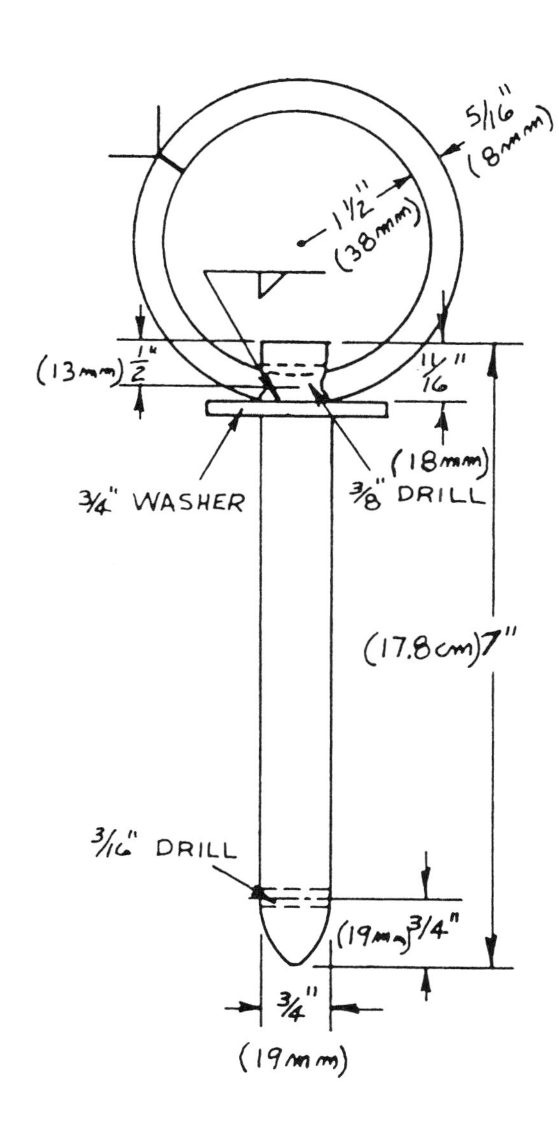

Construction Procedure:

1. Select and shape material according to plan.
2. Machine: (a) Drill holes for pull ring and safety catch pin. (b) Taper end for easy insertion. (c) Remove oil and oxide if present.
3. Heat Treatment: Suspend TDBHP in molten salt. CAUTION - Do not allow water to contact molten salt - result violent explosion. Allow TDBHP to remain for about 1to 2 hours, depending upon depth of case desired. Remove and oil quench at end of heat period.
4. Case hardening may be done using "Kasenit No. 1", a commercial surface hardening compound. Heat part uniformity to a bright red (about 1650 degrees F.) Dip or roll in the compound to form a fused shell. Reheat stock to 1650° F. Quench stock in water. Process may be repeated to increase depth of case hardness. Kasenit Co., Mahwah, New Jersey.
5. Fabrication: Arc weld washer and ring according to plan.
6. Check hardness with edge or corner of file.

Questions:

1. Spark test before heat treatment, circle condition.
   Low C Medium C High C
2. File test before and after heat treatment of TDBHP Ring and Washer
3. Rockwell hardness on TDBHP test before heat treatment B_______________ C________________
4. Rockwell hardness after heat treatment B__________________ C__________________
5. How can you visually distinguish hot-rolled from cold-rolled steel? ____________________
6. How can you use a piece of angle iron to align holes in round rod or pipe? ___________________
7. What tools do you need for locating and marking locations of holes? _______________________
8. What drill speed should be used in drilling the holes? ____________________________
9. What grinding wheel safety precautions should be observed in grinding this pin? ______________
10. How may an arc striking plate be used in weld ing this pin? ______________________________

Material:

1 - 5/16" x 10-1/4" M1020 hot rolled round stock
1 - 3/4" flat washer
1 - 3/4" x 7" round C1020 cold rolled steel

Name________________________

Date_________________________

Grade________________________

Evaluation Score Sheet:

Item	Points Possible	Points Earned
1. Length of pin	5	______
2. Tapered area smooth	10	______
3. Washer square to pin	10	______
4. Washer spaced correctly	10	______
5. Weld on washer	15	______
6. Pull ring round	10	______
7. Pull ring centered	10	______
8. Weld on pull ring	15	______
9. Holes centered and aligned	5	______
10. Attitude and work habits	10	______
Total Points	100	______

Note: This activity could be completed using 1" round, cold rolled stock, however the pull ring hole will need to be drilled at 7/16" rather than 3/8".

LAB EX-5 - CASE HARDENING OF MILD STEEL

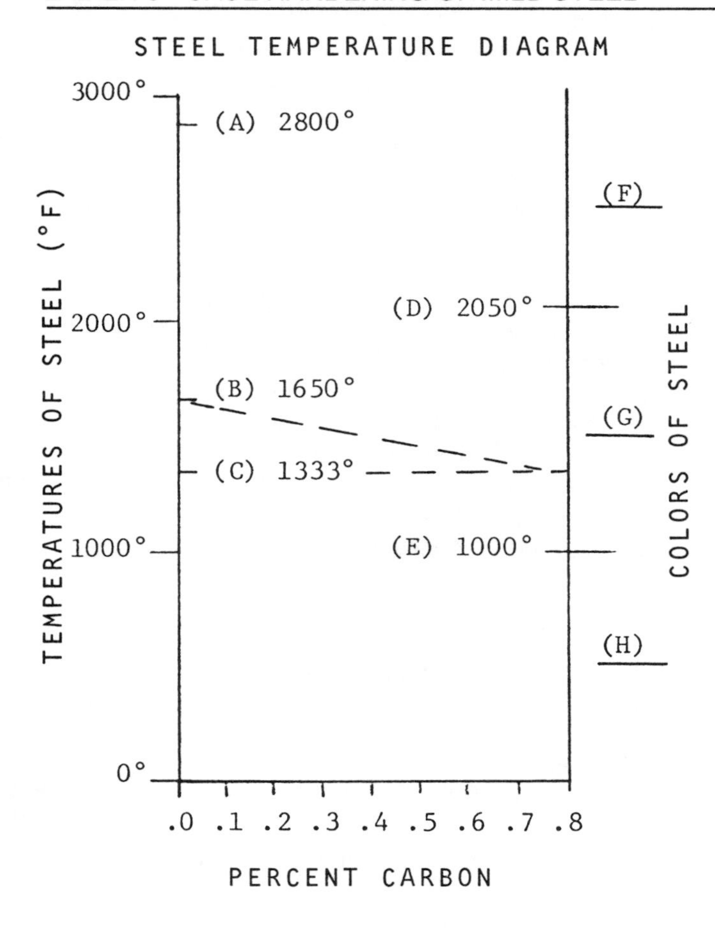

Identification of parts of diagram:

A. ______________________

B. ______________________

C. ______________________

D. ______________________

E. ______________________

F. ______________________

G. ______________________

H. ______________________

Case Hardening Procedure:

1. Clean the stock to be hardened.
2. Place carburizing compound in flat metal pan.
3. Heat stock with oxy-acetylene torch to 1650 degrees F (bright red).
4. Submerge stock in carburizing compound until a fused shell is formed on stock.
5. Re-heat stock to 1650 degrees F.
6. Quench stock in water using a scrubbing motion.
7. Spark test surface of stock to check carbon content.
8. Use file to check surface for hardness.

Operation Teaches Ability to:

1. understand the properties of case hardened steel.
2. use oxy-acetylene torch for heating.
3. identify steel temperatures by color.
4. add carbon to surface of steel using carburizing compound.
5. harden carbon steel by quenching.
6. identify carbon content of steel by spark testing.
7. identify hardness of steel by the use of a file.

Materials:

Mild steel stock
Kasenit No. 1, Surface-hardening compound
Oxy-acetylene torch with heating tip
Flat metal tray
Container of cold water
Metal tongs
Safety glasses and gloves

Evaluation Score Sheet:

	Item	Points Possible	Points Earned
1.	Identification (5 each)	40	______
2.	Carbon content of surface	20	______
3.	Hardness of surface	20	______
4.	Evidence of clean stock	10	______
5.	Safety and work habits	10	______
	Total Points	100	______

Name______________________________

Date________________ Grade________________

LAB EX-6 - TOOL SHARPENING GAGE

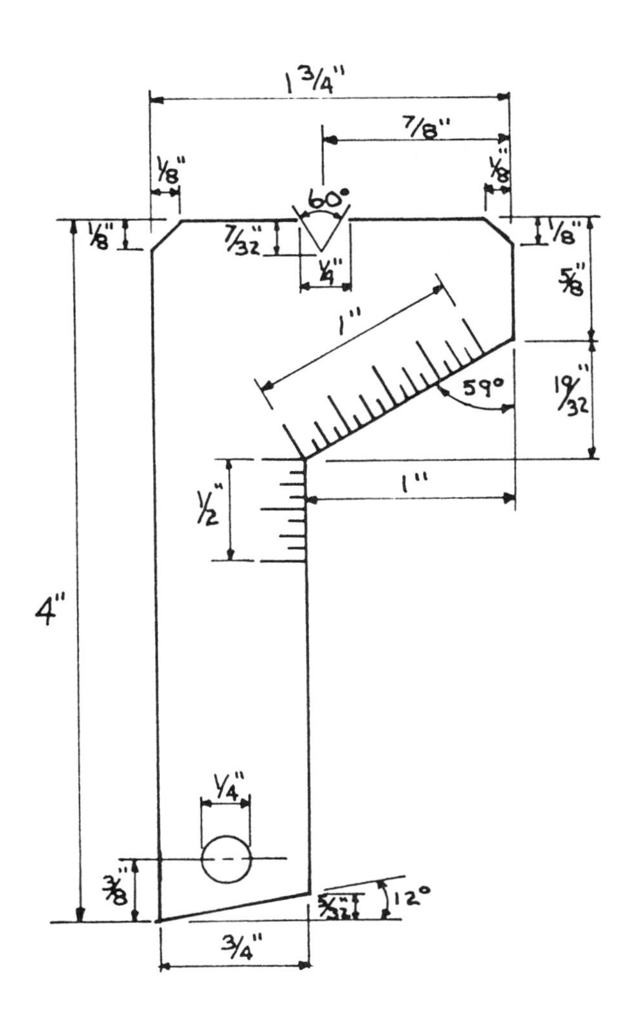

Gage can be used to measure:

1. Cold chisel cutting angle
2. Twist drill, cutting edge
3. Twist drill, lip clearance

Material:

1-7/8" x 4-1/16" x 3/32"
aluminum or brass
(aluminum is easier to work)

Construction procedure:

1. Square one corner of aluminum or brass stock.
2. Measure and scribe outline on stock with awl.
3. Mark graduations with awl.
4. Cut out tool gage with hacksaw. (Use protective blocks on each side when cutting in vise.)
5. Dress to the lines with file, bevel corners.
6. Use three-square file to cut chisel vee.
7. Polish with steel wool.
8. Drill size holes 1/4", 3/16", 1/8", and 1/6" may be drilled in the body of the gage.
9. Submit to instructor for evaluation.

Questions:

1. Why is aluminum easier to work than brass?
   ____
2. Why is it important to square or make a corner 90 degrees? ______________________
3. Which hacksaw blade did you use; 18-24-32? Why?
   ____
4. When should you use draw filing? Cross filling?
   ____
5. Why should files always be fitted with handles.
   ____

Evaluation score sheet:

	Item	Points Possible	Points Earned
1.	Length of gage	10	______
2.	Width of narrow end	10	______
3.	Width of wide end	10	______
4.	Vee position for cold chisel angle	10	______
5.	Angle for drill cutting edge	15	______
6.	Accuracy of the 1" and 1/2" rules	10	______
7.	Angle for drill lip clearance	10	______
8.	Correct angle and size of bevel corners	5	______
9.	Finish	10	______
10.	Attitude and work habits	10	______
	Total Points	100	______

Name______________________________

Date__________________ Grade__________

LAB EX-7 - SHARPENING THE TWIST DRILL BIT

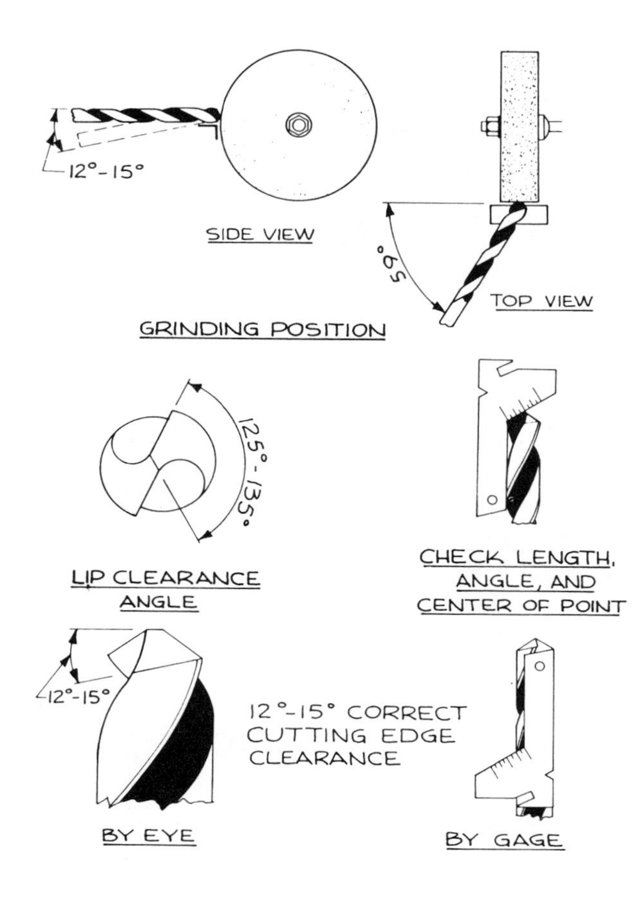

Operational Procedures:

1. Dress grinding wheel with dressing tool.
2. Hold drill bit against face of wheel at 59 degree angle on cutting lip.
3. Carry drill bit up the wheel face by dropping end and rotating very slightly in clockwise direction.
4. Make slow deliberate strokes, the full width of the cutting lip.
5. Do not lower cutting lip below the horizontal position as this will round the cutting edge.
6. When one lip is ground, rotate the drill one-half turn and grind the other lip.
7. Use tool gage to check equal lengths of lips, 59 degree angle cutting lip and 12-15 degree lip clearance.
8. Test bit by drilling hole in mild steel plate.
9. Stop while drilling, turn drill press in reverse direction to release drill bit from hole. Note depth of cut of each lip.
10. Make grinding corrections on drill bit as indicated by hole.
11. Submit drill bit, cuttings and metal for evaluation.

Evaluation Score Sheet

Item	Points Possible	Points Earned
1. Cutting lip angle, 59 degrees	20	______
2. Cutting lips equal length	15	______
3. Lip clearance, 12-15 degrees	15	______
4. Correct angle between dead center & cutting lip, chisel edge angle, 120-135 degrees	10	______
5. Smoothness of grinding surface, lip to heel	5	______
6. Bit cuts spiral chips	10	______
7. Hole drilled is correct size	10	______
8. Work habits and attitude	15	______
Total Points	100	______

Material:

Dull drill bit 3/8"-1/2"
Tool grinder with correct grinding wheel

Questions:

1. Check the specifications of the grinding wheel and determine if it is proper for this job.
2. What is the purpose of dressing a grinding wheel? ____________________
3. What kinds of wheel dressers are available? ____________________
4. How should the tool rest be positioned? ____________________
5. Describe the chip formation during the testing of the sharpened drill bit.

____

Name____________________

Date____________ Grade____________

LAB EX-8 - TOOL SHARPENING GAGE

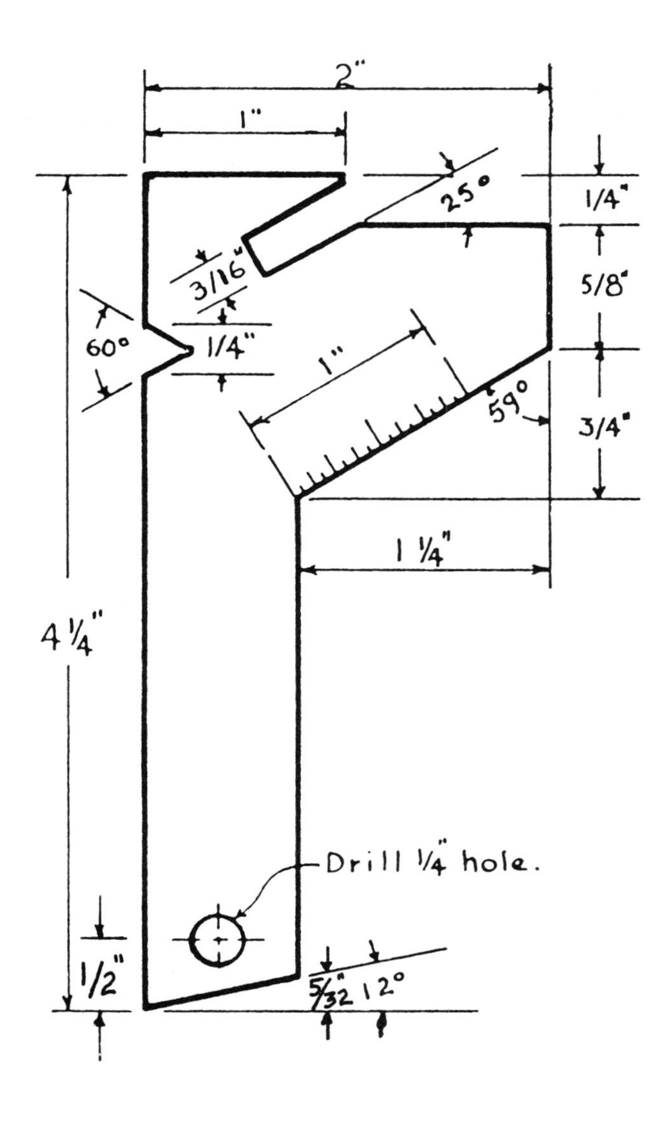

Construction Procedure:

1. Measure and scribe outline on stock with awl.
2. Cut out tool gage with hacksaw, use protective blocks when holding in vise.
3. Use flat file to cut wood chisel slot.
4. Use taper file to cut cold chisel vee.
5. Position and drill 1/4" hole at narrow end of gage.
6. Measure and scribe a 1" rule by 1/16" graduations.
7. Cut rule indicator marks with cold chisel.
8. Polish with steel wool for final finish.
9. Submit to instructor for evaluation.

Construction Teaches: Ability to:

1. understand selection of the flat stock.
2. measure and transfer outline of gage to stock.
3. scribe outline on metal with awl.
4. understand the correct hacksaw blade to select for cutting stock.
5. to use the hacksaw.
6. understand the correct files to select for filing slot and vee.
7. cut the slot with the flat file.
8. cut a vee with the taper file.
9. understand the need for center punching before drilling.
10. use the center punch.
11. understand fastening metal before drilling.
12. adjust speed and use of the drill press.
13. calibrate and indicate rule marks on metal.
14. set rule marks in metal with the cold chisel.
15. understand correct grinding wheels and setting of tool rest.
16. understand correct method of finishing and polishing flat stock.
17. use abrasive wheel and emery cloth to polishgage.

Gages:

A) Plane iron or wood chisel
B) Cold chisel or center punch
C) Twist drill, cutting edge angle and length
D) Twist drill, lip clearance

Material:

1 - 2/-1/4 x 4-1/2 - 3/32" stainless steel, mild steel, brass or 1/8" aluminum

Evaluation Score Sheet:

	Item	Points Possible	Points Earned
1.	Length of gage	5	______
2.	Width of narrow end	5	______
3.	Width of wide end	5	______
4.	Angle of wood chisel slot	15	______
5.	Vee position for cold chisel angle	15	______
6.	Angle for drill cutting edge	15	______
7.	Accuracy of the 1" rules	10	______
8.	Angle for drill lip clearance	10	______
9.	Hole centered	5	______
10.	Finish	5	______
11.	Attitude and work habits	10	______
	Total Points	100	______

Name________________________________ Date________________ Grade___________

LAB EX-9 - DRILL BIT, BOLT AND WASHER GAGE

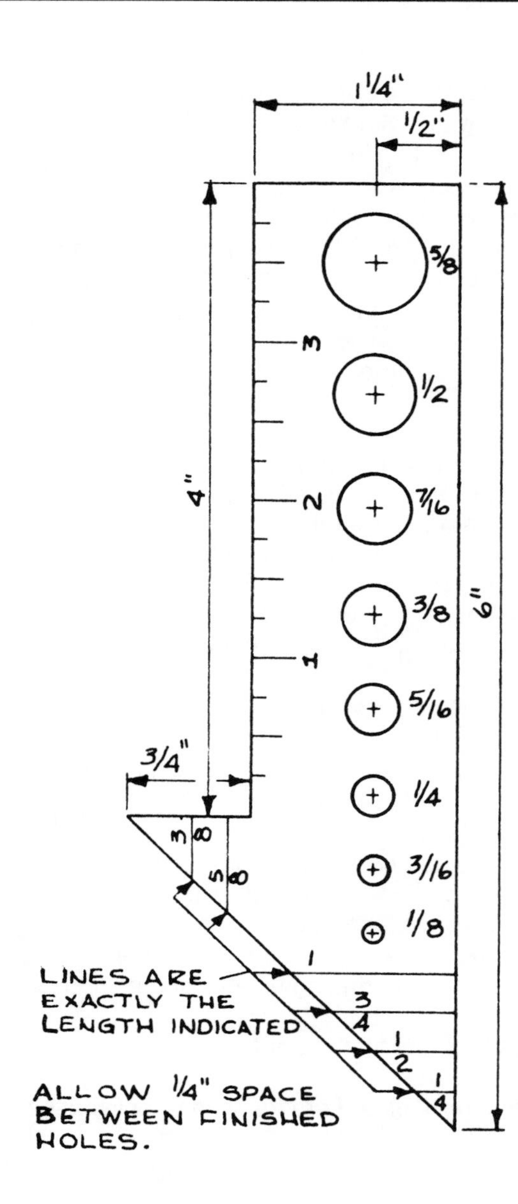

Construction Procedure:

1. Measure and scribe outline on stock.
2. Cut out with hacksaw, use protective blocks.
3. Use flat file to file all edges.
4. Measure and center punch for holes.
5. Select bit, position and drill holes.
6. Measure and scribe lines for inside diameter gages.
7. Cut marks with cold chisel.
8. Measure and scribe 4" rule in 1/4" graduations.
9. Cut marks with cold chisel or other sharp tools.
10. Polish surfaces.
11. Mark in numbers identifying size of gages and rule. Use engraving, metal stamp set, or paint on numbers.
12. Submit to instructor for evaluation.

Construction Teaches Ability to:

1. understand selection of flat stock
2. read plans, measure and scribe
3. use awl
4. understand hacksaw blade selection
5. use hacksaw
6. understand correct file selection and operation
7. file correctly
8. understand need for center punching and securing metal for drill
9. selection of proper bit sizes
10. adjust and operate drill press
11. measure and scribe gage and rule marks
12. understand correct method of finishing and polishing stock
13. use abrasive wheel and emery cloth for finishing
14. use engraving tool, metal stamps, or fine painting

Evaluation Score Sheet:

	Item	Points Possible	Points Earned
1.	Length of right side	5	______
2.	Length of left side and lip	10	______
3.	Angle of bottom	5	______
4.	Width of top	5	______
5.	Correct placement of holes	15	______
6.	Accuracy of inside of diameter gage marks	15	______
7.	Accuracy of rule marks	15	______
8.	Number identification	10	______
9.	Finish	5	______
10.	Attitude and work habits	15	______
	Total Points	100	______

Gage Applications:

1. Drill bit size. (series of holes) Used to determine the size of abit or bolt.
2. Inside diameter (points on bottom and left side). To measure inside diameter of hole to determine size of bolt and size of given washer.
3. Bolt length (graduated rule on left). To determine length of bolt required, a given bolt, and the threads on a given bolt.

Material:

1 - 2-1/4" x 6-1/4" - 3/32" stainless steel, mild steel, brass or 1/8" aluminum

Name______________________________

Date______________________________

Grade______________________________

LAB EX-10 - TOOL SHARPENING GAGE

Material:
1 - 5 cm x 10.5 cm x 2 mm
stainless steel, mild steel, brass or aluminum

Construction Procedure:
1. Square one corner of the selected stock.
2. Measure and scribe outline on stock with awl.
3. Mark graduations with awl.
4. Cut out tool gage with hacksaw. (Use protective blocks on each side when cutting in vise.)
5. Dress to the lines with the file, bevel corners.
6. Use three-square file to cut chisel vee.
7. Polish with steel wool.
8. Submit to instructor for evaluation.

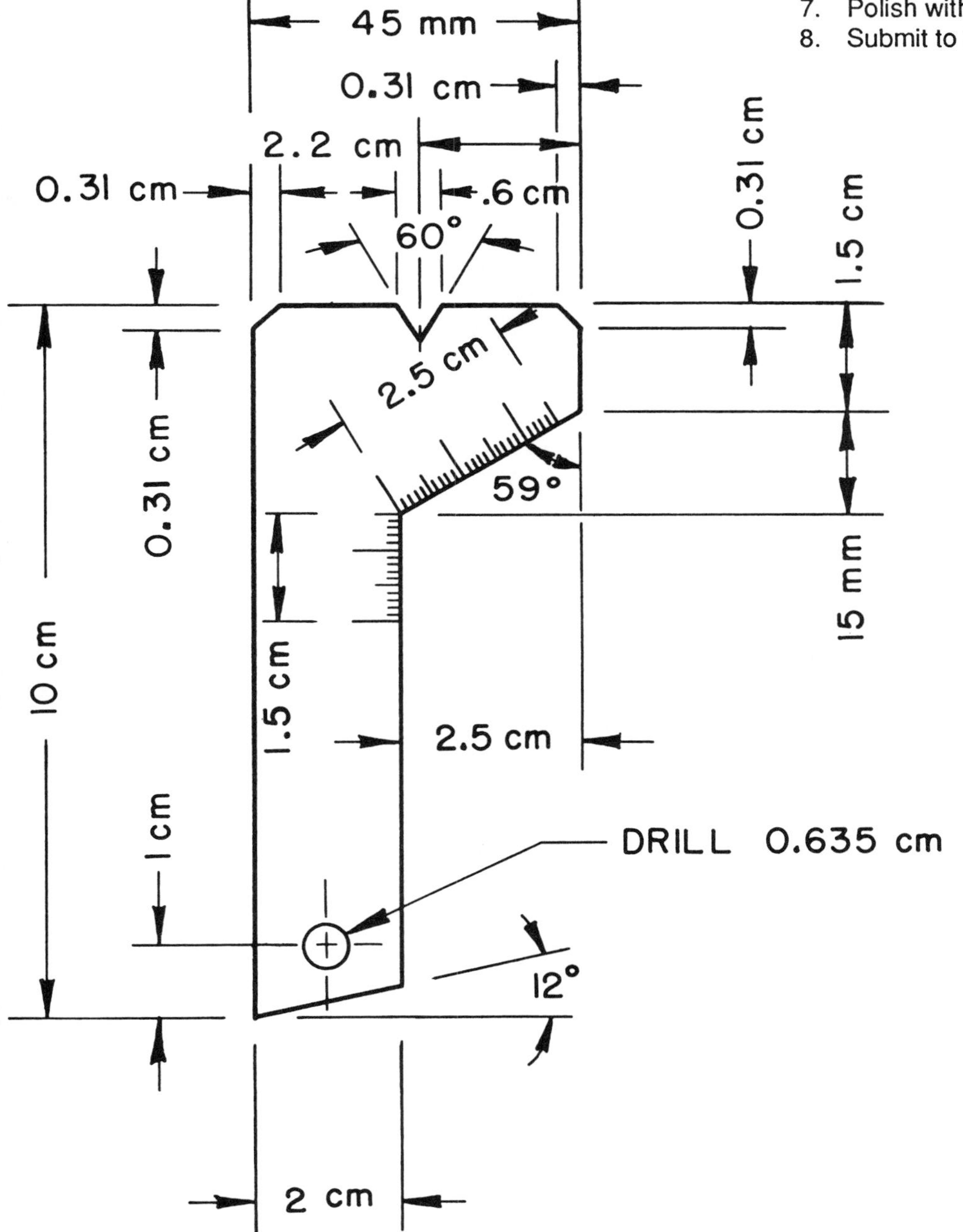

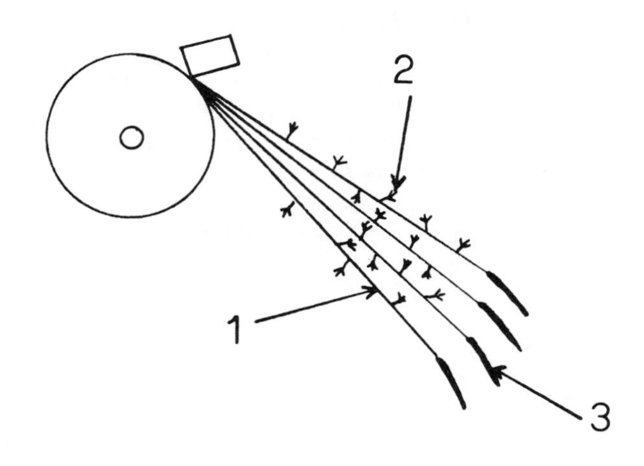

Identification:

1. ____________________
2. ____________________
3. ____________________

Methods of Identifying Metals:

Spark test
File test
Color
Weight
Surface texture
Texture of a break
Shape
Use
Sound when tapped
Forge marks
Cast marks

Exercise Teaches: Ability to:

1. understand characteristics which lead to metal identification.
2. identify non-ferrous metals.
3. identify ferrous metals.
4. identify steels.
5. identify the carbon content of steels.
6. identify the various cast irons.

Metal Identification Exercise:

Identify the metal samples provided and give the carbon content where applicable.

	Type of Metal	% C.
A.	____________________	______
B.	____________________	______
C.	____________________	______
D.	____________________	______
E.	____________________	______
F.	____________________	______
G.	____________________	______
H.	____________________	______
I.	____________________	______
J.	____________________	______

Spark Testing:

Hold the metal sample with moderate pressure against the grinding wheel and observe the sparks given off.

No sparks = non-ferrous metal
Long iron lines = low C steel
Short iron lines = high C steel
Few explosions = low C content
Straw colored iron lines = steel
Red iron lines = cast iron
Curved appendages on iron lines = cast iron

Evaluation Score Sheet:

	Item	Points Possible	Points Earned
1.	Identification (5 each)	15	______
2.	Identification of samples (7 each)	70	______
3.	Attitude and work habits	15	______
	Total Points	100	______

Name____________________________

Date________________ Grade____________

UNIT IV
METAL WORKING BY WELDING

Welding is a fast, efficient and economical method of joining metal. Anchored in the basic sciences, the practical applications are seemingly endless. Welded vehicles have taken man to the moon and to the extreme depths of the ocean.

Although welding is only about 100 years old, more than forty different processes have been developed. Figure 1, Master Chart of Welding Processes, shows six basic processes from which more exotic processes have developed. It is beyond the scope of this book to cover all the processes. The more common open-arc welding, gas welding and brazing will be covered in detail.

Figure 2, a chart, shows related processes of cutting and soldering. Again, only the more common processes will be covered.

There are two general areas of welding applications. **Production welding**, using assembly-line procedures accounts for the greatest tonnage of weld metal deposited. Chemists, metallurgists and engineers form a team where efficiency and quality control results in a competitive product. The weldor is a factory worker often working on the piece-work of job-related repetitive welding stations regu-

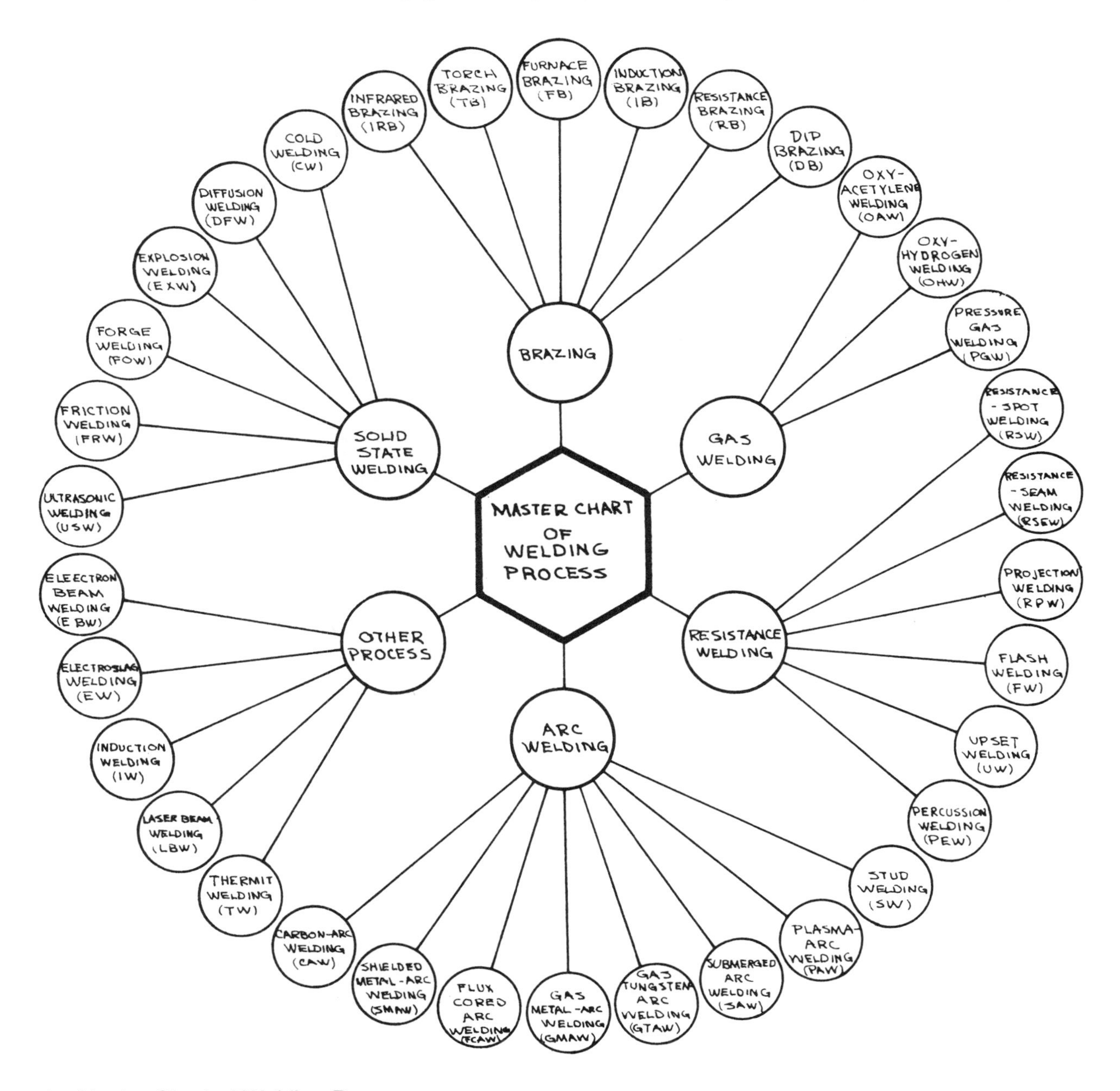

Figure 1. Master Chart of Welding Processes.

lated by management and labor unions. Extent of use of the various welding processes on different types of steel is given in Table 1.

Production welding requires skill and speed in application but little knowledge of different welding processes, metals and design.

By contrast, successful **maintenance welding** is a highly skilled occupation requiring ingenuity, confidence, imagination and knowledge of physics, chemistry, metallurgy and engineering. Obviously, there are many levels of skill and ability in maintenance welding ranging from home or hobby shop to pressure vessel maintenance. Factories would stop if maintenance welding were neglected. The entire field of hard surfacing is largely a maintenance operation. A great amount of welding done in farm shops involves maintaining productive machines and equipment.

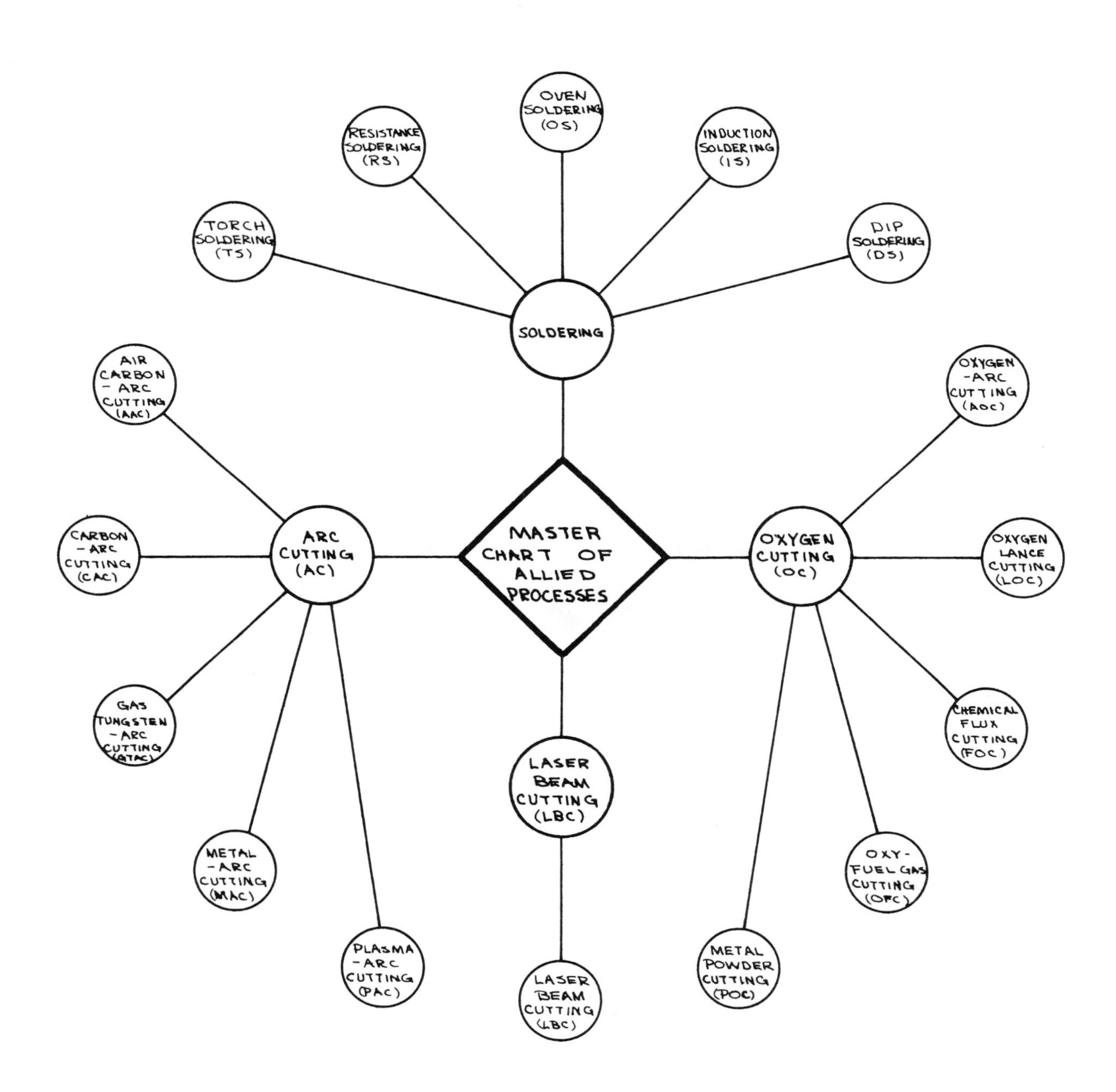

Figure 2. Master Chart of Allied Processes.

TABLE 1. EXTENT OF THE USE OF WELDING PROCESSES TO JOIN VARIOUS STEEL.

	Shielded Metal Arc Welding	Submerged Arc Welding	Gas Shielded Metal Arc Welding	Oxy-Acetylene Welding	Spot, Seam and Projection Welding
Low carbon:					
Sheet	Common	Common	Common	Common	Commom
(plate or bar)	Common	Common	Common	Occasional	Occasional
Medium carbon:					
Sheet	Common	Common	Occasional	Common	Occasional
(plate or bar)	Common	Common	Occasional	Occasional	Rare
High carbon:					
Sheet	Common	Occasional	Rare	Common	Rare
(Plate or bar)	Common	Occasional	Not used	Common	Not used
Low alloy:					
Sheet	Common	Common	Occasional	Common	Occasional
(Plate or bar)	Common	Common	Occasional	Occasional	Rare
Stainless:					
Sheet	Common	Occasional	Common	Occasional	Common
(Plate or bar)	Common	Common	Occasional	Occasional	Occasional

NOTES

NOTES

UNIT V
OXY-ACETYLENE WELDING AND CUTTING

PRECAUTIONS AND SAFE PRACTICES

The hazards affecting the health of operators in welding processes are burns, metallic dust, action of injurious light or heat rays generated by the flame, and toxic gases or fumes from the mixture of gases used in the torch or the action of heat upon the metals and fluxes used. Welding or cutting is not particularily hazardous, but even as with steam, gas or electricity, common sense precautions must be taken and enforced. Weldors should be thoroughly instructed in the performance of their work regarding the following safe practices and precautions for protection of themselves and others who may be nearby.

1. **Never use oil on oxygen connections** or any equipment through which oxygen passes. Oxygen under pressure coming in contact with oils or grease may explode. No lubrication of apparatus is necessary, but beeswax, soap or gylcerine may be applied to connections if desired.

2. **Never force connections** that do not fit.

3. **Fasten acetylene and oxygen cylinders securely** to the wall, post or rack before "cracking" the valves. Cracking the valves cleans dirt and dust from the cylinder orifices.

4. **Acetylene hose connections have left-handed threads** and are attached to red hoses. This insures against connecting an oxygen regulator to a cylinder containing a combustible gas. Use regulators only with the gases for which they are intended.

5. **Always see that connections are secure** before using equipment. Test for leaks with soapy water and a brush. The packing nut on the acetylene valve spindle often becomes loose. Do not hang a torch by the hose on the cylinder or regulator valves. All hose should be examined carefully at frequent intervals for leaks, worn places and loose connections.

6. **Always wear goggles or a cellulose acetate shield** when working with a lighted torch. Lens or shield should be shade no. 4 or 5 for fusion welding. Shade no. 2 or 3 filter lenses are satisfactory for brazing and for protection from stray light from welding and cutting operations for a radius of approximately 100 feet.

7. **Always use friction lighters** or satisfactory pilot flames. Never use matches for lighting a torch as bad burns or explosions may result.

8. **Check pressures carefully**. Always use the proper size tip and the proper gas pressure. Using acetylene pressures above 15 lbs. is hazardous. Acetylene is highly inflammable and might cause fire or explosion if released in quantity.

9. **Keep cylinders away from fire**. Never use leaky cylinders. Before starting to weld or cut, make certain there is no material or opening leading to material that sparks and hot slag will ignite. See that there is clear space between cylinders and work so regulators can be reached quickly. Sparks and hot pieces of iron can be sources of fire. Watch them!

10. **Do not open the acetylene valve more than one and one-half turns**. One- half to three-quarters of a turn is usually adequate. Leave cylinder valve wrench in place so the acetylene cylinder can be shut off quickly in an emergency.

11. **Turn pressures into gages gradually**. Close valves and release gage pressure when not in use.

12. **Use care that hoses are not damaged** in cutting or welding. Do not repair leaky hoses with tape.

13. **Never attempt to transfer acetylene** from one cylinder to another.

14. **Use particular caution when welding or cutting in dusty or gassy locations** or any place where sparks or an open flame might be hazardous.

15. **Never do any welding or cutting on containers that have held an inflammable substance** until they have been cleaned and safeguarded with carbon dioxide or nitrogen gas.

16. **Do not weld or cut range boilers** or other heavily galvanized, painted or lead coated material inside the shop as the fumes are highly toxic. Arrange for good ventilation or individual respiratory protection against nitrous and carbon monoxide gases, toxic fumes, metallic oxides and dusts. Lead poisoning is by far the most important of the metallic poisonings from the standpoint of frequency. Zinc and cadmium oxides are common. Flourides are present in the fluxes for stainless steel welding.

17. **Guard against possible exhaustion of oxygen in the air breathed** when welding in unventilated spaces.

18. **Do not relight flames (on hot metal**) in a pocket or small confined space.

19. All oxy-acetylene **operations should be under the supervision of trained and qualified personnel.**

20. **Wear clothing suitable for the kind of work** to be done to keep out flying sparks and slag. (Sweat shirts inside out are dangerous.)

21. **Store extra cylinders away from working areas.** Always keep cylinders in the upright position to avoid releasing the acetone.

22. **Be ready to put out any fire promptly** with fire extinguishers, pails of water, water hose or sand.

23. If there is a possibility that a smoldering fire may have been started, keep a person at the scene of the work for at least a half hour after the job is completed.

24. When the welding or cutting is finished, **extinguish the flame by closing the acetylene valve** (red hose and gage), **then turn off oxygen valve**.

25. **Stand erect and relaxed**. Injuries due to difficult and awkward postures maintained for long periods must be considered.

26. **Keep flashbacks or backfires at a minimum** by proper maintenance of equipment. Backfires are caused by (1) dirty tips, (2) improper gas pressures, usually too low, (3) leaks or loose connections, (4) overheating the metal and tip and (5) luminous cone in pool. Flashbacks (burning behind mixing chamber) are caused by (1) loose tip or nozzle, (2) dirty or damaged seat, and (3) cracked or distorted torch head.

OXY-ACETYLENE EQUIPMENT

A complete oxy-acetylene welding and cutting unit is shown in Figure 1. **Oxygen and acetylene**, the basic gases for this welding process, are stored in the two cylinders. As the gases are under pressures, **regulators** are needed to regulate the correct pressures for the welding and cutting operation. The **oxygen** (green) and **acetylene** (red) hoses have the primary purpose of delivering the gases from the regulators to the welding and cutting torch. The **welding or cutting torch** receives and again regulates the gases through valves so that the correct amount of each gas is delivered to the **torch head** or **tip** where the gases burn to provide the heat for the welding or cutting operation.

Other basic parts of the system include (1) the **off-on valves** of the cylinders, (2) the **cylinder pressure gage** and **outlet pressure gage** as found on both the oxygen and acetylene regulators, (3) the **regulator adjusting screws** or cross bars used for adjusting the pressure delivered from the regulators, (4) the **cylinder caps** for protection of the cylinder valves during storage or transporting the cylinders, (5) the **torch valves** to regulate the amount of each gas entering the welding torch, (6) the **welding blowpipe** or **mixing chamber** of the welding torch and (7) the **welding or cutting tip**. All of the basic parts and their specific function will be discussed in more detail in this unit.

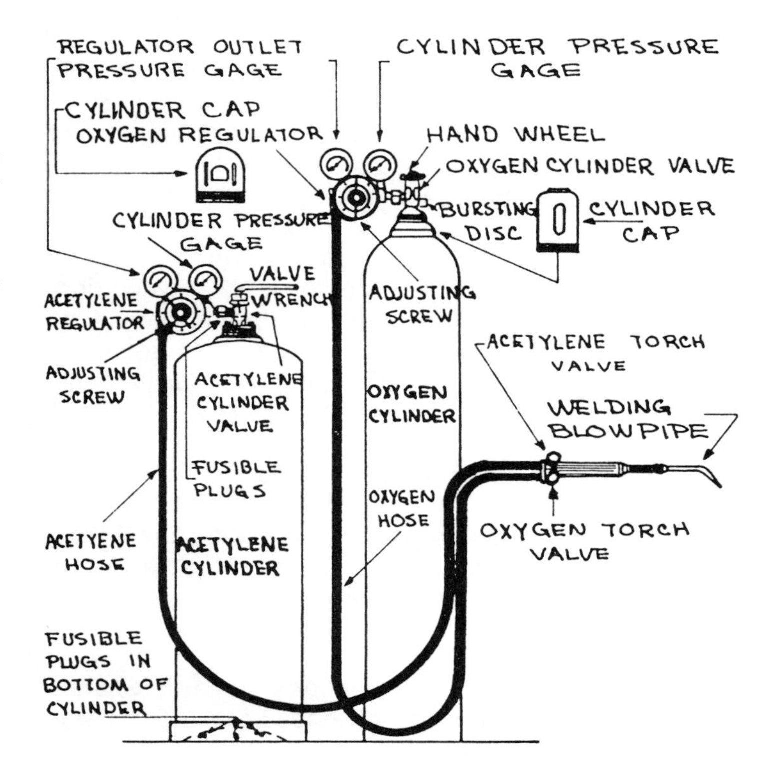

Figure 1. Oxy-Acetylene Welding and Cutting Unit.

There are two types of oxy-acetylene welding setups; these being the **stationary** and **portable** units. The stationary welding setup, oftentimes referred to as a **manifold system**, is used when the equipment need not be portable or need to be moved around. The manifold system is common in school shops or production lines where a number of stations are required and where the work can be brought to the welding equipment. The manifold system usually has two or more cylinders of each of the gases manifolded or connected together. The system is equipped with master regulators to control the pressure and gas flow. A greater volume of gas is required when a number of individuals are welding at one time. The cylinders are generally located in other rooms or outside when connected to the manifold system. The gases are then piped to the individual welding stations. The gases are regulated at the cylinders and then again at each individual station before being used for the actual welding operaton. A simple stationary welding setup is shown in Figure 2. Notice the two pipes delivering the two gases from the manifold sysem. Each gas is then regulated

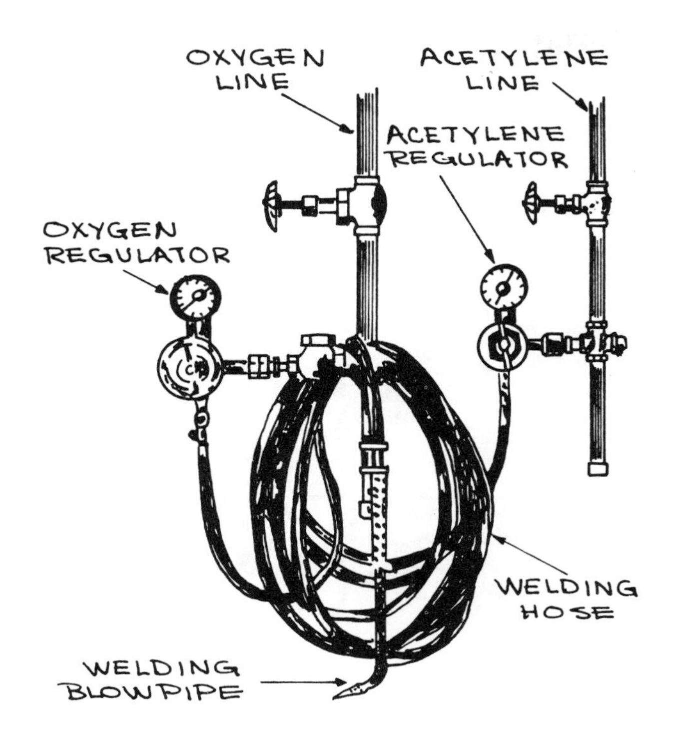

Figure 2. Stationary Oxy-Acetylene Welding Setup.

from the line pressure to the desired welding pressure before being delivered to the welding torch.

A portable welding unit as shown in Figure 3 has the main advantage of being portable so the total unit can be taken directly to the job. Normally, the unit is mounted on a set of wheels and can easily be wheeled throughout the shop or to do any job. The portable unit as shown has the two gas cylinders, an oxygen and acetylene regulator, hoses leading from the regulator to the welding torch and either a welding or a cutting torch.

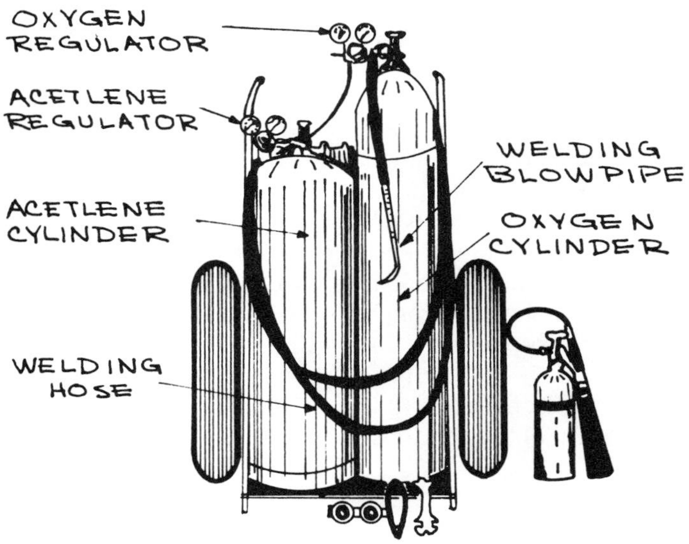

Figure 3. Portable Oxy-Acetylene Welding Setup.

Although most portable units are commonly used for cutting, there are cases where the portable unit would be needed for welding operations. To avoid changing the welding and cutting torch heads, portable units are often equipped with a double hose system in which a Y-connection is used at each regulator outlet so that double hoses, two oxygen and two acetylene, can be connected from each regulator. This way two torches, one for welding and one for cutting, are always available on the portable oxy-acetylene unit. Of course, either two cutting torches or two welding torches could be mounted on the same unit at one time. In this sense, the portable unit is similar to a manifold system in that two welding or cutting stations could be served from one unit.

Safety is generally more of a problem with the portable unit as it is being moved from place to place. In transporting, keep cylinders in upright position and make certain the unit is secure to avoid tipping and possible damage to parts of the unit. Parts of the portable unit are usually more subject to damage because it is moved throughout the shop or work area. One other disadvantage of the portable unit is the fact that if any one part is damaged, the total unit is out of commission, whereas with the manifold system, one station could be out of commission but the rest of the system would be operable.

The portable system with the double hose setup would probably be best for the farm and other shop applications. In a school or production situation, the manifold system with one or two portable units available would best meet the requirements for oxy-acetylene welding and cutting needs.

THE GASES

In the late 1800s, LeChatelier, a French chemist, discovered that the combustion of acetylene with oxygen produced a flame having a temperature far higher than that of any gas previously known. Oxy-acetylene welding and cutting is one of the more important processes in our present day industrial complex. This process plays a very important role in almost all industry. When oxygen and acetylene gases are mixed in the correct proportions and ignited, one of the hottest flames known is produced. The flame reaches a temperature of 6300° F. and melts metals so completely that they actually flow together to form a complete bond without an application of any mechanical pressure.

With the oxy-acetylene flame, such metals as iron, steel, cast iron, brass, aluminum, copper, and stainless steel, may be welded as well as some dissimilar metals such as steel and cast iron, copper and cast iron, and brass and cast iron can be joined. The oxy-acetylene flame can be used to melt the edges of pieces of metal and a filler metal called a welding rod is then melted into the molten pool of the base metal to complete the welding operation. Since the welding rod is similar to the metal being joined, it unites to form a single homogeneous piece of metal.

Welds in which the base metals are melted and allowed to flow together with or without a filler rod are referred to as **fusion welds**. The other common weld made with oxy-acetylene flame is referred to as brazing or braze welding; this type of weld is also referred to as an **adhesion weld**. In brazing, the base metals are not melted but are heated to a dull red or tinning temperature at which time a filler rod having a lower melting temperature than the base metal is added to the surface of the pieces being joined. The filler rod flows by capillary action over the surface of the joint adhering to the base metals, Upon cooling, the braze weld has strengths comparable to that of a fusion weld.

Another common use of the oxy-acetylene flame is that of **cutting**. In the cutting process, a stream of oxygen is directed against a piece of heated metal, causing the metal to oxidize or burn away thus cutting the metal to the desired shape.

In addition to fusion welding, braze welding and cutting, the oxy-acetylene flame has many other uses such as tempering, case hardening, annealing, hard surfacing as well as being an excellent heat source for heating metals for bending, shaping, forming, forging or hot cutting. Now that we have an idea of the many uses of the oxy-acetylene flame, we should have an understanding of the sources of the two gases used in this very important process.

OXYGEN

Oxygen (O_2) is a colorless, odorless and tasteless gas that is slightly heavier than air. It will not burn by itself but will support the rapid combustion of most materials. Oils, paints, greases or other similar materials must never come in contact with oxygen valves or regulators because of its explosive nature of these materials.

The air we breathe is a mixture of a number of gases. Air is approximately 20% oxygen, 79% nitrogen with the remainder composed of water vapor and gases auch as argon, neon, helium, and carbon dioxide. To obtain oxygen for welding purposes, it is necessary to separate it from the other gases. Two methods are used to produce oxygen, these being the **electrolytic** and the **liquid-air** methods.

The electrolytic method, the most expensive and also the least common method, utilizes the fact that water is made up of oxygen and hydrogen. By sending a current through a solution of water containing caustic soda, oxygen is isolated on one pole while hydrogen collects on the other. This process makes it possible to separate and collect the two gases.

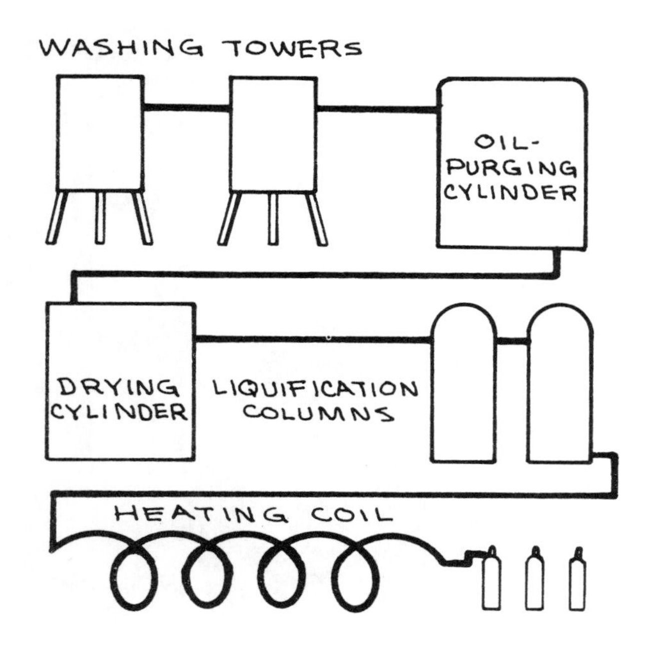

Figure 4. Schematic of Liquid-Air Process for Production of Oxygen.

The liquid-air process is most commonly used in the production of commercial oxygen. Shown in Figure 4 is a schematic of the complete process of the liquid-air method of producing oxygen. Briefly and simply stated, this method consists of separating the various gases in air by cooling them to a very low temperature. The air is drawn from the outside into huge containers called washing towers. The air is washed using a solution of caustic soda. As the air leaves the washing tower, it is compressed and passed through oil-purging cylinders in which oil particles and water vapor are removed. Next, the air passes into cylinders containing dry caustic potash in which the air is dried and the remaining carbon dioxide and water vapor are removed. The dry, clean, compressed air is then taken into the liquification columns where the air is cooled and then expanded to approximately atmopsheric pressure. This process of changing the extremely cold air under pressure to the lower atmospheric presure causes the air to liquify. Because of their different boiling points, the various gases are easily separated. Oxygen boils at -296°F whereas nitrogen boils at -320°F. The nitrogen, having the lower temperature, evaporates first, leaving the liquid oxygen in the bottom of the condenser.

The liquid oxygen next passes through a heated coil which changes the liquid into a gaseous form. Once the oxygen has been separated from the other gases, it is compressed and stored in the pure state in a gaseous form in steel containers at a pressure of 2200 psi at 70°F. As gases expand when heated and contract when cooled, the oxygen pressure in the tank will change as the ambient temperature changes.

ACETYLENE

Acetylene (C_2H_2) is colorless with a very distinctive odor. It is highly flammable and has a wide explosive range when it is mixed with air or oxygen. A mixture containing only one (1) part of acetylene to forty (40) parts of air will explode violently if it comes in contact with fire. Although pure acetylene is not a health hazard, the impurities in acetylene are toxic and produce a harmful anesthetic effect when breathed. Acetylene is lighter than air and because of its unstable nature at pressures greater than 15 psi, it is mixed with **acetone** before being placed under pressure in the cylinder.

Acetylene is produced from **calcium carbide** and **water**. The commercial generator consists of a large water tank. When calcium carbide is dropped into the water, a chemical reaction takes place and bubbles of acetylene gas are given off and rise to the surface. The gas is then collected, purified, cooled and slowly compressed into cylinders at approximately 250 psi.

The chemical properties of acetylene first involve the production of calcium carbide. Calcium carbide is produced by burning coke and limestone in an electric furnace. The chemical formula is :

$$\underset{\text{(coke)}}{3C} + \underset{\text{(limestone)}}{CaO} + \text{heat} = \underset{\text{(Calcium carbide)}}{CaC_2} + \underset{\text{(Carbon monoxide)}}{CO}$$

Next, the calcium carbide is mixed with water in the acetylene generator and acetylene gas is produced as:

$$\underset{\text{(Calcium carbide)}}{CaC_2} + \underset{\text{(water)}}{H_2O} = \underset{\text{(Acetylene)}}{C_2H_2} + \underset{\text{(Calcium hydroxide)}}{Ca(OH)_2}$$

Where large amounts of acetylene are required such as in a production welding shop, it is most economical to make acetylene on the spot with the acetylene generator. A commercial generator is pictured in Figure 5. Small portable acetylene generators are also available; however, they would not be recommended for a school or home shop because of their danger, problem with the water freezing and also getting rid of the calcium hydroxide residue left over from the production of acetylene.

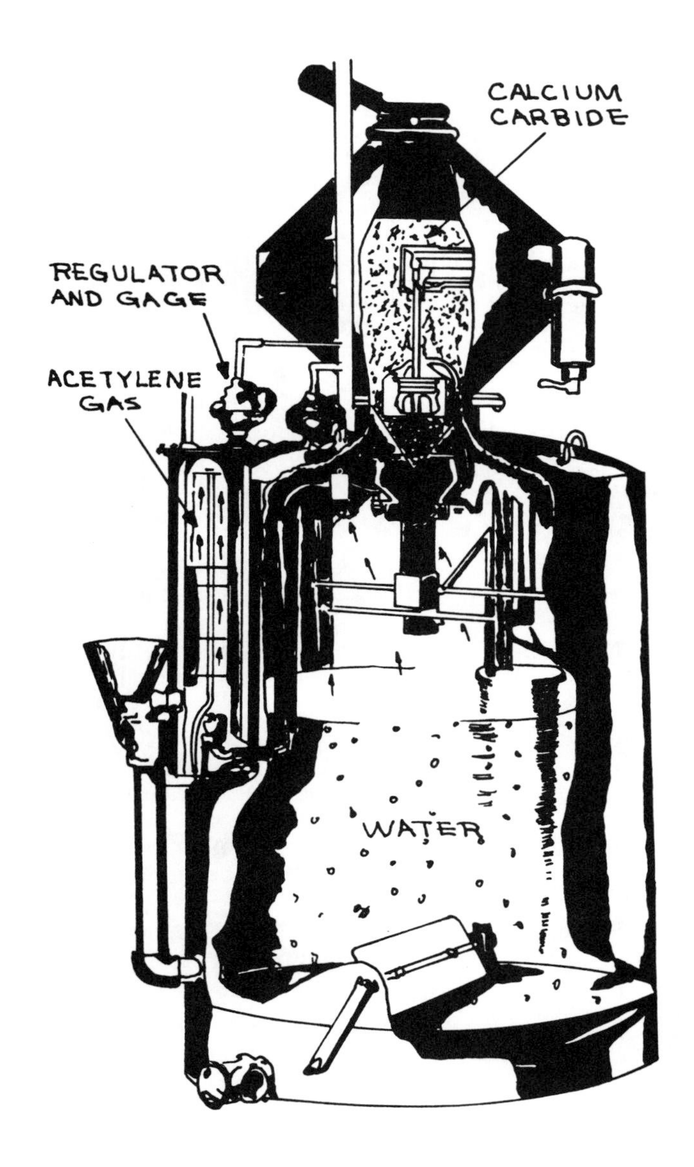

Figure 5. Commercial Acetylene Generator.

CYLINDERS

The standard gas cylinders and capacities are shown in Figure 6. As noted, the gases, oxygen, nitrogen, hydrogen, argon, helium and carbon dioxide are all stored in similar cylinders. The main difference is color of the cylinders and the amount of gas stored in the various sizes of cylinders. The amount stored is almost always given in cubic feet; for example, the common oxygen cylinder when at 2200 psi and 70°F contains 244 cubic feet of oxygen. Of the gas cylinders illustrated, the most common gases used in welding are oxygen, argon, helium, carbon dioxide and acetylene. The acetylene cylinder, as noted, is somewhat different from the other cylinders. Further explanation will follow during the discussion of oxygen and acetylene cylinders.

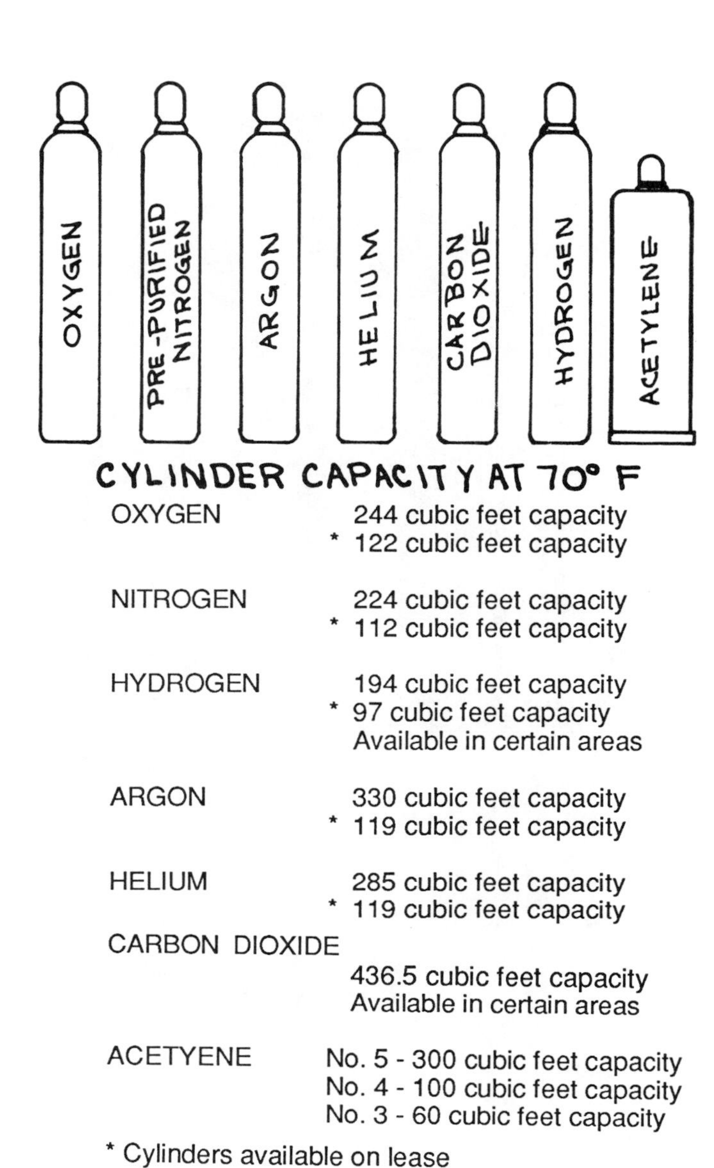

CYLINDER CAPACITY AT 70° F

OXYGEN	244 cubic feet capacity
	* 122 cubic feet capacity
NITROGEN	224 cubic feet capacity
	* 112 cubic feet capacity
HYDROGEN	194 cubic feet capacity
	* 97 cubic feet capacity
	Available in certain areas
ARGON	330 cubic feet capacity
	* 119 cubic feet capacity
HELIUM	285 cubic feet capacity
	* 119 cubic feet capacity
CARBON DIOXIDE	436.5 cubic feet capacity
	Available in certain areas
ACETYENE	No. 5 - 300 cubic feet capacity
	No. 4 - 100 cubic feet capacity
	No. 3 - 60 cubic feet capacity

* Cylinders available on lease

Figure 6. Standard Gas Cylinders and Cubic Feet Capacity of Various Gases Used in Welding.

OXYGEN CYLINDERS

Oxygen cylinders are made from a single plate of seamless, drawn steel. The cylinder is tested with water pressure at 3360 pounds per square inch. As shown in Figure 7, the cylinder is equipped with a **high pressure valve** made of bronze which can be opened by turning the hand wheel on top of the cylinder. The valve should always be operated by hand and turned slowly to avoid damage to the regulator. Oxygen valves should always be turned completely open as the high pressure could cause the valve to leak if not opened completely. Shown in Figure 8 is the top of the oxygen cylinder and the major parts.

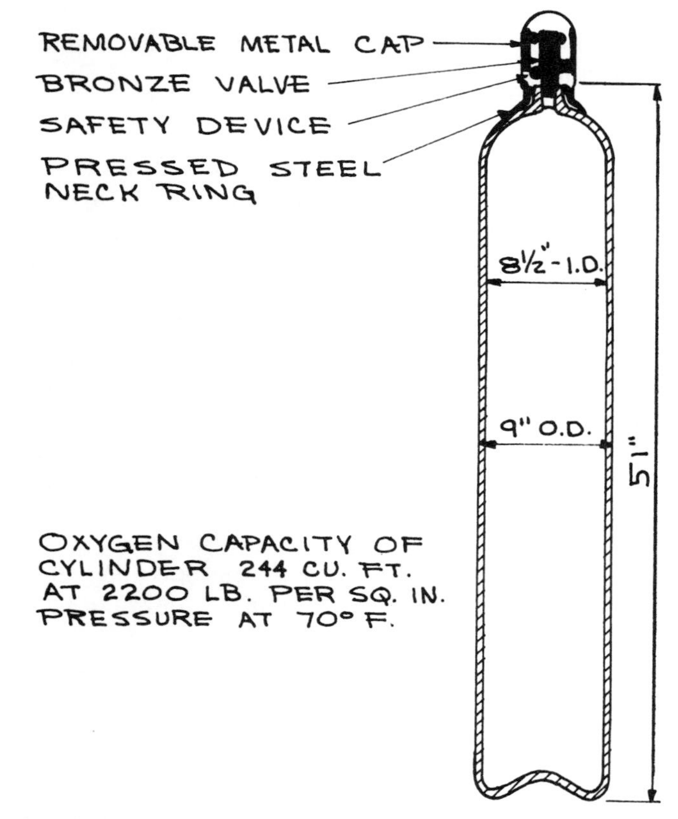

Figure 7. 244 Cubic-Foot Oxygen Cylinder.

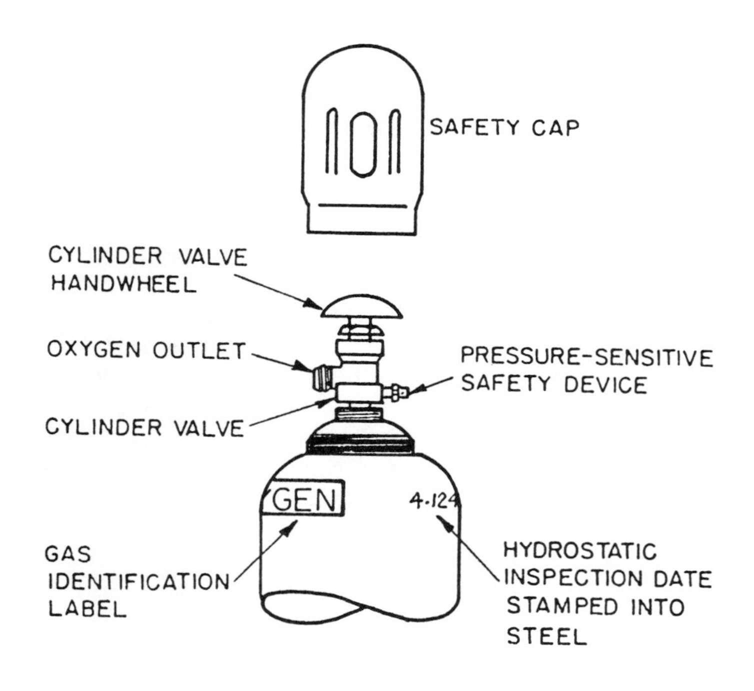

Figure 8. Oxygen Cylinder Top.

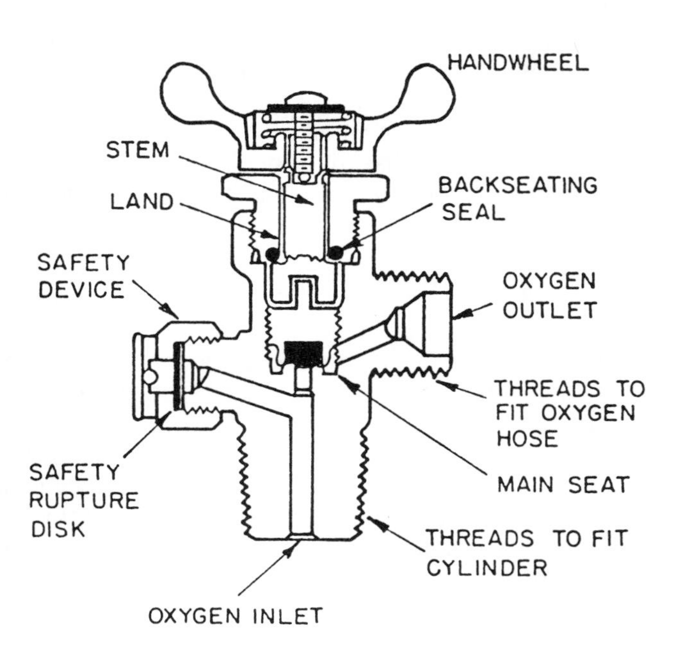

Figure 9. Oxygen Cylinder Valve.

As shown in Figure 9, the cylinder contains a safety device that would permit the oxygen to drain off slowly in the event the temperature would increase to a point which would cause the pressure in the cylinder to go beyond the safety limit of the cylinder. Thus, if a cylnder were exposed to a hot flame as in a fire, the safety device would relieve the pressure before the cylinder could reach the point where it would explode. Note Figure 10 which reveals the cubic feet of gas in a cylinder by temperature and cylinder pressure. For example, if the tank pressure gage reads 1550 psi and the temperature is 70°F, the cylinder would contain 170 cubic feet of oxygen.

Other parts of the standard cylinder are the removable protector cap and the pressed steel neck ring. The protector cap must always be in place to protect the valve from damage when the cylinder is not in use.

There are three common sizes of oxygen cylinders, these being **244 cubic feet**, **122 cubic feet**, and **80 cubic feet**. The larae size is considered the standard size cylinder as illustrated in Figure 7. All oxygen cylinders are filled to 2200 psi at 70° F. Oxygen cylinders are not normally sold but remain the property of the oxygen manufacturer. This way, the cylinders can be inspected regularly and kept in good condition at all times. The cylinders are loaned to customers, usually free for a period of 30 days, after which they are subject to a demurrage charge.

VARIATIONS IN OXYGEN PRESSURES WITH TEMPERATURE CHANGES

Gage pressures indicated for varying temperature conditions on a full cylinder initially charged to 2200 psi at 70°F. Values identical for 244 cu. ft. and 122 cu. ft. cylinder.

Temperature Degrees F.	Pressure psi approx.	Temperature Degrees F.	Pressure psi approx.
120	2500	30	1960
100	2380	20	1900
80	2260	10	1840
70	2200	0	1780
60	2140	-10	1720
50	2080	-20	1660
40	2020		

OXYGEN CYLINDER CONTENT

Indicated by Gage Pressure at 70° F. 244 Cu. Ft. Cylinder

Gage Pressure psi	Content Cu. Ft.	Gage Pressure psi	Content Cu. Ft.
190	20	1200	130
285	30	1285	140
380	40	1375	150
475	50	1465	160
565	60	1550	170
655	70	1640	180
745	80	1820	200
840	90	1910	210
930	100	2000	220
1020	110	2090	230
1110	120	2200	244

122 cu. ft. cylinder content one-half above volumes

Figure 10. Effect of Temperature on Gage Pressure and Content of Oxygen Cylinders.

ACETYLENE CYLINDER

As discused earlier, the properites of acetylene are such that special precautions must be taken on order to keep acetylene under pressure. **Free acetylene can never be stored under high pressure**. Acetylene cylinders, as illustrated in Figure 11, are therefore packed with a porous material. The porous material, usually asbestos and balsa wood, is saturated with **acetone**. Acetone is a chemical having the ability to **absorb many times its own volume** of acetylene without changing the nature of the acetylene.

At normal temperatures and pressure, one volume of acetone has the ability to absorb 20 volumes of acetylene gas. Thus, it is possible to increase the pressure of stored acetylene to a maximum of 250 psi in a full cylinder. As the acetone is slightly heavier than acetylene, it remains in the cylinder unless the cylinder would be used in a horizontal rather than an upright position. If the cylinder is transported in a horizontal position, which it should not be unless completely necessary, it should be allowed to sit in the upright position at least two (2) hours before using or the acetone could be lost from the cylinder. As illustrated in Figure 11, the cylinder has a steel valve usually operated by a T-wrench. This valve should never be opened more than one and one-half turns, usually one-half to three-quarters of a turn is adequate. The slight opening is a safety factor as it permits the quick closing of the valve in case of an emergency. Also, the T-wrench or valve turning device should always be left in place on the cylinder. The acetylene cylinder, depending upon the size and type of cylinder,

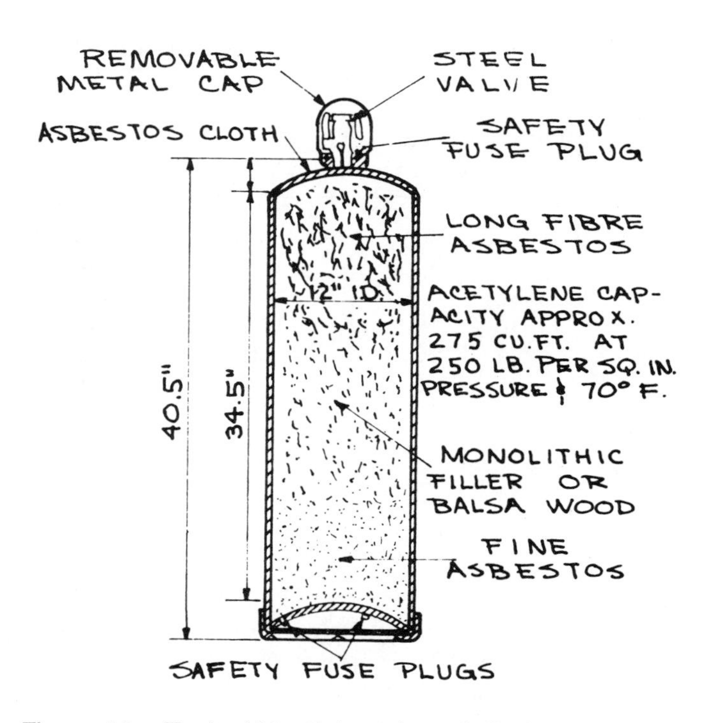

Figure 11. Typical No. 5 Acetylene Cylinder.

has a removable steel cap which, like oxygen, should be kept in place to protect the valve when not using the cylinder.

The acetylene cylinder has a number of safety fuse plugs usually at the top around the valve and in the bottom which have the purpose of relieving excess pressure should the cylinder be subjected to undue heat or other mechanical pressure.

Acetylene is sold in three common sizes, these being 60, 100 and 300 cubic feet, note Figure 6. The actual acetylene content of the cylinder is usually slightly below the rated capacity; therefore, each cylinder is weighed and stamped with the actual amount before shipping, and the customer pays only for the actual amount of acetylene in the cylinder. Like oxygen, acetylene cylinders are owned by the manufacturer and are available either on a lease or month free basis. It is important that the manufacturer checks the cylinder and recharges the acetone when needel, for safety reasons.

REGULATORS

The regulator or reducing valve is a mechanical device which reduces the high pressure of the gases as they flow from their respective cylinders down to the pressure used for welding or cutting. In addition to pressure reduction, the regulator must produce a steady flow of gas under varying cylinder pressures. For example, an oxygen cylinder might have a cylinder pressure of 2200 psi; however, a pressure of only 7 pounds is needed for welding. The regulator must maintain a constant pressure of 7 pounds even if the cylinder pressure drops down to 800 psi. Note Figure 12

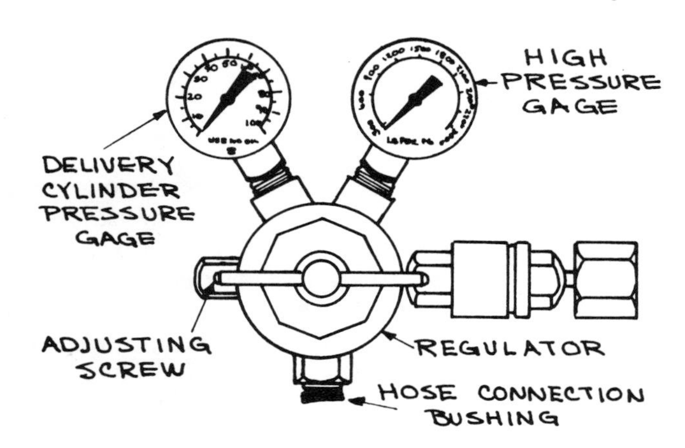

Figure 12. Gas Pressure Gages and Regulator.

which illustrates the high pressure (tank pressure gage) and the delivery pressure (working pressure gage).

There are two common types of regulators, (1) **single-stage** and (2) **two-stage**. The single-stage regulator reduces the pressure of the gas from the cylinder to working pressure in one stage. Note Figure 13 which illustrates a schematic drawing of the operating principles of the oxygen and acetylene single-stage regulator. As shown, as the pressure-adjusting screw (P) is turned in, tension is applied to the spring (S) and the pressure is regulated. The two-stage regulator reduces the pressure in two stages. As the pressure drop in each stage is less than a two-stage regulator, the pressure regulation is more accurate and less fluctuating. A cross sectional view of a two-stage regulator is shown in Figure 14. In the first stage, the gas flows from the cylinder into a high pressure chamber. A spring and diaphragm keeps a predetermined gas pressure in this chamber. For oxygen, this pressure is usually set around 200 pounds and for acetylene regulators, the first stage reduces the pressure to approximately 50 pounds. In other words, for oxygen the pressure is reduced from 2200 pounds for a full tank to 200 pounds while for acetylene, the pressure is reduced from 250 pounds for a full tank to 50 pounds.

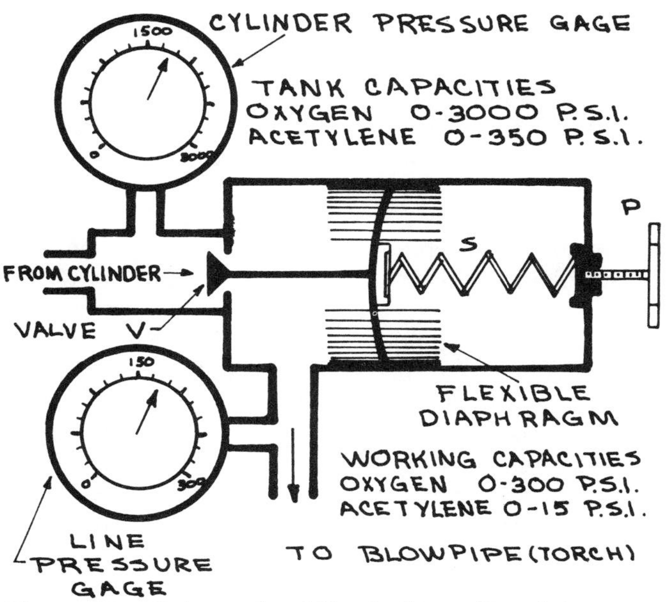

Figure 13. Schematic of Single-Stage Regulator

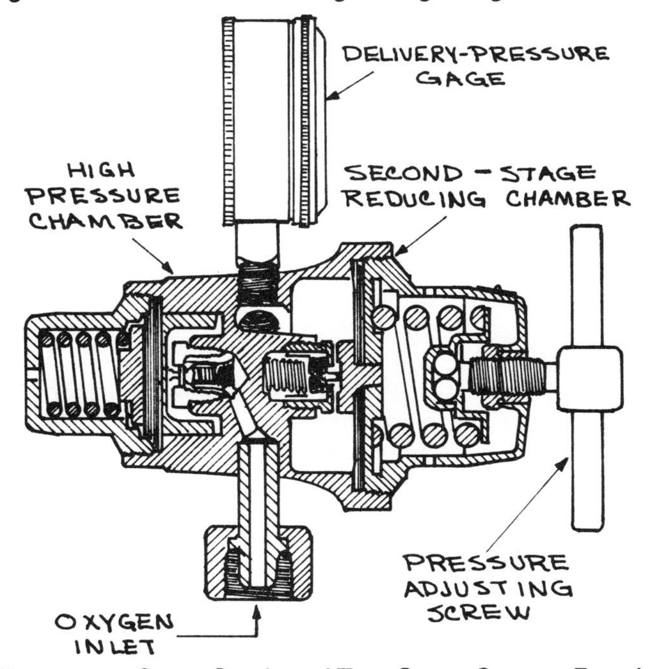

Figure 14. Cross Section of Two-Stage Oxygen Regulator.

From the high pressure chamber, the gas then flows into a second reducing chamber. Control of the pressure in the second reducing chamber is accomplished by a pressure adjusting screw which can be adjusted to the desired pressure according to the welding or cutting job. The main advantage of the two-stage regulator is that it provides for a more uniform flow of gas since the working pressure does not change as the tank pressure decreases.

The single-stage regulator is not as expensive as the two-stage type. The single-stage regulator has no intermediate or high pressure reduction chamber. The gas flows directly from the cylinder to the chamber and diaphragm controlled by the pressure adjusting screw. The main problem with the single-stage regulator is that as the cylinder pressure drops, the working pressure also drops which means that the adjusting screw must be occasionally readjusted to maintain a constant working pressure.

Both types of regulators have two gages as shown in Figure 12. One gage indicates the **tank pressure** while the other reveals the line or **working pressure**. Shown in Figure 15 are the typical tank pressure and working pressure for (A) oxygen and (B) acetylene regulators. As noted, the tank

A

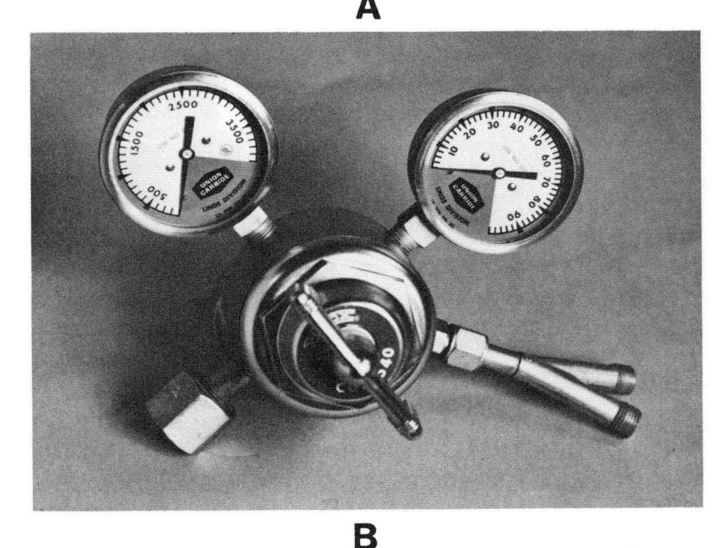

B

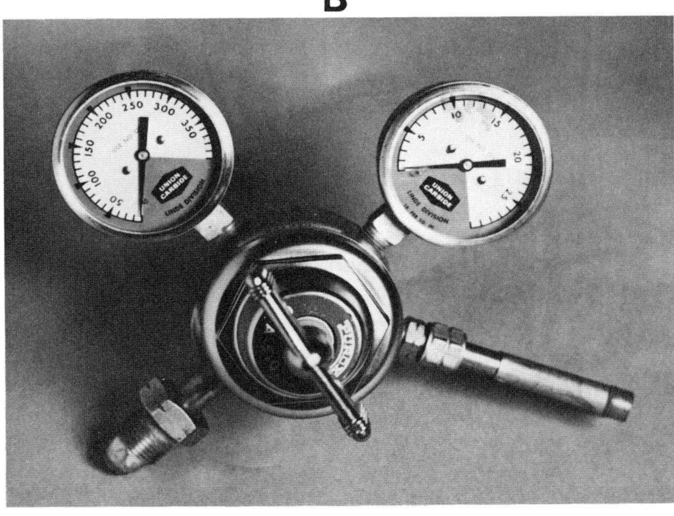

Figure 15. Typical Tank Pressure and Working Pressure Gages on Oxygen (A) and Acetylene (B).

pressure gage for oxygen is graduated from **0-4000 pounds**, whereas, the typical acetylene high pressure gage goes from **0-400 pounds**. The working pressure for the oxygen gage is usually calibrated from **0-100 pounds** while the acetylene working gage goes from **0-30 pounds**. Oftentimes the high pressure oxygen gage has a second scale which registers the content of the cylinder in cubic feet at 70°F.

The most important point to remember when using a regulator is to make certain that the adjusting screw is turned out before the cylinder valve is opened. If the adjusting screw is not turned out before the high pressure from the tank is released, the sudden force of pressure on the working gage could cause damage to the regulator. Further, pressures over 15 pounds in the acetylene working gage could create a possible explosive situation as free acetylene is very unstable over 15 psi. Some acetylene regulator working gages have a **red danger band** for pressure **above 15 pounds** calibrated on the gage. The regulators are the most expensive part of the oxy-acetylene unit; therefore, they should be handled with care at all times. The following rules for handling and using regulators should be kept in mind:

1. Be extremely careful while removing regulators from the cylinder. Never allow a regulator to remain on a bench top for any length of time.
2. Always check the adjusting screw before releasing cylinder valve.
3. Never use oil on a regulator.
4. Do not attempt to interchange the oxygen and acetylene regulators.
5. Have a qualified repairperson check the regulator if it does not function properly.
6. If a regulator creeps(does not remain at set pressure) have it repaired immediately.
7. If the gage pointer fails to go back to zero when the pressure is released, have it repaired.
8. Always keep a tight connection between the regulator and the cylinder. Be sure to crack the valves before attaching the regulator to the cylinder. Cracking is done by turning the valve on slightly before attaching the regulator to blow out all foreign particles in the valve.

TORCHES

WELDING TORCHES

The **oxyacetylene torch** illustrated in Figure 16 mixes oxygen and acetylene gases in exact proportions and controls the size and quality of the flame. Torches, sometimes referred to as the blowpipe, are made in several different sizes. As illustrated, the blowpipe consists of a handle equipped with two needle valves, one for adjusting the flow of acetylene and one for adjusting the flow of oxygen. Also shown are the **hose connection threads**, the **mixing head**, **torch head** and **welding tip**.

Oxy-acetylene torches for welding are classified as **low-pressure injector** and **equal pressure types**. The low-pressure injector blowpipe as illustrated in Figure 17 is designed to operate an acetylene pressure of less than one (1) pound per square inch. Originally, the injector blowpipe was designed to be used with the acetylene generator which produced acetylene at very low pressure. As shown, the blowpipe consists of an oxygen and acetylene valve and passages for both gases; however, as noted, the acetylene passage is much larger than the oxygen. Also note the injector nozzle in which oxygen comes out through the center of the nozzle with acetylene all around the outside of the injector nozzle. The acetylene is drawn into the mixing head through the injector nozzle under pressures from 10 to 40 psi. Through this method, the required amount of acetylene is drawn into the torch head and mixed with the oxygen for welding. One advantage of the injector blowpipe is that small fluctuations in the oxygen supplied will produce a corresponding change in the amount of acetylene drawn into the torch head, thereby keeping the proportions of the two gases constant while the torch is operating. One major disadvantage of the injector type blowpipe is the fact that mixed gases, because of the low pressure in the acetylene passage could travel back through this passage and create a possible flashback.

Shown in Figure 18 is the **equal pressure blowpipe**. The equal pressure type operates on acetylene pressures of 1 to 15 psi depending upon the tip size. In this type, the oxygen and acetylene are fed through the blowpipe needle valves, which are equal in size, at equal pressures and delivered in equal volumes to the mixing chamber. The mixing chamber serves only to mix the gases. The equal pressure blowpipe has two advantages over the injector type: (1) **the flame is easier to adjust** and (2) **there is less chance of a flashback**. Both kinds have two needle valves, one oxygen and one acetylene valve. On the rear end of the blowpipe, there are two fittings for connecting the hoses, the **oxygen fitting** has **right-hand threads**, whereas the **acetylene connection** has **left-hand threads**. As shown on Figure 19, the acetylene connection nut is grooved while the oxygen nut is plain. A special

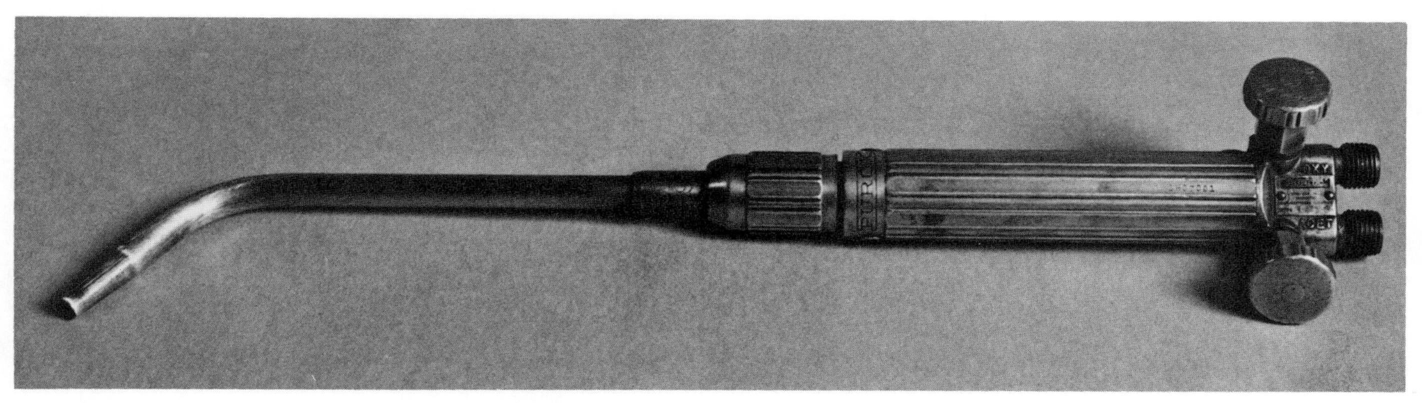

Figure 16. The Oxy-acetylene Welding Blowpipe.

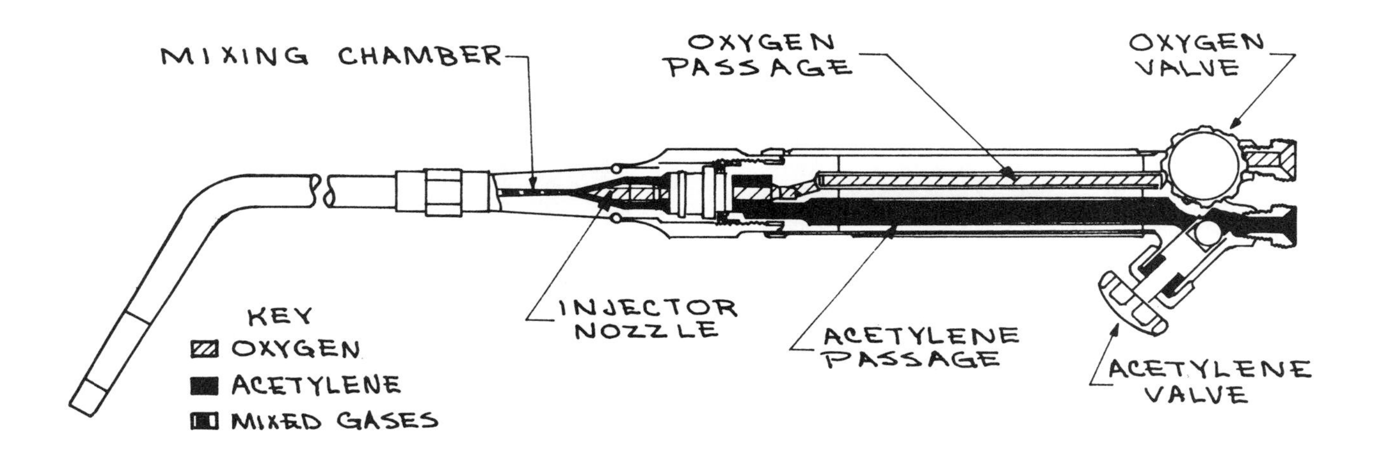

Figure 17. Low-Pressure Injector Blowpipe.

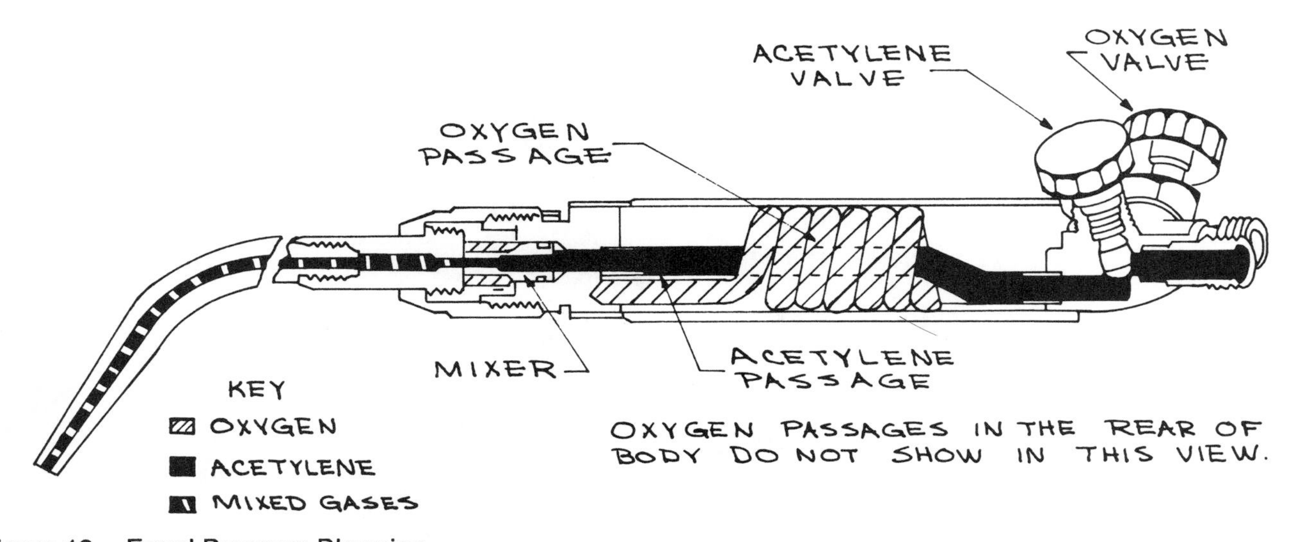

Figure 18. Equal Pressure Blowpipe.

nonporous hose is used for welding. See Figure 20. To avoid interchanging, a color system is used. The **oxygen hose** is usually **green** and the **acetylene hose is red**. Note Figure 19. Shown in Figure 20 is the web-joined welding hoses used on most welding units.

Blowpipe tips are interchangable for a given torch and are available in various sizes for welding different thicknesses of metal. The size of the tip is governed by the diameter of its opening which is marked on the tip. The number system used depends largely on the manufacturer. The most common system consists of numbers ranging from 0-15 with the larger number being the greater tip diameter and capacity. Tips should be kept clean by brushing the end of the tip with fine sandpaper and the use of soft wire or specified tip drill or tip cleaner to clean the passage opening. **Never use a twist drill** as the opening could be enlarged.

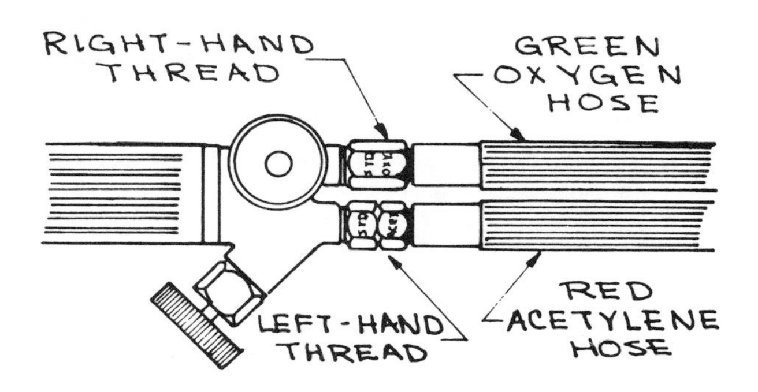

Figure 19. Oxy-Acetylene Blowpipe Needle Valves and Hose Connecions.

CUTTING TORCHES

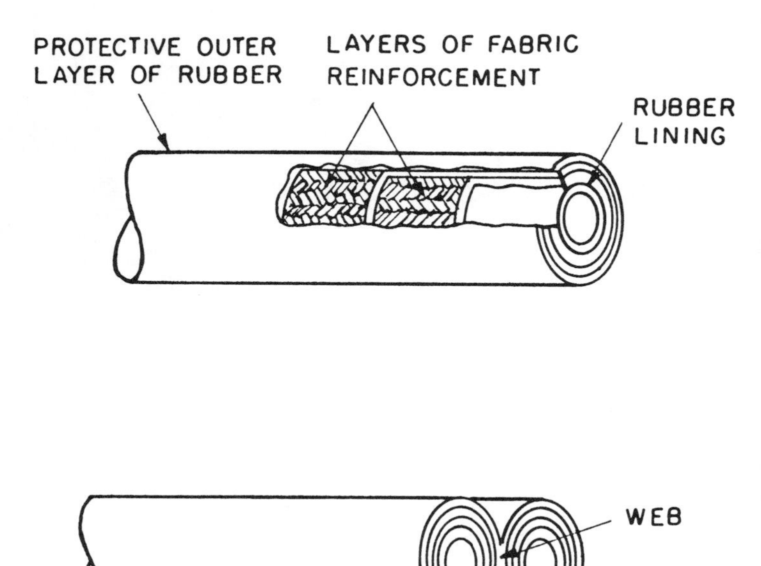

Figure 20. Oxy-Acetylene Hoses.

Shown in Figure 21 is a typical **oxy-acetylene flame cutting blowpipe**. This blowpipe utilizes the equal-pressure mixing system similar to the equal-pressure welding blowpipe. The main difference as shown is that the cutting blowpipe has an additional **lever** for the control of the oxygen used to burn the metal. The blowpipe has the oxygen valve near the oxygen lever and the acetylene valve in the position as found on the welding blowpipe. A cut-away of the cutting blowpipe is shown in Figure 22. As illustrated, the blowpipe has a large passage way for oxygen as controlled by the lever to pass directly and in large amounts through the center of the cutting tip. Also passing to the tip is a mixture of oxygen and acetylene which serves the preheating flame. To better understand the makeup of the cutting blowpipe, note Figure 23, which illustrates the typical cutting tip showing the **four preheat orifices** and the **cutting oxygen orifice**. The center opening permits the flow of the cutting oxygen and the smaller holes are for the heating flame.

The blowpipe as shown in Figure 22 uses the balanced or equal-pressure mixer for mixing and delivering the oxygen and acetylene used in the preheating phase of the blowpipe. The two valves in this type are used to control oxygen and acetylene for the preheat flame. The **lever arm** is the only control over the cutting oxygen.

Shown in Figure 24 is a cutting attachment that can be used with the regular welding blowpipe. The attachment has an oxygen valve which is used with the acetylene valve on the welding blowpipe to balance the flame for preheating. The oxygen valve on the blowpipe must be turned wide open to allow large amounts of oxygen as controlled by the oxygen

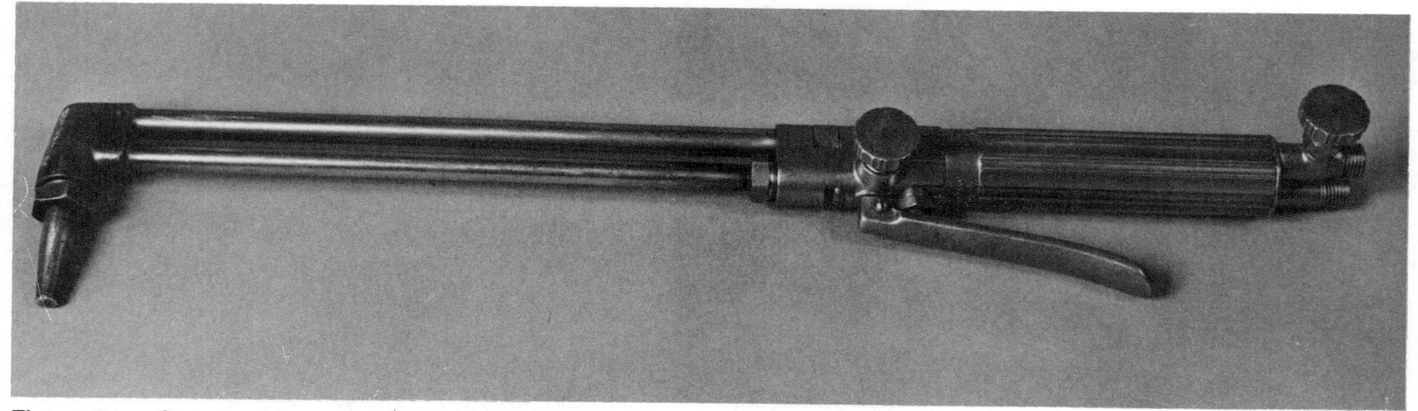

Figure 21. Oxy-Acetylene Flame Cutting Blowpipe.

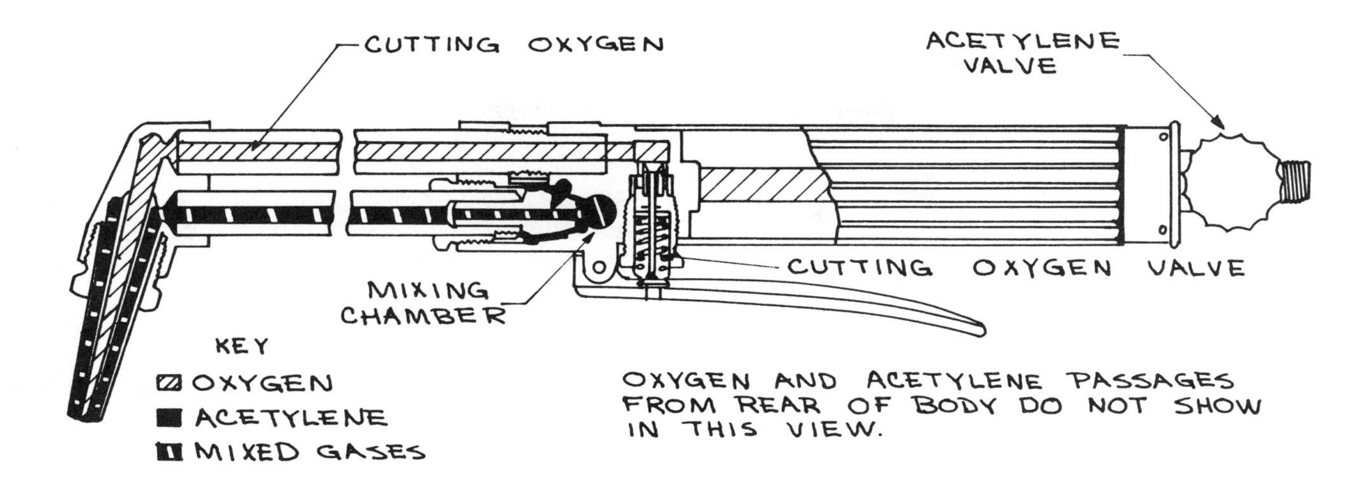

Figure 22. Equal-Pressure Oxy-Acetylene Cutting Blowpipe.

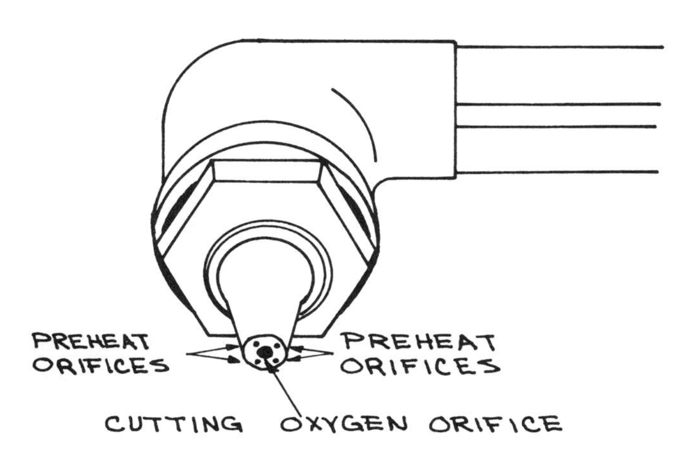

Figure 23. Oxy-Acetylene Cutting Tip.

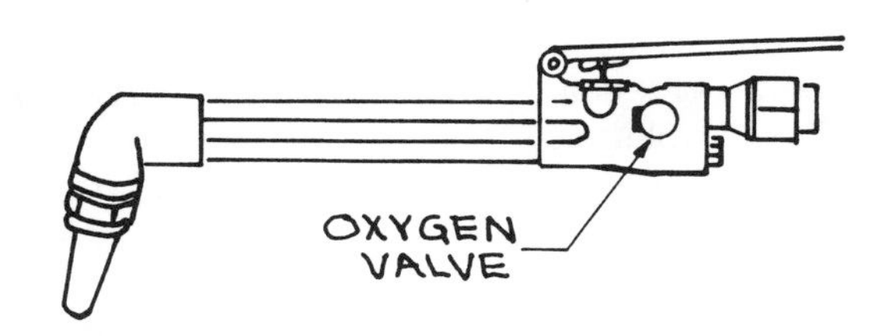

Figure 24. Oxy-Acetylene Cutting Attachment.

Figure 25. Flashback Arresters Stop Explosions.

lever on the attachment for the cutting oxygen.

As with the welding blowpipe, a number of tips for cutting are available for different thicknesses of metal. Similar procedures should be followed in cleaning and caring for cutting tips as discussed for the welding tip.

PREVENTING BACKFIRES AND FLASHBACKS

A **backfire** is an explosion usually accompanied by a popping sound, confined to the torch head. It occurs when the heating head or cutting tip is held too close to the work. As the velocity of mixed gas is slowed when it leaves the tip, it drops below the rate needed to feed the flame. The flame then burns back inside the torch. If the torch head heats to above 600°F, spontaneous ignition of the gases in the torch may occur, leading to a flashback. A **flashback** is an explosion that progresses through the torch and gas supply equipment. Its consequences can be a damaged torch, a burst hose, or a violent explosion of regulator or cylinder, and fire started by the flame and supported by unrestricted flow of fuel gas and oxygen. Note Figure 25. Backfires can rapidly turn into flashback or fire in the head of the torch. The popping sound turns to a whistle as the gases burn inside the torch head. If this happens, close the oxygen torch valve immediately, shut down the equipment, and check for the cause of the problem. Loose connections or leaks in the gas hose can reduce gas flow below flame requirements, causing the flame to burn back into the torch.

CHECK VALVES AND ARRESTERS

Check valves as shown in Figure 26 prevent reverse flow of gases. They are one-way valves, consisting of a ball or plate forced against a sealing surface by a spring. Gas

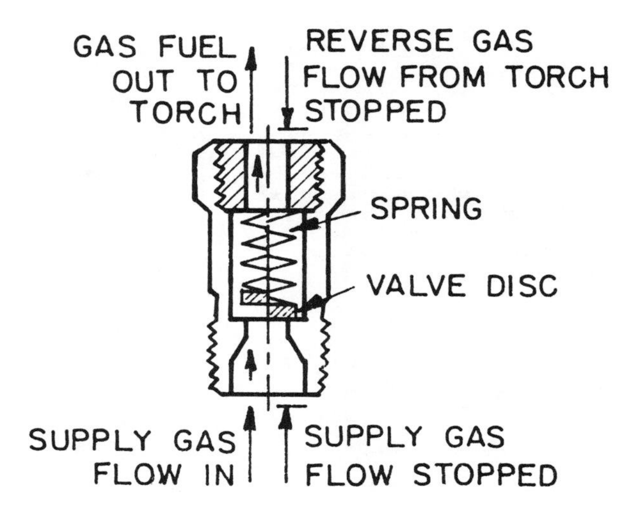

Figure 26. The Check Valve.

pressure opens the valve and allows gas to flow. When flow stops, the valve closes and does not allow gas to flow in the opposite direction. Check valves should be placed on torch inlets. Sensitive to dirt and to damage to the seating surface, they should be checked at least once every six months to make sure that they are in working condition.

Flashback arresters, see Figure 27, eliminate the risk of an explosion in the regulator and cylinder. Small, lightweight flashback arresters fit on the regulator to halt flash-

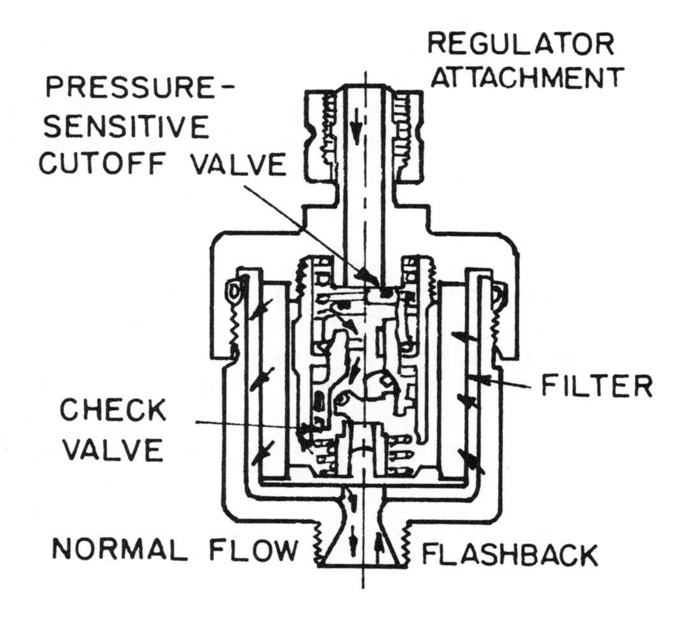

Figure 27. The Flashback Arresters.

back at that point, and on pipelines at drop points. In normal operation, left, gas flows through the open cut-off valves and check valve through the flame arrester filter into the hose. Study Figure 27. In the event of a flashback, right, the stainless steel filter stops the flame and the pressure wave activates the cut-off valve, stopping the flow of gas to extinguish the flame. The check valve operates when gas flows towards the cylinder. If the arrester is exposed to fire, the thermal cut-off valve shuts the gas supply. The arrester is reusable after a flashback.

Modern flashback arresters incorporate many safety functions:

1. A heavy-duty reverse-flow check valve in the arrester to stop flow of gas in the wrong direction.
2. A pressure-sensitive cut-off valve that cuts off gas flow if an explosion occurs.
3. A thermal cut-off valve that stops gas flow when the arrester heats above 220°F. Check valves and flashback arresters are essential to protect people, equipment, and factories from dangers caused by improper use of oxyfuel equipment.

THE OXY-ACETYLENE FLAME

Now that we have a basic understanding of the materials and equipment involved in oxy-acetylene welding and cutting, we must next look at the most important product of the total process, that of the oxy-acetylene flame. Earlier, we discussed that the oxy-acetylene flame is one of the hottest flames on earth, producing temperatures upward to 6300 degrees Fahrenheit depending upon the proportions of oxygen and acetylene supplied to the blowpipe tip.

TYPES OF FLAMES

The complete combustion of one volume of acetylene requires 2 1/2 volumes of oxygen. However, equal volumes of oxygen and acetylene are supplied from the cylinders through the blowpipe tip; therefore, the remaining 1 1/2 volumes of oxygen come from the surrounding atmosphere. The parts of the oxy-acetylene flame are illustrated in Figure 28. Note the flame has three distinct parts, **the inner cone**, the **acetylene feather** and the envelope.

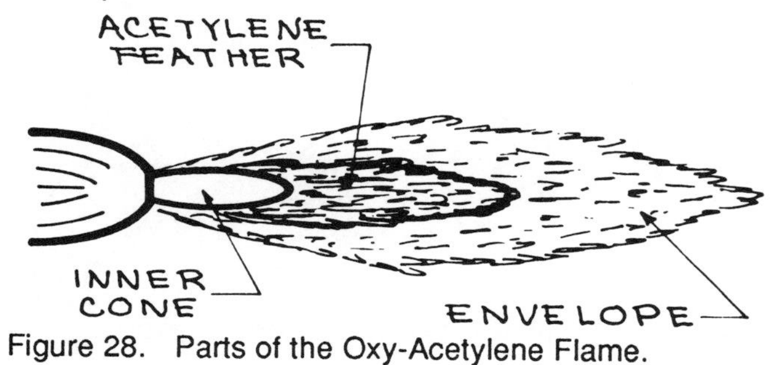

Figure 28. Parts of the Oxy-Acetylene Flame.

NEUTRAL FLAME

5,850 - 6,000
DEGREES FAHRENHEIT

OXIDIZING FLAME

6,000 - 6,300
DEGREES FAHRENHEIT

CARBURIZING FLAME

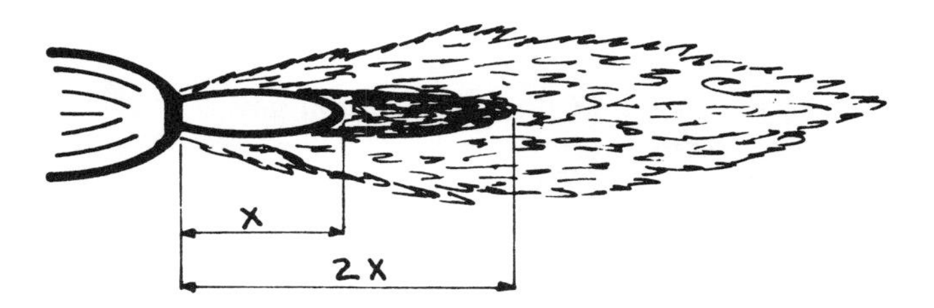

APPROXIMATELY 5.500
DEGREES FAHRENHEIT

Figure 29. Neutral, Oxidizing and Carburizing Flames.

There are three distict types of flames that you will need for various oxy-acetylene welding and cutting jobs. Shown In Figure 29 are the three flames. As noted, the **neutral flame** has only the inner cone and the envelope. Actually, the acetylene feather is present except with the neutral flame, the feather and the inner cone should be exactly the same length. The neutral flame produces approximately 6000 ° F of heat at the end of the inner cone. Another characteristic is the shape of the inner cone, note the cone has a rounded blunt shape.

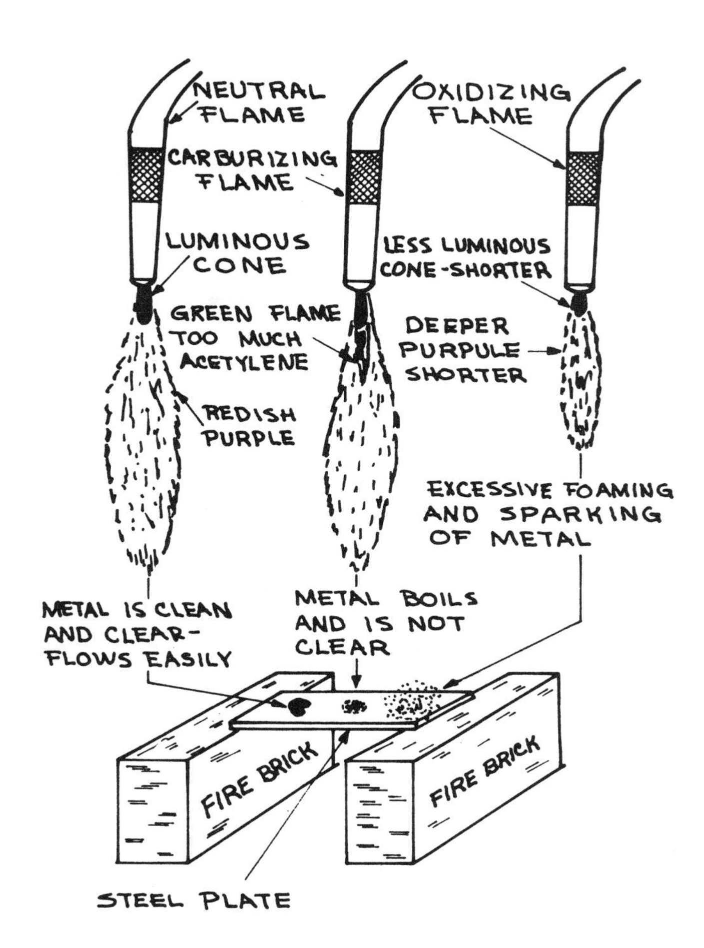

Figure 30. Effect on Steel of the Neutral, Carburizing and Oxidizing Flames.

The **oxidizing flame** is the second flame illustrated in Figure 29. The parts of the flame as with the neutral flame are the inner cone and the envelope; however, this flame is much different from the neutral flame. First, note the shape of the inner cone; it is shorter and appears to be quite pointed or sharp. Also, as will be seen in the laboratory, the inner cone is a very bright white and the envelope has a purplish cast. Also, the flame burns with a decided roar or hissing sound, The oxidizing flame uses excess oxygen and burns at approximately 6300 ° F.

The third flame shown in Figure 29 is the **carburizing flame**. As noted, the acetylene feather is now visible beyond the inner cone. The carburizing flame is the result of excessive acetylene in proportion to oxygen. The carburizing flame is somewhat cooler than either the neutral or oxidizing flame burning at approximately 5500° F. The length of the acetylene feather can be varied by turning on more acetylene. For example, a 2x or 3x flame would indicate that the acetylene feather is 2 times or 3 times the length of the inner cone.

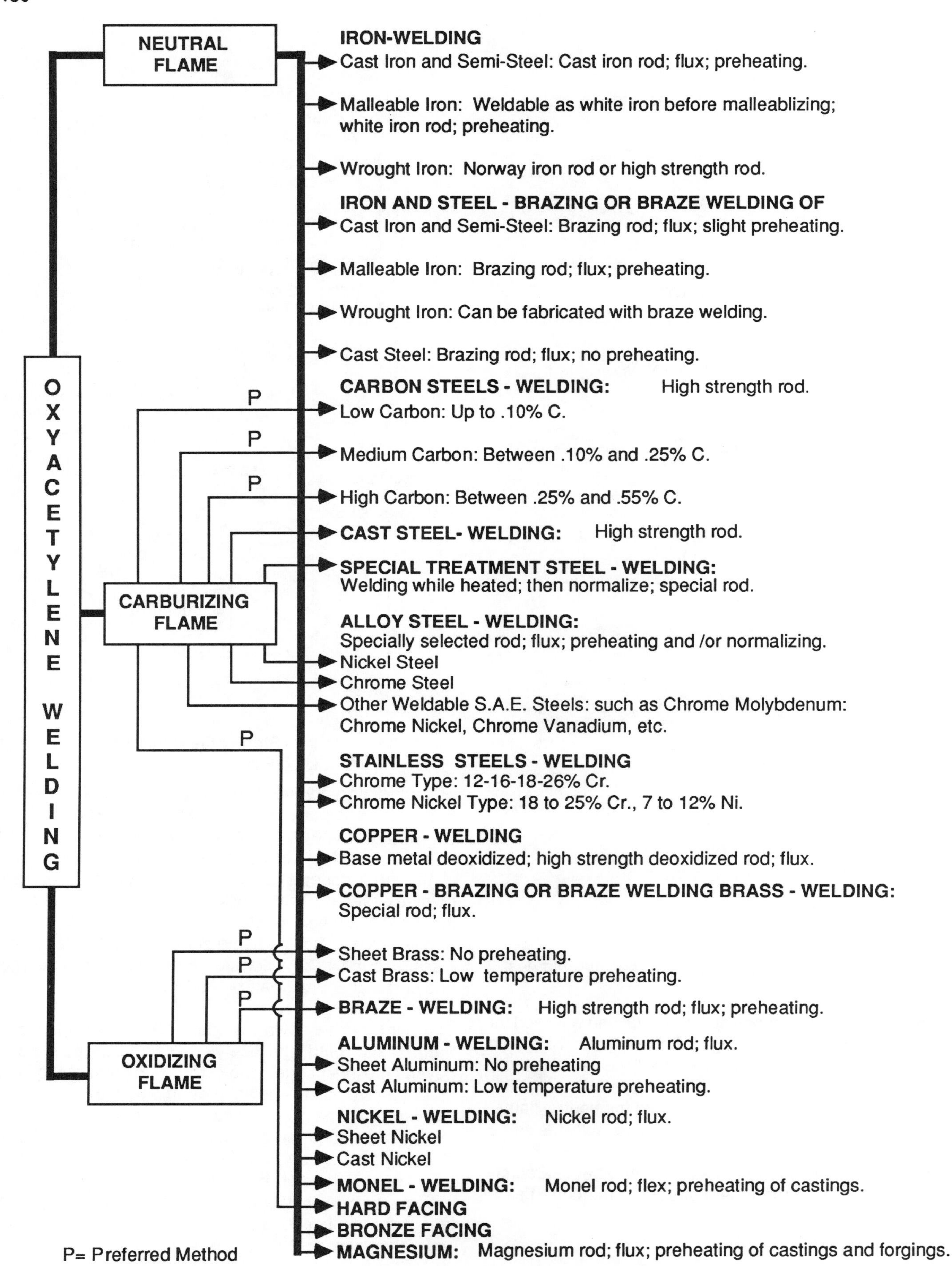

Figure 31. Preferred Oxy-Acetylene Flame for Various Welding Applications.

Now that we have a basic idea of the three flames and how they are adjusted, we must next understand how the three flames affect steel during the welding process. Illustrated in Figure 30 is the effect of the three flames on steel. As noted, the neutral flame causes the metal to flow smoothly, with very few sparks, and is clean and clear. The carburizing flame, excessive acetylene, causes the metal to boil, and it is not clear and clean. Upon cooling, the surface of the metal is pitted and appears to be brittle. The carburizing flame is sometimes known as a reducing flame because it has a tendency to remove the oxygen from the iron oxides in the metal.

The third flame in Figure 30, the oxidizing flame, causes the metal to foam and numerous sparks are given off. The cooled metal has a shiny surface.

As there are many welding applications, not all requiring a neutral flame, through experience it is known that certain flames do work better for specific applications. Shown in Figure 31 are the various welding applications and the preferred method or flame recommended for different metals. As illustrated, the neutral flame is the major process used for the most metals; however, there are certain metals and applications where either the carburizing or oxidizing flame has an advantage over the neutral flame. The carburizing flame is best for carbon steels using a high strength rod, alloy steel welding, and hard-facing. The oxidizing flame is the preferred method for sheet brass, cast brass and braze welding using high strength rod. The major process used for the other applications is the neutral flame.

CHEMISTRY OF THE FLAME

It is important that one has a basic understanding of the chemistry of the oxy-acetylene flame. There are a number of gases and chemicals involved with the production of the flame. The most common and their chemical formulae are:

1. Carbon - C
2. Hydrogen - H
3. Oxygen - O_2
4. Acetylene - C_2H_2
5. Carbon Monoxide - CO
6. Carbon Dioxide - CO_2
7. Water - H_2O

Shown in Figure 32 is the chemical composition of the various parts of the oxy-acetylene flame. As noted, the chemical reaction for the inner cone is:

$$C_2H_2 + O_2 = H_2 + 2\ CO$$

This equation is referred to as the **primary combustion** and takes place at the inner cone. As noted, one volume

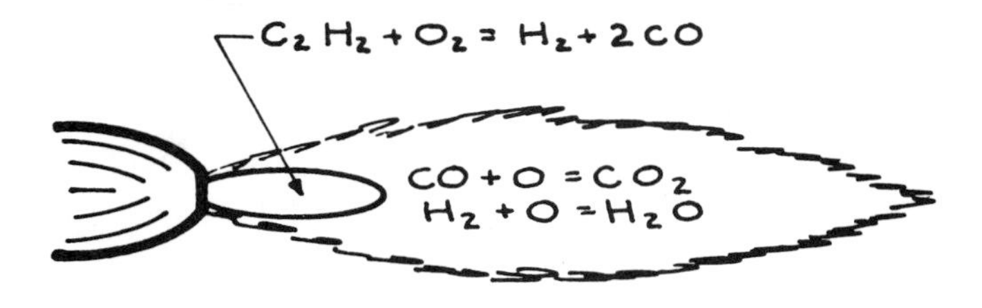

Torch Flame - Complete Reaction

1 volume of acetylene combines with 2 .5 volumes of oxygen and burns to form 2 volumes carbon dioxide and 1 volume water vapor plus heat

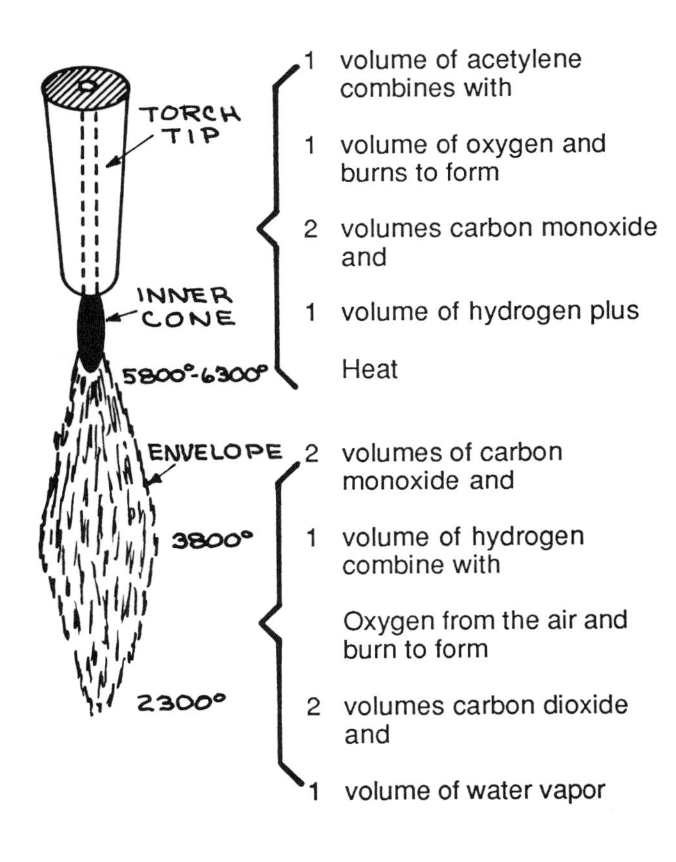

Figure 32. The Chemistry of the Oxy-Acetylene Flame.

of acetylene (C_2H_2) is united with an equal volume of oxygen (O_2). From this, two volumes of carbon monoxide (CO) and one volume of hydrogen (H_2) are formed. In addition, heat at a temperature of approximately 6000° F is produced.

The carbon monoxide and hydrogen are both combustible gases and are burned in the envelope in the presence of additional oxygen provided by the atmosphere as the

secondary stage of combustion. The chemical equation for the secondary stage is:

$$2\,CO_2 + O = 2\,CO_2$$

and

$$2\,H_2 + O_2 = 2\,H_2O$$

As noted in Figure 32, the carbon monoxide (CO) combined with oxygen (O_2) to produce carbon dioxide (CO_2); whereas, the hydrogen (H_2) is combined with oxygen (O_2) to produce water (H_2O). Heat is also produced in the secondary stage; however, as noted, the heat produced in the secondary stage of combustion is somewhat lower and ranges from 2300° to 3800° F.

By combining the chemical equations for the primary and secondary stages, complete combustion of acetylene in oxygen is accomplished and the chemical equation is as follows:

$$\underset{\text{(Acetylene)}}{2\,C_2H_2} + \underset{\text{(Oxygen)}}{5\,O_2} = \underset{\text{(Water)}}{2\,H_2O} + \underset{\text{(Carbon Dioxide)}}{4\,CO_2} + \text{Heat}$$

As noted, one volume of acetylene will combine with two and one-half volumes of oxygen However, writing the equation with fractions of oxygen, 2 1/2 voulmes, is more difficult; therefore, the acetylene volume is doubled to arrive at a whole number for oxygen in the complete combustion equation.

Two problems which were discussed earlier in this chapter therefore sometimes experienced with the oxy-acetylene flame are (1) **backfires** and (2) **flashbacks**. When the flame goes out with a loud snap or "pop", it is called a backfire. A backfire can be caused by:

1. Operating the torch at lower pressures than required for the tip.
2. Touching the tip against the work or in the molten pool.
3. Overheating the tip.
4. Unequal pressure when using the equal pressure blowpipe.
5. Loose or dirty tip.
6. Obstruction in the tip.

A flashback, somewhat more dangerous than the backfire, occurs when the flame burns back inside the blowpipe. It makes a shrill hissing or squealing sound. A flashback could be caused by:

1. Clogged orifice in blowpipe.
2. Clogged tip.
3. Improper functioning of the blowpipe needle valves.
4. Incorrect gas pressure.
5. Kinked hoses.

When a flashback occurs, close the oxygen valve immediately, then close the acetylene valve and correct the problem before relighting the torch. As discussed earlier, the use of check valves or a flashback arrester would be good safety devices to add to the oxy-acetylene welding unit.

SETTING UP AND OPERATING OXY-ACETYLENE EQUIPMENT

Before you can learn how to operate the oxy-acetylene welding equipment, it is important that you know how to properly set up the oxy-acetylene unit. Before assembling the welding equipment, be sure that a fire extinguisher is available and that you know how to use the extinguisher.

To set up the oxy-acetylene welding unit, follow these steps:

1. **Secure Cylinders**- Place the acetylene and oxygen cylinders on the portable carrier, attach them securely and remove the protective caps. Inspect the outlet nozzles making sure the connection seat and screw threads are not damaged.

2. **Crack cylinder valves**- As shown in View 1 of Figure 33, each valve should be opened slightly to blow out any dirt or foreign material that may be lodged in the outlet nozzle. Dirt lodged in the outlet nozzle could damage the regulator when the pressure is turned on.

3. **Attach regulators**- Connect the oxygen and acetylene regulators to their respective cylinders and tighten the union nut. Use a close-fitting wrench to tighten the nuts and avoid stripping the threads. Note view 2 of Figure 33.

4. **Connect hoses to regulators**- Attach the red hose to the acetylene regulator outlet (left-hand threads). Also note the groove on the connection nut of the acetylene hose. Attach the green hose to the oxygen regulator outlet (right-hand threads). Turn the connection nuts tightly as shown in view 3 of Figure 33.

5. **Blow out hoses**- Next release the regulator adjusting screws by turning them counterclockwise until they are loose. Open the cylinder valves slowly 1/4 to 1/2 turn, then open the acetylene cylinder 3/4 to one turn and the oxygen full on, note view 4 of Figure 27. Now turn the regulator screws clockwise just enough to blow out the hoses, clearing out any dirt or foreign material as shown in view 5 of Figure 33.

6. **Connect hoses to blowpipe**- Before attaching hoses to the blowpipe, blow out the interior of the blowpipe and welding tip as shown in view 6 of Figure 27. After cleaning the blowpipe, release the regulator screws.

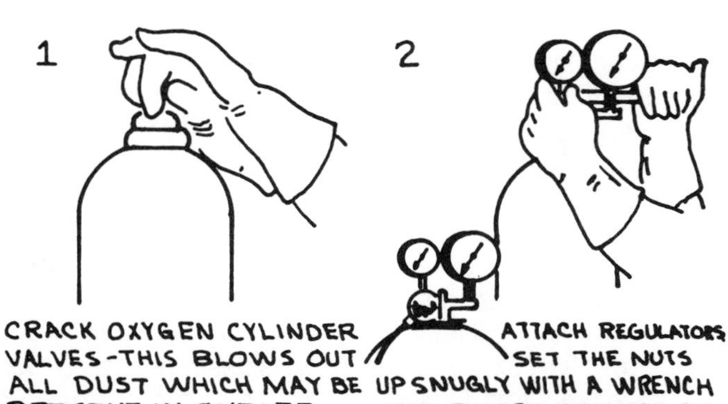

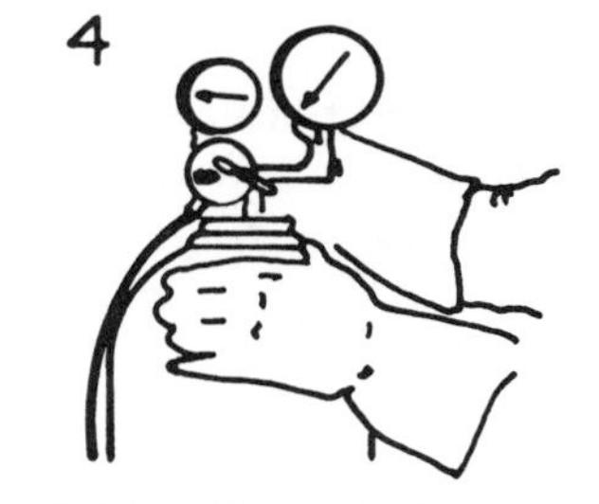

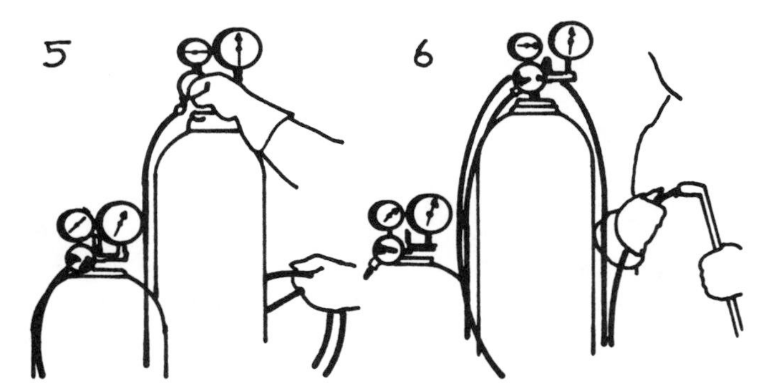

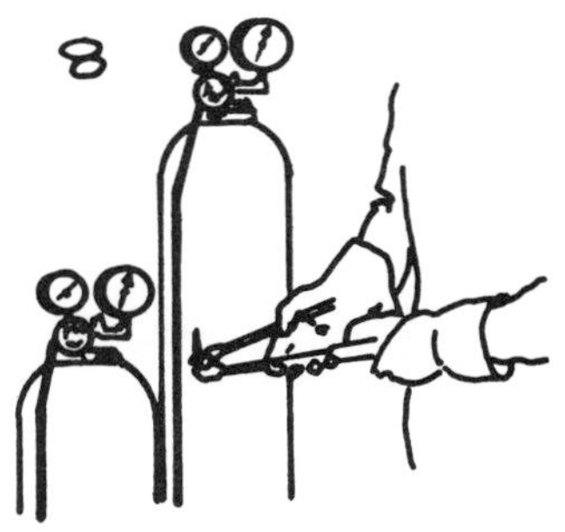

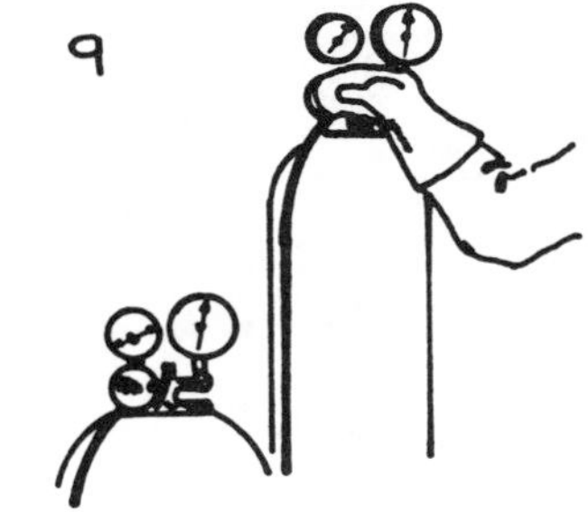

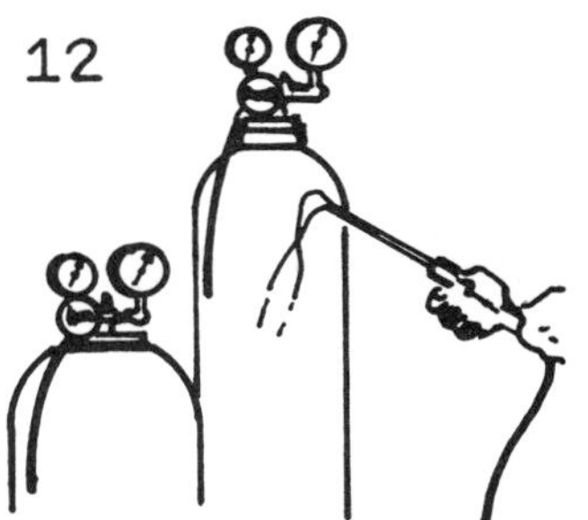

Figure 33. Setting Up and Starting Oxy-Acetylene Unit.

Attach the red hose to the acetylene connection, normally marked ACET, on the blowpipe with left-hand threads and the green hose to the connection marked OXY on the blowpipe. Note view 7 of Figure 33.

7. **Select and attach tip**- Select the welding or cutting tip and attach to the blowpipe. Tighten the cutting tip with properly fitting wrench and welding tip with hand. Never use a pliers as it will slip on the nut and round corners. Also, welding tips will be too tight and could lock in place. Note view 8 of Figure 33.

8. **Turning on oxy-acetylene unit**- Turn the oxygen cylinder valve wide open and the acetylene cylinder 3/4 to one turn. As shown in view 9 of Figure 33, next open the needle valves of the torch. Adjust the oxygen regulator screw until the desired working presssure appears on the working pressure gage, then close the oxygen needle valve on the blowpipe. Note a similar procedure in view 10 of Figure 33. Open the acetylene screw to the desired working pressure. Close the acetylene needle valve at the blowpipe.

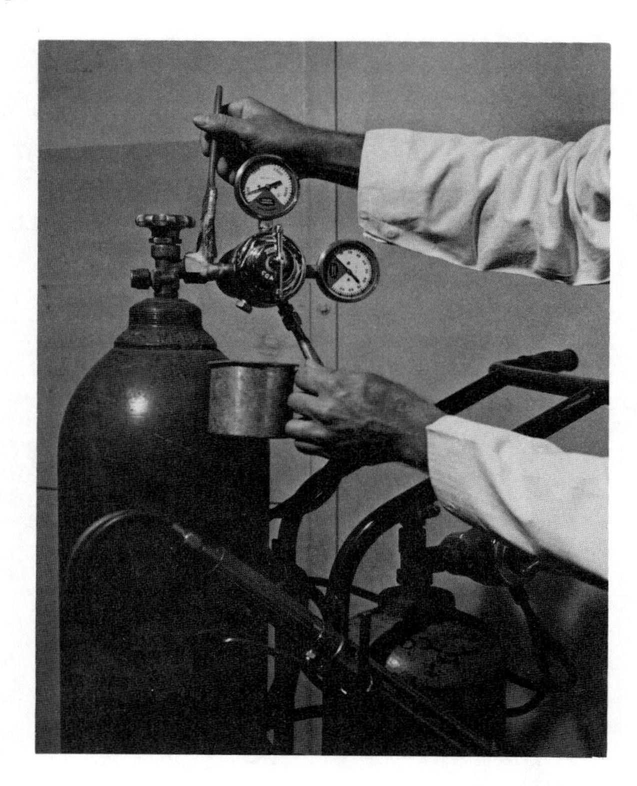

Figure 34. Using a Paint Brush and Soap Suds to Check for Leaks.

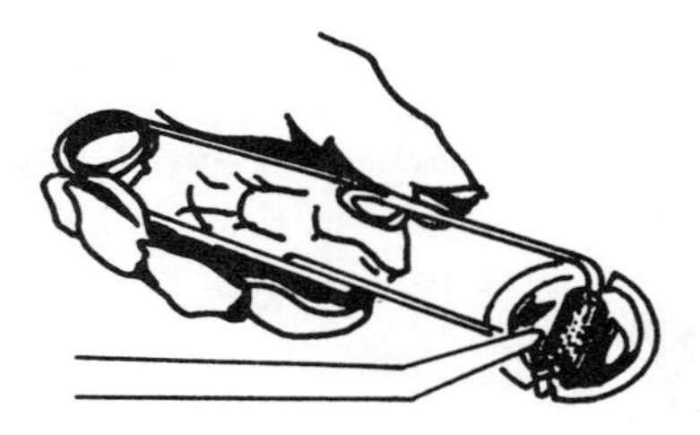

Figure 35. Use the Friction Lighter to Light the Oxy Acetylene Flame.

9. **Test for leaks**- It is a must to check for leaks before attempting to light the unit, note Figure 34. Leaks are not only dangerous but also waste gas. To test for leaks, apply soapy water with a paint brush to the following valves:
   A - oxygen cylinder valve
   B - oyxgen regulator valve
   C - acetylene cylinder valve
   D - acetylene regulator valve
   E - hose connection of regulators and torch
   F - oxygen and acetylene needle valves

10. **Lighting the torch**- Always use a friction lighter as shown in Figure 35 to light the oxy-acetylene flame. The correct procedure for lighting is illustrated in views 11 and 12 of Figure 33. First, open the acetylene valve on the blowpipe approximately three-quarters of a turn. With the friction lighter held about one inch away, ignite the acetylene gas at the tip. If not enough acetylene is turned on, you will note considerable sooting; therefore, quickly turn on more acetylene until the flame has a slight tendency to jump away from the end of the tip. Now open the oxygen needle valve and adjust to a neutral flame or other flame desired. If you desire a smaller flame, turn the oxygen needle valve off slightly and then balance the flame by turning the acetylene needle valve off until you have again reached a neutral flame. If a larger volume flame is desired, turn on more acetylene and balance by turning on more oxygen. Always light the flame down and away from other people or flammable materials.

After the proper flame is achieved, recheck the working pressure gages to see that the desired pressures are available for welding or cutting. Note changing the regulator pressure will change the flame at the blowpipe tip; therefore, it will be necessary to readjust the flame.

SHUTTING OFF THE OXY-ACETYLENE UNIT

It is equally important that you follow the proper procedure in shutting off the acetylene unit. The following sequence of operations should be followed in shutting off the oxy-acetylene unit.

1. Close the acetylene needle valve.

2. Close the oxygen needle valve.

3. If you are shutting the unit down such as at the end of the day, next close the acetylene cylinder valve.

4. Close the oxygen valve.

5. Open the oxygen needle valve on the blowpipe to release all pressure on the working and tank pressure gages of the oxygen regulator.

6. Turn the oxygen pressure adjusting screw out counter-clockwise on the oxygen regulator.

7. Close the oxygen needle valve at the blowpipe.

8. Open the acetylene needle valve releasing all pressure on the acetylene regulator.

9. Turn out the pressure adjusting screw at the acetylene regulator.

10. Close the acetylene needle valve at the blowpipe.

Shutting off the flame and releasing all pressures on the welding unit is often referred to as **bleeding the lines**. Bleeding the lines at the end of the work day does waste gas; however, for safety purposes, it is a must that the lines be bled at the end of each work day. Further, releasing the pressure on the gages, regulators, hoses and total unit adds greatly to the life of the unit.

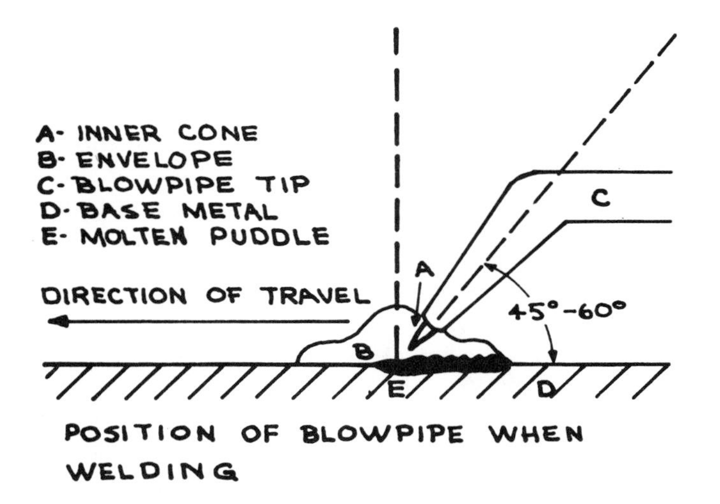

Figure 36. Proper Position of Blowpipe Tip and Base Metal for Running a Bead Without a Filler Rod.

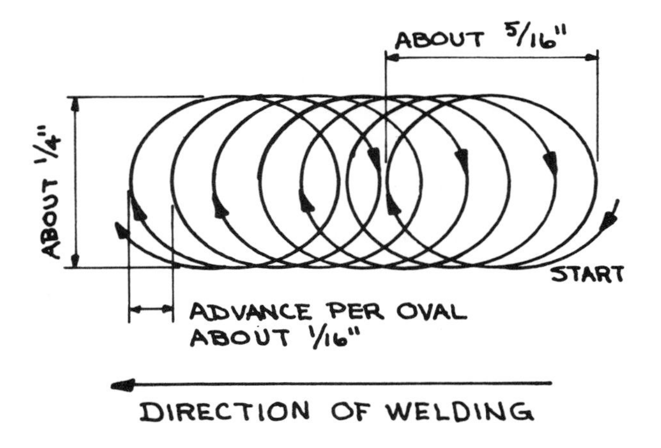

Figure 37. Proper Motion of Blowpipe Tip for Running a Bead Without a Filler Rod.

FLAT POSITION WELDING

The first technique that must be mastered before anyone can become proficient in oxy-acetylene welding is **holding the blowpipe**. There are two accepted methods; the first method is commonly used when welding in the sitting position. In this method, the blowpipe is held like a pencil by grasping the handle of the blowpipe with the thumb and first two fingers with the hose over the outside of the wrist. The second method is commonly used when welding in the standing position. Here the blowpipe is held like a hammer with the fingers lightly curled around and underneath. It is suggested that the oxygen and acetylene valves should be pointed down to prevent the operator's forearm from interfering with the adjustment of the blowpipe flame. In either holding method, it is important that the blowpipe be balanced in the operator's hand to avoid tiring the hand or arm.

The next technique that must be mastered is the **proper position and motion of the blowpipe** in respect to the metal. This discussion will pertain to flat position welding. The proper position and motion of the blowpipe for carrying a puddle without a welding rod is shown in Figure 36. Note the welding tip is slanted at approximately 45° toward the direction of travel with the inner cone of the neutral flame approximately 1/8 inch from the surface of the metal. In looking toward or directly away from the direction of travel, the angle of the tip should always be 90° so as not to create a lopsided, molten puddle by directing more heat to one side or the other.

The diameter of the puddle will vary with the thickness of the metal normally 1/4 to 1/2 inch in diameter. To carry a puddle, oftentimes referred to as a running bead, move the blowpipe in a clockwise, circular motion as shown in Figure 37. If you are right handed, start the weld at the right edge and move right to left; if you are left handed, reverse your direction. Do not move the blowpipe ahead of the puddle. The forward speed should allow for the puddle to form and result in a uniform bead having even, half-moon ripples. If too slow of travel, a wide bead or possible burn through will result, whereas too fast of travel results in a narrow, high bead with little or no penetration into the base metal.

Once the basic skill of manipulation of the welding blowpipe in running beads or carrying a puddle without a rod is developed, the next step is to add a filler rod to the puddle. The purpose of the **filler rod** is to build up the surface of the base metal by adding a filler material generally consisting of mild steel or material similar to the base metal.

Filler rods come in various sizes, ranging from 1/16" to 3/8" in diameter. The general rule is to select a filler rod having a diameter equal to the thickness of the base metal. If the rod is too large, the heat of the molten puddle will be insufficient to melt the rod and if too small, the heat cannot be absorbed by the rod resulting in a hole being burned through the base metal.

The filler rod is commonly used to weld two pieces of metal together. Attempting to fuse two pieces of metal together without a filler rod would certainly result in a weld somewhat thinner and probably weaker than the original metal. The strength of a weld where a filler rod is used depends largely upon the skill of the operator to blend or interfuse the rod with the edges of the base metal.

As with running a bead without a rod, the width or size of the molten puddle is controlled with the amount of heat and speed of travel. If travel is too slow, the puddle becomes too large and the base metal may be burned away. Further, the modular structure of the metal is affected causing it to become quite brittle resulting in a hard brittle weld that will probably break.

If the puddle is carried too rapidly, the rod will not have a chance to fuse properly with the base metal resulting in a narrow, poorly fused bead with uneven ripples.

When you first attempt to weld with a filler rod, the rod may tend to stick or freeze to the base metal. This is because the rod is not in the middle of the molten puddle. When the rod is allowed to get to the edge of the puddle or out of the puddle, there is not sufficient heat to melt the rod; therefore it freezes to the base metal. To loosen the rod, simply direct the flame to the tip of the rod freeing it from the base metal.

The use of the filler rod requires coordination of both hands, one holding the blowpipe and one manipulating the rod. The position of the filler rod and welding blowpipe in making a butt weld is shown in Figure 38. The rod should be held at approximately the same angle as the blowpipe tip except it is slanted away from the tip. Before starting the weld, tack the left hand end of the two pieces to keep them from distorting apart. To begin to weld, melt a small puddle using only the flame to melt the base metal. As soon as the metal becomes molten, insert the rod. As the rod melts in the puddle, advance the torch forward concentrating the heat on the base metal and not on the rod. The rod must be melted by the molten puddle and not allowed to melt and drop into the puddle as oxygen from the air will combine with molten filler rod burning part of it away causing a weak, porous weld.

DIRECTION OF TRAVEL

Figure 38. Welding With a Filler Rod Using the Forehand Technique.

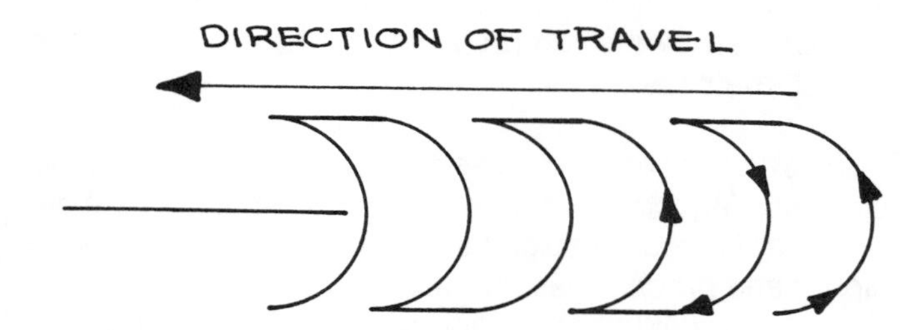

Figure 39. Semicircular Tip Movement Used In Running a Bead With a Filler Rod.

To run a bead, manipulate the blowpipe tip in a semicircular motion as shown in Figure 39. Continue to raise and lower the rod as the molten puddle is moved forward across the metal. When the rod is not in the puddle, keep the tip of the rod just inside the outer envelope of the neutral flame. The circular motion illustrated in Figure 37 can be used as an alternative movement in welding with a filler rod.

There are two specific welding techniques in using the filler rod. The **forehand technique** is shown in Figure 38. As noted, the direction of travel is from right to left with the flame directed ahead of the bead or toward the direction of travel. This is the most common method for butt welds or running beads using the filler rod.

The **backhand method** is illustrated in Figure 40. The weld progresses from left to right and the welding flame is directed backward toward the completed portion of the weld. The rod is held between the completed weld and the flame. Complete fusion between the edges of the metal and the weld puddle is more difficult to obtain because there is less heat on the edges of the base metal. When the operator makes the first (maybe the only) pass, they should be careful to watch the leading edge of the molted pool for a "key-hole" to form. This is vital for obtaining a full-penetrated weld. The backhand technique is used for vertical down welds and has advantages over the forehand technique for pipe welding. Specific directions for welding various joints will be given in the Laboratory Exercise section of this unit.

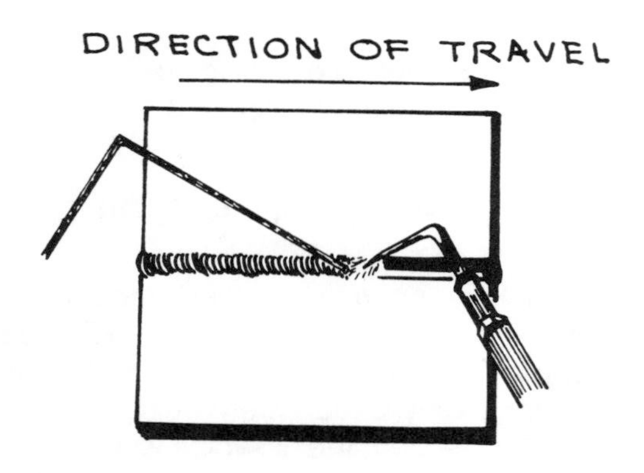

Figure 40. Welding With a Filler Rod Using the Backhand Technique.

FUSION WELDING CAST IRON

Cast iron may be fusion welded using a cast iron rod and special cast iron flux. The main difference in ths operation is that more heat is applied as the rod and base metal are fused or puddled together; therefore, preheating of the total part is essential to avoid fractures at other locations. The edges of the parts are beveled, the surfaces cleaned, and the parts positioned similar to braze welding of cast iron.

After preheating the total area begin welding on the right hand end of the groove by bringing the edges of the vee to a molten form. Heat the end of the rod while melting the edges of the vee. Dip the rod into the flux and insert the rod into the molten puddle. The heat of the puddle should melt the rod as in fusion welding of steel. Build the molten puddle up so that the weld is slightly higher than the metal surface. When built up, move the puddle forward. When white spots or gas bubbles appear, add more flux and play the flame around the surface bringing the impurities to the top. Proceed across the surface completing the weld.

Reheat the entire surface after the weld is completed and allow to cool evenly and slowly. In testing the weld, the fracture should occur along the edge of the completed weld.

BRAZE WELDING

Brazing is defined by the American Welding Society as a group of welding processes using a filler rod of a non-ferrous metal or alloy whose melting point is higher than 800° F, but lower than that of the base metals being joined. The filler metal is distributed between closely fitted surfaces of the joint by capillary action. **Capillary action** would be defined as the power of a heated surface to draw and spread molten metal. Brazing differs from fusion in that the base metal is **not** brought to a molten state. Further, there is little or no alloying or mixing of the filler rod with the base metal. The strength of the braze weld depends entirely upon the uniform flow of a thin layer of filler material between closely fitted joints of the base metal. The strength is comparable to that of a fusion weld providing proper welding procedures are followed.

BRAZE WELDING MILD STEEL

Braze welding, one of the brazing processes, is carried out much as in fusion welding except that the base metal is not melted. The base metal is brought up to what is known as a tinning temperature (dull red color) and a bead is deposited over a joint with a braze welding rod. Even though the base metal is not melted, the unique characteristics of the bond when using the brazing rod is such that the strength is comparable to joints made through fusion welding. A microscopic view of a braze weld would show the small pores in the base metal opening when heated, the braze filler rod flowing into the pores and the pores closing like small vises on the filler material during the cooling process.

The braze filler rod is composed mainly of **copper** and **zinc**; it has high corrosion resistance and high strength and ductility. Small amounts of tin, iron, silicon and manganese are added to the rod to improve its ability to flow as well as reduce the tendency to fume. The filler rod is approximately 60% copper and 40% zinc and has a melting temperature of approximately 1600°F. This temperature is somewhat below the melting temperature of steel (2700°F) and cast iron (2200°F). The 1600°F melting temperature of the rod is arrived at as copper melts at approximately 2000°F whereas zinc melts at approximately 800°F.

As the joint for braze welding must be clean, the use of a chemical cleaning material or flux is essential for braze welding. The joint is first cleaned mechanically by grinding, buffing, sand blasting or chipping. Even after mechanical cleaning, numerous oxides and metal impurities can remain on the surface of the metal. A flux must be applied to remove these impurities. The impurities are dissolved and brought to the surface of the weld not allowing them to mix with the filler material. One special precaution for using fluxes is that they oftentimes contain flourides which produce harmful toxic fumes. Avoid breathing the fumes and weld only in well ventilated areas.

Flux can be applied in two manners (1) **powdered flux** applied to the joint by the heated rod and (2) **pre-fluxed brazing rods** manufactured with a thin coat of flux which melts from the rod as heat is applied during the welding operation. Complete and effective operations may be performed with either method of applying flux.

There are a number of advantages and disadvantages of braze welding. Advantages to keep in mind are:

1. Welds can be made more rapidly than fusion welds because less heat is required.
2. Less stress is locked-up in the weld because of the low melting temperature of the brazing rod plus the ability of the filler material to stretch during cooling.
3. Breaks in cast iron can be restored to original strength.
4. Dissimilar metals can be joined such as copper to steel and cast iron to steel.
5. Thin materials will have less distortion because of the lower heating temperatures.
6. Brazing filler material has high corrosion resistance.

There are some disadvantages of braze welding, these being:

1. Braze welding cannot be used at temperatures above the melting point of the filler rod.

2. Braze welding should not be used on parts or welds which would be exposed to temperatures above 500°F as the brazing filler material will weaken.

3. In some cases, the brazing filler may not be as corrosion resistant as the metals being joined.

4. Color may not match the materials being joined.

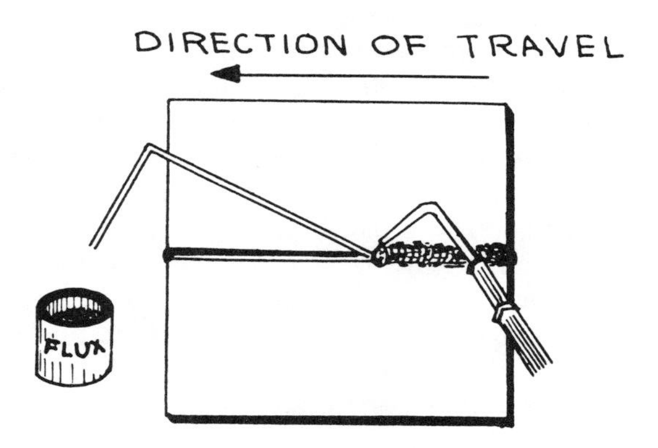

Figure 41. Braze Welding Two Pieces of Mild Steel Together Using the Brazing Welding Rod.

Braze welding of steel is illustrated in Figure 41. Note the angle of the blowpipe tip and filler rod. The position of the rod and blowpipe is comparable to welding steel with a steel filler rod as discussed earlier. This method would be referred to as the forehand technique. To make quality braze welds, the following procedure must be used:

1. Mechanically clean the joints to be welded. Remove all oil, grease, paint, rust, dirt, and other foreign material or the brazing material will not stick.

2. Prepare the joints - Vee parts to a 90° groove if over 1/8" in thickness. Clean or grind 1/4" to 1/2" along sides and edges of vee.

3. Position parts - Clamp or tack weld in place.

4. Preheat the base metal to cherry red, approximately 1600°F. Hold the flame about 1" from joint. A carbon arc torch can be used instead of the oxy-acetylene flame.

5. Chemically clean the joint - If powdered flux, apply by dipping heated rod into flux. Coat the area to be welded or wherever the bronze material is to be applied. If using a flux coated rod, melt the flux from the rod on to the joint surface.

6. Tin the surface of the weld. Apply a thin coating of the brazing filler material to the surface to be welded. Tinning will indicate if the surface is properly cleaned as the filler rod should flow evenly over the joint surface.

7. Keep the temperature of the weld at approximately 1600°F. If too cold, the brazing material will form balls and not stick to the surface.

8. Do not overheat the weld, as the zinc will burn away at a low temperature; overheating will result in white fumes and a white deposit on the weld, If the zinc is burned away, a copper colored weld will result. Direct the flame ahead of the molten puddle adding the filler rod to approximately 1/16" thickness. Be sure complete fusion is made between additional layers of the brazing rod.

9. Cool braze welds slowly. Quenching anneals the weld causing it to be weak. The cooled weld should have color similar to the filler rod. A cooled, copper colored weld indicates the zinc has been burned away. Clean excess flux from cooled weld.

BRAZE WELDING CAST IRON

Illustrated in Figure 42 is braze welding of cast iron. Braze welding is well accepted for repairing gray cast iron. As braze welding requires less heat than fusion welding, the malleability and grain structure of the cast iron is not changed during braze welding.

As in braze welding of mild steel, chemical cleaning or the use of a flux is essential. The brazing rod as used in welding steel is used. The following procedure should be used during the welding operation:

1. Bevel edges to an angle of 90°.

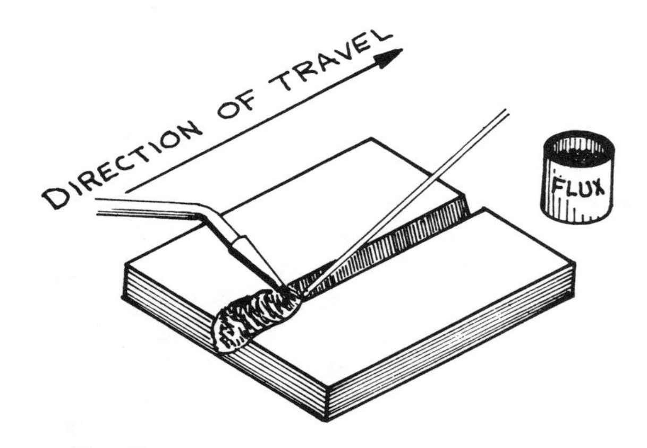

Figure 42. Braze Welding of Cast Iron.

2. Clean the metal mechanically. Grind back 1/2" from each beveled edge where brazing filler material is to flow.

3. Position parts by tacking or clamping.

4. Adjust blowpipe flame to slightly oxidizing.

5. Preheat part to be welded to a dull red color.

6. Apply flux to entire vee and flat surface beside the vee.

7. Tin the surface to be welded. Tinning is very important, the total surface must be tinned to be assured that the brazing material will stick to the cast iron.

8. Add layers as needed moving from left to right to fill the vee so that the final pass provides the correct degree of reinforcement. It should show a distinct ripple effect and be slightly convex over the surface of the cast iron joint.

9. Cool slowly. Remove flux by buffing or with a chisel.

10. Test joint for strength. A successful weld will fracture along the welded joint.

LOW TEMPERATURE WELDING PROCESSES

SILVER BRAZING

Silver brazing alloys or **silver solder** are often called hard solders. This group of solders have a melting point above 800°F. However, at least one manufacturer of silver solder is marketing an alloy composed of 96% tin and 4% silver with a wire flux core and a melting point of 440°F. Because of its melting point it is classified as a soft solder.

The making of silver brazed or silver-soldered joints is similar in some respects to braze welding; however, the joint design is often similar to that used for solder type fittings. Most silver brazed joints are closely fit, therefore, the filler material is drawn throughout the joint by capillary action.

Alloys that contain as little as 10 percent silver are known as **silver alloys**. Silver alloys used in brazing are usually composed of silver, copper, and zinc. A common silver brazing rod would be **20% silver**, **45% copper** and **35% zinc**. This rod would have a melting temperature of approximately 1430°F with a flowing point of 1500° F. This particular material could be used to join copper, bronze, brass as well as for steel and for joining dissimilar metals. Some manufacturers of silver brazing rods have used small amounts of cadmium to increase weld strength. If the rod contains cadmium, the operator should take particular caution as cadmium fumes are very toxic. Avoid breathing the fumes and weld at all times in a very well ventilated area. Packages of silver brazing rods which contain cadmium must be clearly marked by the supplier.

The cost of silver brazing alloys seems high in comparison with other brazing or welding rods; however, the relatively small amount required and the special properties of the brazed joints compensate with other welding and brazing materials. However, it is possible to design joints that have tensile strengths of 40,000 to 60,000 psi provided the film of the alloy is kept thin and the stress is placed in shear. The clearance of the joint must be such that capillary action will draw the fluid silver alloy into the joint. Best results have been obtained when parts are separated by about 0.0015 to 0.002 of an inch.

Because of their high electrical conductivity and their resistance to shock and vibration, silver brazing alloys are widely used in making electrical connections. Another important property is the high resistance to corrosion of silver alloys. They are often used in the fabrication of chemical tanks where joints might be chemically attacked if other materials were used. Combination seams are often used in dairy, textile and laundry industries in which silver brazing the inside seam, smoothes the joint and minimizes corrosion. The procedures for silver brazing are quite similar to brazing with the braze welding rod. The joints must first be cleaned thoroughly to remove all oxides and foreign material. Fine steel wool is usually sufficient for cleaning. Joints must be smooth and fit snuggly as described in the section on joint design.

A saturated solution of borax or a special paste is used as a flux. The purpose of the flux is to protect the metal from oxidation by dissolving oxides, promote the capillary action or flow of the filler material and indicate the temperature of the base metal or when to feed the fiiler metal into the joint. It is most effective if applied with a brush to the parts to be joined and to the silver alloy wire. As some fluxes contain flourides producing toxic fumes, always weld in a well ventilated area and avoid breathing fumes.

A neutral oxy-acetylene flame from a small tip is recommended for silver brazing. After the joint has been fluxed, gently heat the metal to or slightly above the melting point of the alloy-1300 to 1500 degrees F. This is well below the melting point of the metals that will be brazed. The color of the flux indicates the metal temperature. The water boils from the flux at 212°F; at 600°F the flux has a white, puffy appearance; at 800°F a clinging, milky appearance; and at 1100°F it appears as a water clear liquid at which time the preheat temperature has been reached. When preheat has been reached, move the flame away and quickly apply the alloy to the joint. The silver alloy should be melted by the heat of the base metal. The flame should never touch the alloy. Flux should be localized to control the flow of the silver solder. After brazing, all excess flux should be removed by thoroughly washing in water.

SOLDERING

Soldering is the process of forming two similar or dissimilar metals by means of an alloy (combination of metals) having a lower melting point than the metals to be joined. Soldering is not a fusion process, but an adhesion process and mild penetration of the alloy with the joined metal.

SOLDER

Solder is often defined as an alloy filler metal with a melting point not exceeding 800°F. Most soft solders are a mixture of tin and lead with melting points ranging from 400°F to about 500°F depending upon the amount of tin and lead in the alloy. Note Table 1. When equal parts (50-50) of tin with a melting point of 450°F (232°C), is mixed with lead with a melting point of 620°F (327°C) the melting point of the alloy solder is 361 °F (183°C). Other mixtures of tin and lead are available, as soft solder such as 20-80, 30-70, or 40-60 with a lowering of melting points. The higher the percentage of lead, the higher the melting point. 50-50 solder is most often used for general soldering. It can be used on galvanized sheet steel and copper tubing with any available source of heat. Solders with flux cores of rosin or acid are often used because of the limited need for added flux. If the solder alloy contains zinc or cadmium avoid breathing the toxic fumes given off when soldering.

Silver Solder - An alloy of 96% tin and 4% silver with a melting point of 440°F can be classified as a soft solder. This silver solder is used to join stainless steel and stainless steel to other metals. Since it is lead and cadmium free it meets the requirements of the food and drug laws. It is used to some degree in jewelry repair and fabrication, as well as by food processing, dairy industries, air conditioning and refrigeration industries. It may be used on silver, copper, brass, bronze and nickel alloys.

SOLDERING EQUIPMENT

Soldering Iron - Forged soldering coppers (Irons) are illustrated in Figure 43. They are purchased in pairs and are pure copper. Usually pointed and in weights from 1 to 2 pounds per pair. The heavier weights hold the heat longer and reduce the heat loss from the heat source to the work.

Electric Soldering Iron - Electric Soldering Coppers (Irons) are usually available in 100-200 and 300 watt sizes. Note Figure 43. A variety of tips with chisel or diamond points may be purchased.

Soldering Gun - Electric Soldering Gun - the gun as shown in Figure 44, produces intense heat at the point of application to the item being soldered. It is used widely in electrical equipment repairs and in crafts and hobbies. The soldering gun is often sold as a part of a kit with additional tips and accessories included. Sizes range from 100 to 200 watts. Some have dual-heat depending upon trigger position.

TABLE 1. SOFT SOLDERS.

Solder (Form)	Percent Tin-Lead,*1	Approximate Melting Point°F,*3	Suggested Uses
Solid Wire	50-50	4.14	Use with zinc chloride paste flux. Copper plumbing and general purpose uses.
Rosin Core Wire-1/16 or 3/32 inch	40-60	460	Non-corrosive or conductive flux. Household and automotive wiring. Motors, generators and electrical appliances.
Acid Core Wire-1/8 inch	40-60	460	Auto radiators, gutters galvanized sheet steel, non-electrical repairs.
Bar-l lb.	30-70 to 60-40,*2	Variable	Industrial uses.

*1. The tin content is always shown first.
*2. 60/40 is an excellent solder due to its 60% tin content. This alloy is very fluid and is used where a low melting point is required such as in electronics. It becomes molten at 374°F.
*3. Solder melts at a lower melting point but becomes fluid at about these temperatures.

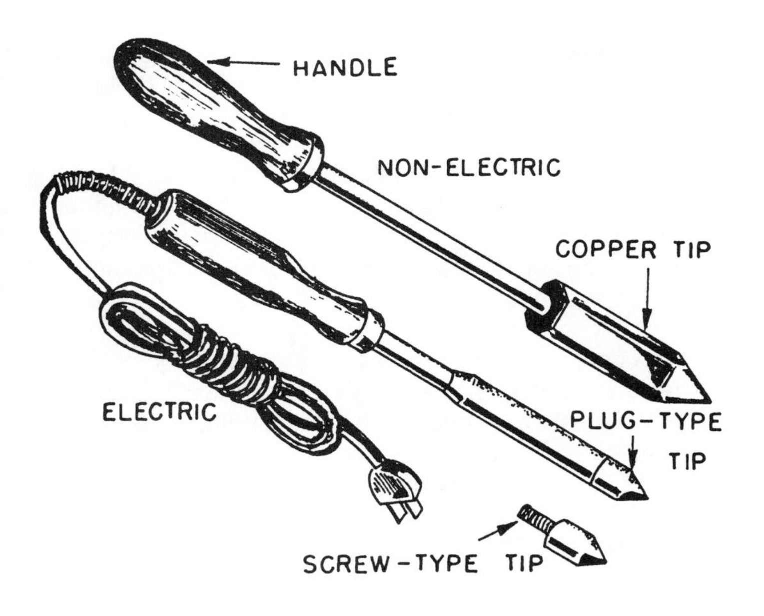

Figure 43. Soldering Irons.

Soldering Aids - In addition to soldering fluxes and solder, many items are needed to expedite the process such as: pliers, clamps, flux brushes, wet cellulose sponge, ammoniac block, wire brush, steel wool, file, zinc chloride flux solution, ceramic soldering block, and snips.

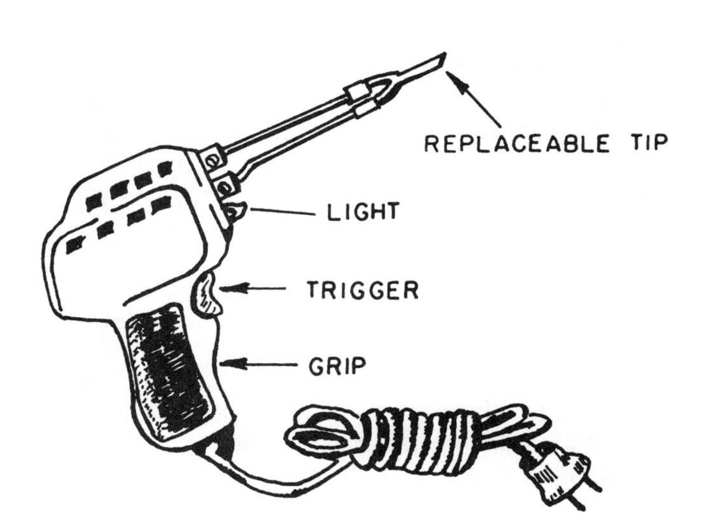

Figure 44. Electric Soldering Gun

Tinning the Soldering Copper - A soldering copper must be tinned before it will pick up and run the solder. Tinning is the coating of the faces of the copper and the soldering gun tip with solder. The copper tip is heated, filed and shaped before tinning. To tin the soldering copper,heat copper, rub on sal ammoniac block, as illustrated in Figure 45, or dip in solution or resin flux, melt solder on ceramic soldering block, rub heated copper in solder and wipe off excess solder with wet cellulose sponge or cloth.

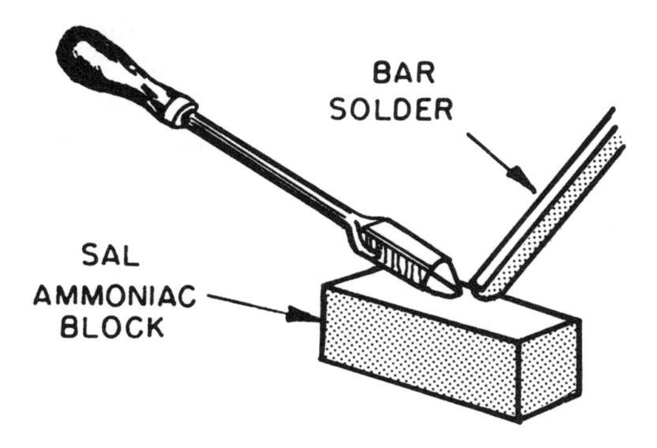

Figure 45. Rubbing Soldering Copper on Sal Ammoniac Block.

FLUXES

Flux materials are used to remove the oxides that result from exposure of the metal to the air. Removal of these oxides decreases surface tension allowing direct contact between the base metal and the solder, the filler metal alloy. The heating of the flux dissolves the oxides and the other impurities, suspends them, while the heavier solder flows under the suspension uniting the metals. Fluxes do not remove paint, varnish, rust, dirt or grease. These must be removed mechanically from the base metal before the flux is applied.

Corrosive Fluxes - Zinc Chloride Paste or Liquid. This paste, available in tube or can, is very convenient to use. The liquid is made by adding zinc flakes to dilute hydrochloric acid (Muriatic Acid) until it is "killed" or "cut." This flux is used in many soldering situations that are non-electrical.

Dilute Hydrochloric Acid (Muriatic Acid) - This acid is often used on galvanized steel before adding the zinc chloride paste or liquid. Hydrochloric acid and zinc chloride are very corrosive. Tools and equipment and the soldered project must be washed thoroughly with water to remove all excess flux material.

Non-Corrosive - Rosin is made from crude turpentine and is corrosive only when it is hot. When cold it is completely inert. It is removed by washing or wiping the soldered area after the solder has set. This flux is used widely in electrical application. Rosin core 50-50 or 40-60 wire is very popular.

EQUIPMENT AND SUPPLIES FOR SOLDERING

The metal to be soldered must be cleaned, fluxed, held securely and heated to the melting point of the solder. The following equipment and supply items are used to achieve these conditions.

Propane Torch - The propane torch as illustrated in Figure 46, is a very convenient source of heat for soldering. It is light in weight, easy to ignite and generally with no objectionable odor. It may be used to heat soldering coppers or the heat from the torch may be applied directly to the metal as in soldering a fitting on a copper pipe. There are numerous torch heads available and cylinders can be recharged or disposable.

Lighting the Torch - (1) Open control valve slowly until a low hiss of gas can be heard. (2) Ignite immediately with match, cigarette lighter or flint spark. (3) Allow torch to warm about 2 minutes, then turn control knob to adjust flame size. Note Figure 47.

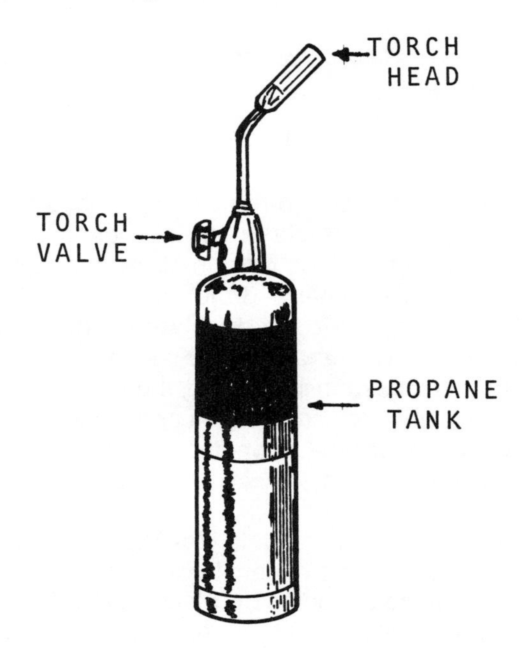

Figure 46. Propane Torch.

SOLDERING SUGGESTIONS

Preparation of the items to be soldered is extremely important. Metals that are coated with rust, oil, dirt, paint and other foreign materials can not be soldered. All pieces should be cleaned thoroughly and fitted closely before soldering is attempted. Muriatic acid is often used in the cleaning the metal but the area must be washed before the flux is applied. If zinc chloride is used as a flux, washing must again take place after soldering. If a rosin flux is used, the soldered joint is wiped with a wet cellulose sponge or cloth after the joint is set.

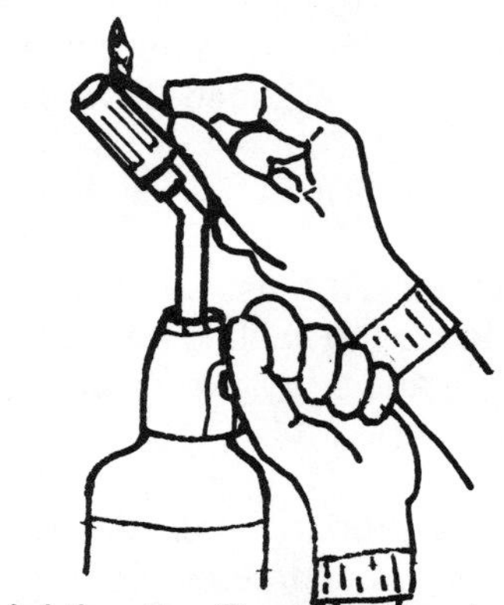

Figure 47. Lighting the Propane Torch.

The first essential in soldering is to heat the metal to the melting point of the solder with a clean hot well tinned soldering copper. The propane torch flame may be applied directly to the item to reach this point as in the soldering of a copper pipe joint.

The hot copper (iron) is moved slowly over the joint to be soldered to transfer the heat to the metal. All of the area to be joined must be heated to the melting point of the solder to assure satisfactory work. Do not attempt to smooth soldered joints mechanically. A glossy finish is assured if sufficient heat is applied and if the heel of the iron is applied evenly to form a straight narrow bead. Solder is fed against the heated iron as the soldering progresses. Well tinned soldering copper and correctly heated work as a result of its application will cause the solder to flow evenly and smoothly. **Always wear gloves** and **eye protection** when soldering.

REQUIREMENTS FOR QUALITY SOLDERING:

1. PROPER FLUX
2. QUALITY SOLDER
3. CLEAN HEAT SOURCE

Soldering Copper Pipe or Tubing - Rigid and flexible copper pipe has been used extensively in home plumbing and industrial applications. If the soldered sweat joints used with this pipe are correctly done there is no opportunity for the fluids carried in the piping systems to come in contact with the solder securing the joints. This eliminates any possibility of the lead in the soft solder to contaminate the water or other fluids carried in the pipe.

The copper pipe is cut square with a pipe cutter or a fine toothed hacksaw. Any exterior burr is removed with a file and the interior is reamed. The area involved in the joint is thoroughly cleaned with fine steel wool. The interior of the copper fitting; coupling union, elbow or tee is cleaned mechanically with fine steel wool or emery cloth. The area of the pipe to be inserted into the fitting as well as the interior of the coupling is coated completely with paste flux containing zinc chloride. The pipe is inserted in the fitting, pressure is applied to insure that the pipe is in contact with the shoulder in the fitting. The fitting is then moved back and forth to insure the complete distribution of the flux and that the end of the pipe is seated against the shoulder in the fitting.

Soldering the Joint - The assembled fitting and pipe are positioned and heat is applied from the flame of the propane torch. 50-50 wire solder is used to solder the joint. When the joint area reaches the melting point of the solder the flame is removed or moved to the other side of the pipe depending on the diameter of the pipe. The solder is then applied to the edge of the fitting as shown in Figure 48. It will be drawn into the joint immediately if the proper temperature is maintained. Solder will be drawn into the joint regardless of its position by capillary. Solder will appear around the edge of the joint when it is completely soldered. The joint is then inspected and wiped with a wet cloth or cellulose sponge after the joint is set.

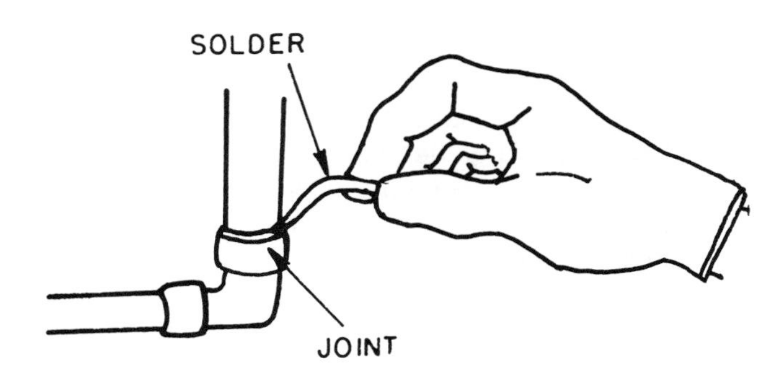

Figure 48. Sweat Soldering a Pipe Joint.

OXY- ACETYLENE FLAME CUTTING

About 1905, through the development of a suitable cutting blowpipe, the cutting of iron and steel with the oxygen and acetylene flame was developed to a useful stage. **Acetylene** has for years been the common fuel for cutting, brazing, and fusion welding. **Propane** and **natural gas** were early competitive fuels for metal cutting. However, advantages for propane and natural gas as cutting fuels are oftentimes found to be disadvantages for acetylene as a cutting fuel and vice-versa. Acetylene has continued over the years as the main industrial flame-cutting fuel. In the early 1960's, the Dow Chemical Company released a new industrial fuel under the trademark of **MAPP**, and its chemical name is methylacetylene propadiene. Although our main discussion will center around the use of acetylene and oxygen for cutting, acetylene will be compared with other industrial fuel gases as a fuel gas for flame cutting.

Before discussing specific cutting methods and procedures, an important question must be answered. Why is it that iron and steel can be cut with the oxy-acetylene flame? It is well known that iron and steel when exposed to air and ordinary temperatures, build up on the surface a layer of iron oxides commonly known as rust. Rusting is the result of the oxygen in the air uniting with the metal causing it to slowly decompose and wear away. If heat is applied to the metal in presence of oxygen, the oxidizing process or rusting speeds up. Basically, this is what takes place during the cutting process. Gas cutting of ferrous metals is a process of preheating the material to its kindling temperature and rapidly oxidizing it with a stream of oxygen. The kindling temperature, point where a metal becomes combustible, for most ferrous metals is 1400°F to 1700°F or a bright cherry red color.

GASES FOR FLAME CUTTING

A basic understanding of the chemistry of flame cutting of the various gases is necessary for a full understanding of the cutting process.

A. **Acetylene**, C_2H_2 is an unsaturated hydrocarbon. This gas gives a higher temperature than other common gases. It burns at about 5900°F with a burning velocity in oxygen at 22.7 feet per second. Acetylene produces 1480 BTU per cubic foot and 21,500 BTU per pound. The reaction for combustion is:

$$C_2H_2 + 2\text{-}1/2\ O_2 \longrightarrow 2\ CO_2 + H_2O$$

A BTU is the amount of required heat to raise the temperature of one pound of water one degree Fahrenheit.

B. **Propane**, C_3H_8. is a saturated hydrocarbon. This gas produces a flame temperature of 4500° F and burns at 12.2 feet per second in oxygen. Propane produces 2520 BTU per cubic foot and 21,800 BTU per second. The chemical reaction for combustion is:

$$C_3H_8 + 5\ O_2 \longrightarrow 4\ H_2O + 3\ CO_2$$

C. **MAPP** (methylacetylene propadiene) gas has a flame temperature in oxygen of 5301°F and a burning velocity of 15.4 cubic feet per second in oxygen. The heat of combustion is 2406 BTU per cubic foot and 21,100 BTU per pound. The combustion reaction is:

$$C_3H_4 + 4\ O_2 \longrightarrow 3\ CO_2 + 2\ H_2O$$

D. **Oxygen** must be used for combustion of the industrial fuel gas and for oxidation of the metals. It must be at least 99.5% pure for efficient cutting. The amount used varies with the type of metal, thickness of metal and width of kerf. The actual cutting of ferrous metals is a chemical process based on the remarkable affinity of oxygen for ferrous metals when heated to or above their kindling temperature. When commercially pure oxygen is brought in contact with most steels or iron at their kindling temperature the following very active chemical reaction occurs:

$$3\ Fe + 2\ O_2 \longrightarrow Fe_3O_4$$

During the normal cutting process, approximately 60% of the metal is washed away by mechanical eroding produced by the kinetic energy of the oxygen stream. This process saves considerable amounts of oxygen. Data in Table 2, Comparison of Properties of Industrial Fuels, reveal further comparison of acetylene and the other industrial gases.

To make rapid cutting of iron and steel possible, it is necessary to have an implement that will heat the iron or steel to a certain temperature and then throw a blast of oxygen on the heated section. The oxy-acetylene cutting tip as illustrated in Figure 23 of this unit reveals that the cutting torch or blowpipe tip provides 4 small oxy-acetylene flames, which supply the heat needed to preheat the steel. The cutting tip as illustrated is also constructed with an oxygen cutting hole to provide the stream of pure oxygen which actually oxidizes the metal producing the cut.

As shown in Figure 22, the cutting blowpipe has two needle valves to regulate the supply of oxygen and acetylene for the four preheat holes and a third valve operated by a lever to control the stream of pure oxygen to the cutting flame.

The pressure of oxygen and acetylene required for a given cut depends upon (1) the tip size and (2) the thickness of the metal to be cut. As shown in Table 4, the oxygen pressure varies greatly with the thickness of metal while the acetylene pressure ranges from 3 to 5 psi. For example, for1/2" thick metal the oxygen working pressure should be approximately 25-30 psi; whereas, the recommended acetylene pressure is 4 psi.

MAKING A CUT

To make a practice cut, select a piece of steel about 1/2" thick. Mark a line with chalk 3/4" from one edge of the steel plate as a guide to follow. To make a complete and efficient cut, study and use the following procedures:

1. To light the blowpipe, turn on the acetylene as for regular welding, lighting the acetylene preheat flame.

2. Turn on the oxygen valve adjusting the preheat flames to neutral.

3. Press the oxygen lever valve. Observe the preheat flames; they should remain neutral. It should be noted that the beginner oftentimes has the metal to be preheated above the kindling temperature discussed earlier which results in an incomplete cut that usually melts back together behind the cut. One rule that should be followed is that it should take a few seconds for the preheat flame to heat the edge of the metal to red hot before the cutting process begins.

4. Place the steel plate so that the cut will clear the edge of the table. A cutting barrel with a grated top is shown in Figure 49. The cut is made directly over an open grate so that oxidized and blown through metal falls into the barrel and not onto the floor. This cutting table provides many safety features as the sparks and metal slag blown through can produce a safety hazard.

5. With the flame properly adjusted, grasp the cutting blowpipe handle with the right hand so that the fingers are in position to operate the oxygen lever.

TABLE 2. COMPARISON OF PROPERTIES OF INDUSTRIAL FUEL GASES

	MAPP GAS	ACETYLENE	PROPANE
Safety:			
Shock sensitivity	Stable	Unstable	Stable
Explosive limits in oxygen, %	2.5 - 60	3.0 - 93	2.4 - 57
Explosive limits in air, %	3.4 - 10.8	2.5 - 80	2.3 - 9.5
Maximum allowable regulator pressure, psi	Cylinder (225 psi at 130°)	15	Cylinder
Burning velocity in oxygen, ft/sec.	15.4	22.7	12.2
Tendency to backfire	Slight	Considerable	Slight
Toxicity	Low	Low	Low
Reactions with common materials	Avoid alloys with more than 67% copper	Avoid alloys with more copper	Few restrictions
Physical Properties:			
Cubic feet per pound of gas at 60°F	8.85	14.6	8.66
Specific gravity of gas(air=1) at 60°F	1.48	0.906	1.52
Boiling range, °F, 760 mm Hg	-36 to -4	-84	-50
Flame temperature in oxygen, °F	5,301	5,589	4,579
Total heating value (after vaporization) BTU/lb.	21,100	21,500	21,800

6. Position yourself as shown in Figure 49 with your left elbow on the table with the forward part of the blowpipe resting in your left hand. This position would be just opposite for the left handed operator. The secret of making clean, smooth, straight cuts is largely dependent upon the operator's ability to hold the blowpipe steady and to be in a comfortable position so that the cutting flame can be evenly carried across the metal being cut.

7. Start the cut at the edge of the plate. A right handed operator normally works from right to left. Hold the blowpipe tip with the inner cone of the preheat flames about 1/16" above the edge of the metal as shown in Figure 50. The tip should be at 90° in all directions to the work piece. Hold the tip in this position until the edge of the metal has been heated to a bright red color.

8. Gradually press down on the oxygen lever and move the torch forward along the chalk line. A shower of sparks will be seen falling from the under side of the cut which tells the operator that the flame is penetrating through the total thickness of the metal.

9. If the cut is not going through the metal, release the oxygen lever and reheat the metal to the bright red color before again pressing the oxygen lever. The cut and **kerf** (area removed by the cutting process) should progress as shown in Figure 51. Note the kerf lines showing a slight amount of drag and a small amount of metal droppings adhering to the bottom of the cut.

10. Progress at a steady rate across the metal just fast enough to insure a fast but continuous cut.

11. When the cut is completed, the metal should fall in two pieces. A slight amount of slag build up may be present at the start and lower edge of the cut. This is not serious and can be easily removed by tapping with a chipping hammer or buffing.

Figure 49. Cutting Barrel With Grated Top.

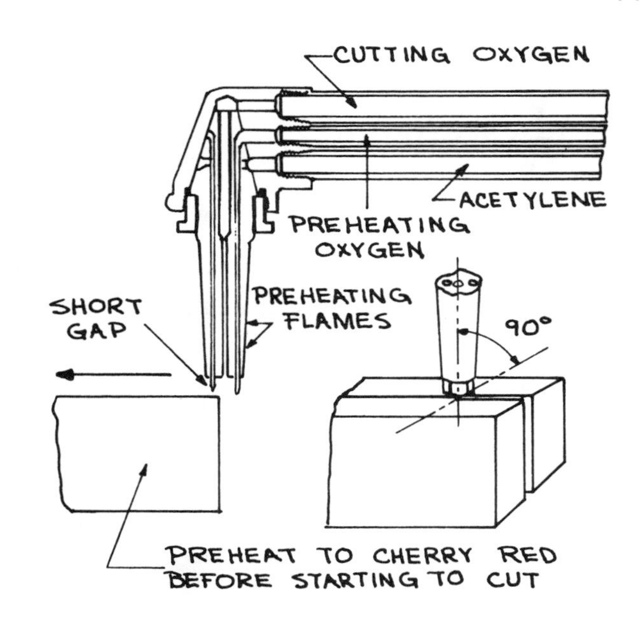

Figure 50. Position of Oxy-Acetylene Blowpipe Tip During Cutting.

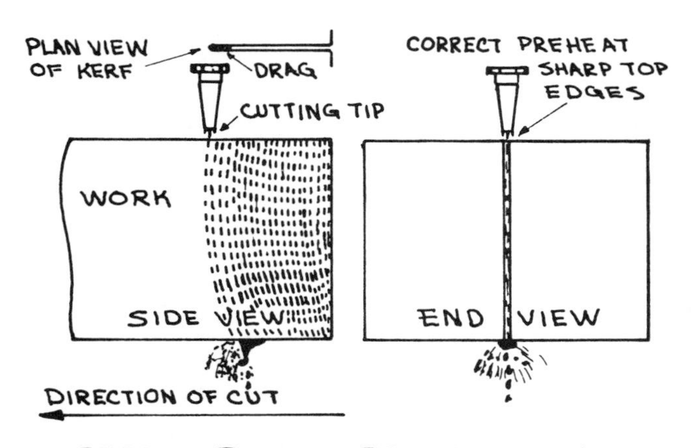

Figure 51. Procedures for Oxy-Acetylene Cutting Showing Kerf and Lines for Correct Cut.

Shown in Figure 52 is the procedure for piercing a hole. As illustrated in view 1, the blowpipe is held in a rigid position with the end of the tip 1/4" to 5/16" above the metal surface until a spot of the plate begins to melt. Next raise the tip 1/2 to 5/8 of an inch above the plate and angle slightly as shown in view 2. Step 3 is to slowly depress the oxygen lever and begin to move the tip in a spiral motion. The purpose of raising and tipping slightly is to avoid blowing slag into the end of the tip which would plug the preheat and oxygen holes. View 4 shows the slag being blown out the opposite side of the puddle and through the thickness of the metal. Lower the cutting tip to the normal height and

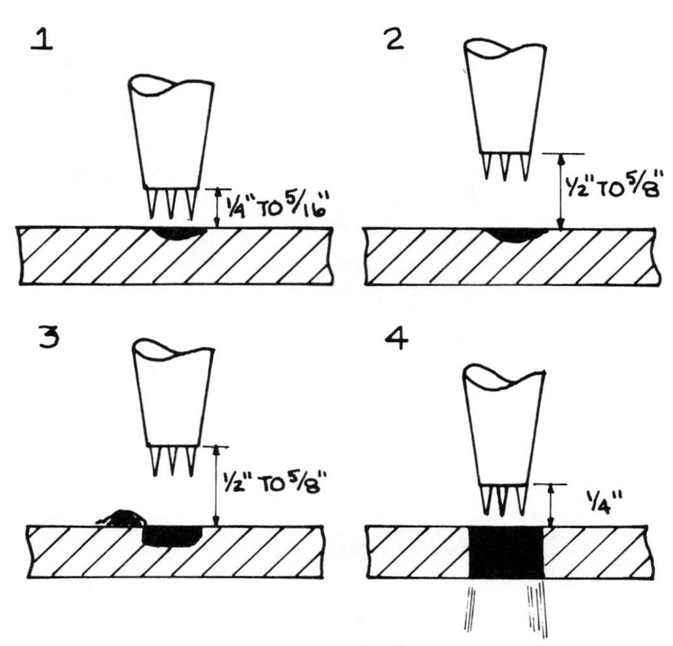

Figure 52. Procedure for Piercing a Hole with Oxy Acetylene Cutting Unit.

continue with a spiral motion until the desired hole is cut through the work piece.

Another common cutting job is making bevel cuts. The position of the preheat holes on the cutting tip in respect to the direction of travel is shown in Figure 53. Note the position of the preheat holes for the bevel cut compared to the straight cut. Cleaner, smoother cuts will result if the tip is positioned as shown.

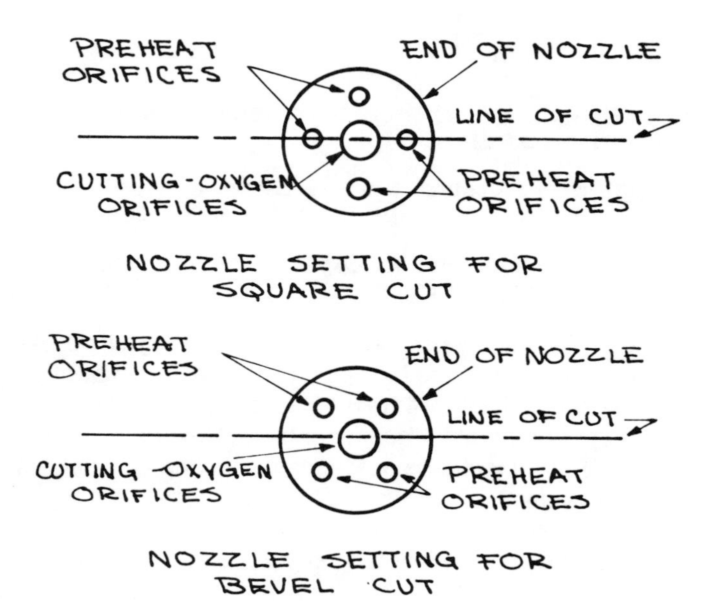

Figure 53. Position of Preheat Holes in Respect to Lines of Cut for Straight and Bevel Cuts.

To make a bevel cut, follow similar procedures as outlined for making the straight cut except the cutting tip should be held at the desired angle instead of the vertical position. An even bevel cut can be made by actually resting the end of the tip on the work as a support and guide for the cut. The angle toward the direction of travel should be at 90° as for the straight cut.

CUTTING CAST IRON

Cast iron can be cut with the oxy-acetylene flame by following a few special procedures. Cast iron is more difficult to cut because of the structure of cast iron and because of the wide range in the quality and chemical composition. One important rule to follow is that once the cut is started, it should be completed. Attempting to restart a cut is very difficult.

More heat, slag and sparks are produced during cutting than when cutting steel. Use the following procedure for cutting cast iron:

1. Use slightly higher pressure for cutting cast iron. For 1/2" material, use 40 psi of oxygen and 7 to 8 psi of acetylene.

2. Light the torch and adjust the flame to a slightly carburizing flame as shown in Figure 54.

3. Approach the edge of the metal at an angle of 40°-50° as shown in Figure 55 for starting the cut. Heat a spot 1/2" in diameter to a molten condition.

4. Begin a circular motion as shown in Figure 54 and at the same time depress the oxygen lever.

5. Slowly move the torch along the line of cut continuing the circular movement of the tip.

6. As progressing forward, straighten the tip to the advanced position 65° to 75° as illustrated in Figure 55. Continue the cut across the work piece using the same circular motion until the cut is completed.

OTHER INDUSTRIAL FUEL GASES

Shown in Figure 56 is a **MAPP cutting unit**. As discussed earlier, MAPP has some special characteristics for cutting that are not present with acetylene cutting. It combines the high heat characteristics of acetylene with the handling safety and convenience of propane. Because of its stability and handling ease, it is distributed in lighter cylinders containing more pounds of actual fuel. However, liquified fuel gases such as MAPP and propane do present vaporiztion problems when used where surrounding temperatures are below the vaporization point of the gas.

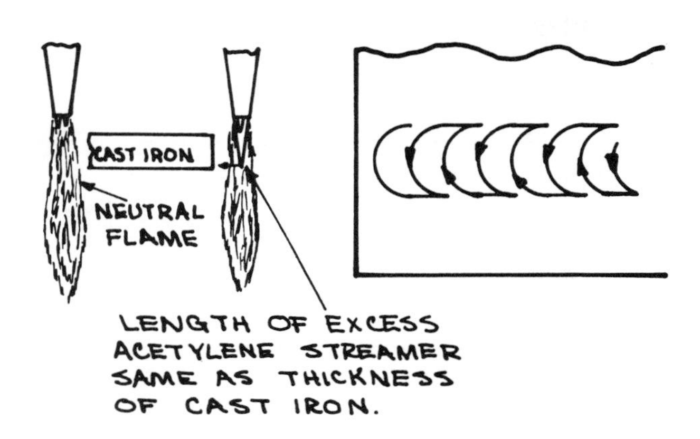

Figure 54. Oxy-Acetylene Flame and Motion for Tip During Cutting of Cast Iron.

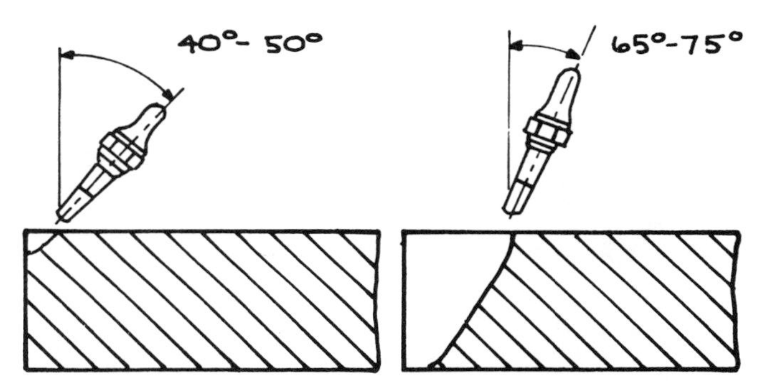

Figure 55. Angle of Cutting Tip for Beginning and Advance Cutting Stages When Cutting Cast Iron.

No special equipment is required for cutting with MAPP gas. The same regulator as used with acetylene can be used. The main difference is in the type of tip. Even though acetylene tips will work, better results are received if special tips fitting the regular blowpipe are used. MAPP and propane cutting tips have **multiple preheat holes** compared to the four commonly found in the acetylene tip. Regular procedures as with acetylene should be followed when cutting with propane and MAPP gas. There are two main problems with most industrial fuel gases when compared to acetylene (1) more oxygen is used when cutting with MAPP and propane because of the longer preheat time required and (2) industrial fuel gases other than acetylene have not proven successful for welding operations because of their lower heating qualities. Due to these two main reasons, acetylene is and will coninue to be the main fuel gas for home shops and agricultural applications. Industrial fuel gases such as MAPP and propane will probably play their main role for industrial cutting where longer cuts and large amounts of gas are used.

Figure 56. MAPP Cutting Unit.

Notes

TABLE 3. COST OF ACETYLENE AND OXYGEN

	Approx. Cap. Cu. ft.	Approx. Retail cost/100 cu.ft.	Approx. Retail cost/cylinder
Large Acetylene Cylinder	297	$18.00	$53.50
Standard Acetylene Cylinder	106	18.75	19.50
*Small Acetylene Cylinder	60	26.60	16.00
Standard Oxygen Cylinder	244	6.55	16.00
*Small Oxygen Cylinder	122	8.20	10.00

* Some companies now sell these cylinders, other companies will lease them on a 10 to 25 year basis; thus, by purchase or lease, all demurrage charges are avoided. One company leases for a 25 year period at $125.00 per pair.

Most standard cylinders are loaned, and a demurrage charge at five cents per day per cylinder is made after a 30 day loan period. Until recently, all gas was sold in loaned cylinders. If acetylene is used in large quantities, it may be produced in a generator from calcium carbide at 1/3 to 1/2 the above cost.

Cost of Oxyacetylene Welding and Cutting Equipment and Supplies

a. **Equipment**

Combination welding and cutting outfits may be purchased for about $200.00. These are quite satisfactory for moderate use. Heavy duty equipment with separate welding and cutting torches may cost up to about $325.00.

b. **Supplies**	**Approx. Retail Cost/lb.**
Mild steel rod	$ 0.70
Cast iron rod	2.70
"Stoody" hard facing alloy	4.15
"Stoody" tube borium	20.00
Stellite, 3/16 rod	21.00
Brazing rod	2.75
Brazing rod, flux coated	3.60
High silver content (45% brazing alloy silver solder)	10.00 Troy ounce
Brazing flux	3.80 per pound can
Cast iron flux	6.50
Silver alloy brazing flux	4.50 per 12 ounce bottle
Goggle lens - shades 3-8	3.50 per pair
Clear cover glasses	3.00 per pair
Medium green cellulose acetate face shield	11.00 each

Note:

Prices quoted on welding rod are for minimum amounts. Variations are largely due to difference in diameter. Treat costs only as approximate, as all are subject to change.

TABLE 4. GENERAL INFORMATION ON OXY-ACETYLENE WELDING AND CUTTING-PLATE THICKNESS, LENGTH OF FLAME, PRESSURES, CONSUMPTION AND COST.

A. **Welding and Brazing (Estimates)**

Gage Pressure 5 psi each, Oxygen and Acetylene

Tip Size *	Thickness	Length of flame (Approximate)	Consumption in cubic ft./hr. Oxygen - Acetylene	Cost dollars Per hour		
				Oxygen	Acetylene	Total
2	28 ga.	1/8	2	.10	.20	.30
4	22 ga.	3/16	4	.20	.40	.60
6	1/16"	1/4	6	.30	.60	.90
9	1/8"	5/16	9	.45	1.00	1.45
12	5/32"	3/8	12	.60	1.20	1.80
15	3/16"	7/16	15	.70	1.50	2.20
20	1/4"	9/16	20	1.00	2.00	3.00
30	3/8"	3/4	30	1.50	3.00	4.50

*"Purox" tips

B. **Cutting (Estimates)**

Plate Thickness In Inches	Tip Size	Regulator Oxygen psi	Pressure psi Acetylene	Cutting Speed in./min.	ConsumptionCost cu. ft./hr.		Cost Total $/hr.
					Oxygen	Acetylene	
1/4	1	15-20	3	18	50	10	8.00
1/2	2	5-30	4	15	83	18	11.90
1	3	35-40	5	10	162	30	22.20
3	4	45-50	5	7	300	40	38.00

Preheating increases consumption of gases in cutting narrow pieces.

CLASSROOM EXERCISE OXY-ACETYLENE WELDING & CUTTING

1. Identify the parts of the welding and cutting unit.

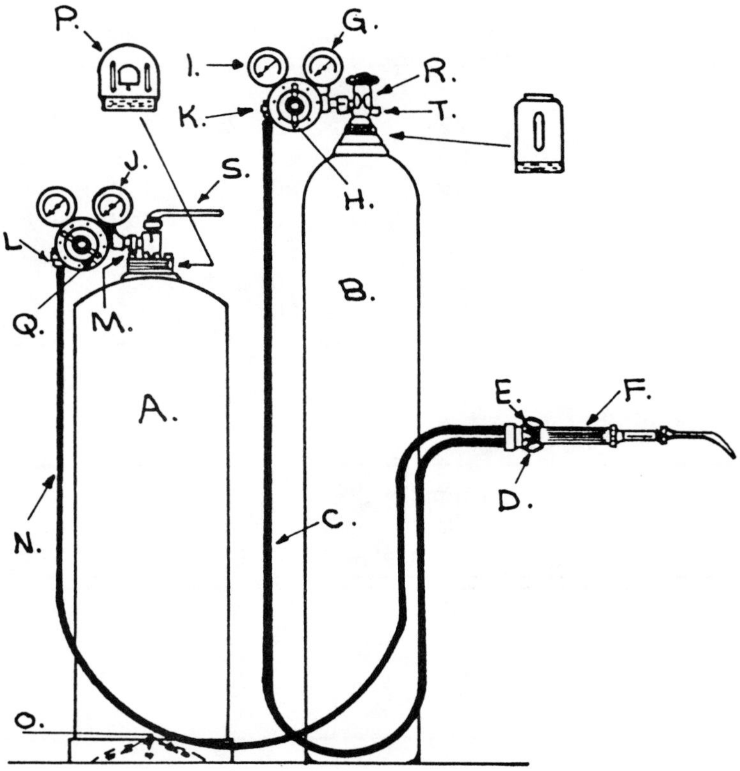

A.____________________

B.____________________

C.____________________

D.____________________

E.____________________

F.____________________

G.____________________

H.____________________

I.____________________

J.____________________

K.____________________

L.____________________

M.____________________

N.____________________

O.____________________

P.____________________

Q.____________________

R.____________________

S.____________________

T.____________________

2. The two methods of production of oxygen are the ____________________ and ____________________ methods.

3. List the chemical formula for the production of acetylene.

____

____

4. The standard oxygen cylinder contains ____________________ cubic feet of oxygen at _______________ psi.

5. At 70°F a full cylinder of oxygen would indicate a pressure of ___________ psi.

6. The three common sizes of acetylene cylinders are ___________, ___________ and ____________ cubic feet.

7. Explain the purpose of acetone in the acetylene cylinder. ____

____

8. Explain the difference between the two-stage and single-stage regulator.

____

9. On the oxygen regulator, the tank pressure gage is normally graduated from _________ to _________ pounds; whereas the working pressure gage goes from ___________ to ___________ pounds.

10. Explain the operation of the low pressure injector type welding blowpipe.

____

11. The most common welding blowpipe is the _______________________ blowpipe.

12. Describe the main differences between the oxygen and acetylene blowpipe hose connections.

a. _______________________________________ b. _______________________________________

13. The oxy-acetylene cutting tip has _____________ preheat holes.

14. Identify the below oxy-acetylene flames and give one specific application for each flame.

Flame Use

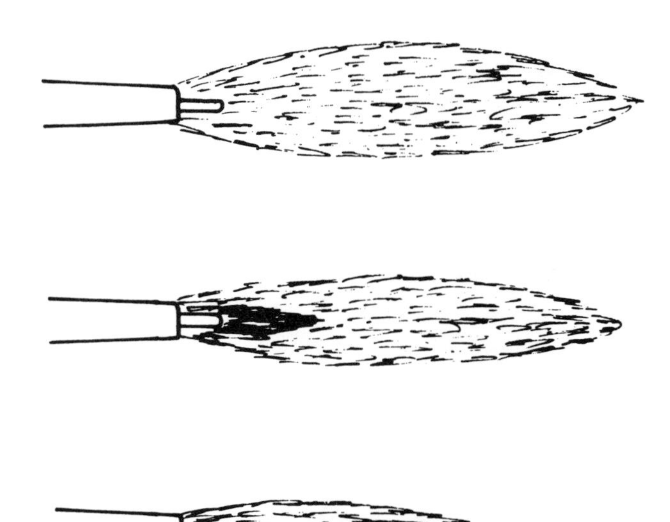

15. The chemical equation for complete combustion of acetylene in oxygen is:

____

16. List 5 causes of a backfire:

a. ____________________ d. ____________________

b. ____________________ e. ____________________

c. ____________________

17. The use of a flashback__________ could eliminate the risk of an explosion in the regulator or cylinder.

18. List the 10 basic steps to follow in setting up and lighting the oxy-acetylene unit.

a. ____________________ f. ____________________

b. ____________________ g. ____________________

c. ____________________ h. ____________________

d. ____________________ i. ____________________

e. ____________________ j. ____________________

19. Explain the difference between the forehand and backhand welding techniques.

____

20. The brazing rod consists of __________ % __________ and __________ % __________

and melts at a temperature of __________ °F.

21. What is the purpose of the flux in brazing? ______________________

____

22. Explain the difference between brazing and braze welding. ______________________

____

23. List the 10 main steps or procedures to follow in braze welding of mild steel.

a. ____________________ f. ____________________

b. ____________________ g. ____________________

c. ____________________ h. ____________________

d. ____________________ i. ____________________

e. ____________________ j. ____________________

24. The silver brazing filler rod is composed of ______________________________, __________________________ and ____.

25. Soldering is the process of forming two________________ or ____________________ metals by means of an alloy having a ____________________melting point than the metals being joined.

26. Soldering is a (a) fusion or (b) adhesion process.

27. Most soft solders are a mixture of 50% _______________ and 50% _______________ and have a melting point of approximately______________________ degrees F.

28. The tip of the soldering iron is forged _________________.

29. Define tinning in the soldering process.____

____

30. The purpose of the flux in soldering is____

____

31. _________________flux is commonly used for soldering electrical application.

32. Requirements for quality soldering are: ____,

____ and ____.

33. List the 5 steps necessary for quality sweat soldering of copper pipe.

a.____

b.____

c.____

d.____

e.____

34. Describe the cutting process of iron or steel. ____

____

35. What is the kerf in oxy-acetylene cutting? ____

____

36. Discuss the basic differences between cutting of iron and steel and cast iron.

a. ____

b. ____

c. ____

LABORATORY EXERCISES OXY-ACETYLENE WELDING AND CUTTING

LAB EX-1 CORNER WELD

Material - Two pieces of 1/8" x 3-1/2" mild steel.
Tip Size - #4 or #6 tip, neutral flame, equal oxygen and acetylene pressures.

Procedure:

1. Set the material together tent-like on the welding table.
2. Tack the one end and begin weld at the other, normally right to left for right handed operator. This is a forehand weld.
3. Weld with a semi-circular motion of the torch tip.
4. Hold tip of inner luminous cone from 1/32 to 1/16 inch from work at 45° angle.
5. Gage forward speed by continued maintenance of puddle or pool of molten metal.
6. The luminous cone should be about 1/4 inch in length for a #6 tip.
7. Weld must have complete and even penetration on under side.
8. Check strength of weld by hammering out flat, weld should not crack or break away at fused edge.
9. Submit completed weld in tent fashion to instructor for evaluation.

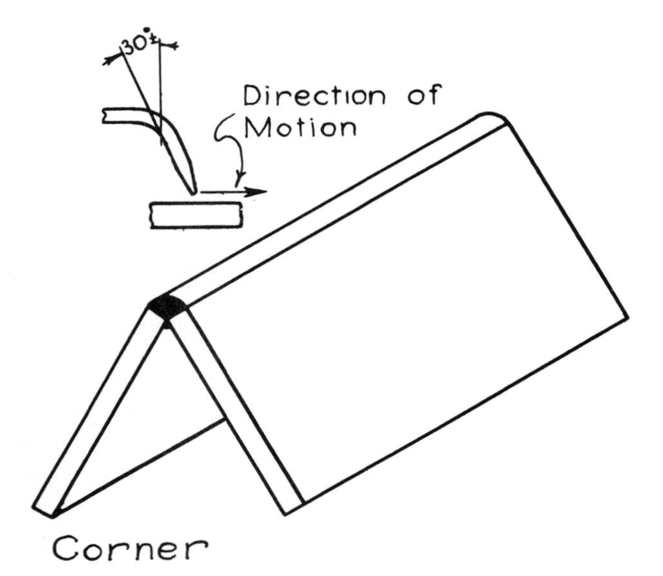

LAB EX-2 BUTT WELD - FOREHAND AND BACKHAND

Material - Two pieces of 1/8" x 1" x 3-1/2" mild steel, and 1/16" or 3/32" steel rod.
Tip Size - #6 or #9 tip, neutral flame, equal regulator pressures.

Procedure:

1. Lay pieces on brick with about 1/16" to 1/8" gap at outer end to avoid overlapping.
2. Weld using a semi-circular movement of the torch tip around the rod using forehand welding technique.
3. Keep rod tip in the molten pool all of the time.
4. Keep molten pool at all times large enough to give complete fusion, 3/8" to 1/2" in diameter.
5. Avoid backfires by keeping lumious cone out of molten pool.
6. Weld from right to left for right handed operation.
7. Travel slowly enough for complete penetration but fast enough to avoid burning through.
8. Too much heat will cause metal structure to become brittle and hard. Cool slowly.
9. Test weld by placing in vise and bend to 90 degrees at edge of weld. Weld should not crack or pull away from fused edges.
10. Submit completed project in flat position for instructor evaluation.
11. Repeat the above weld using the backhand welding technique.

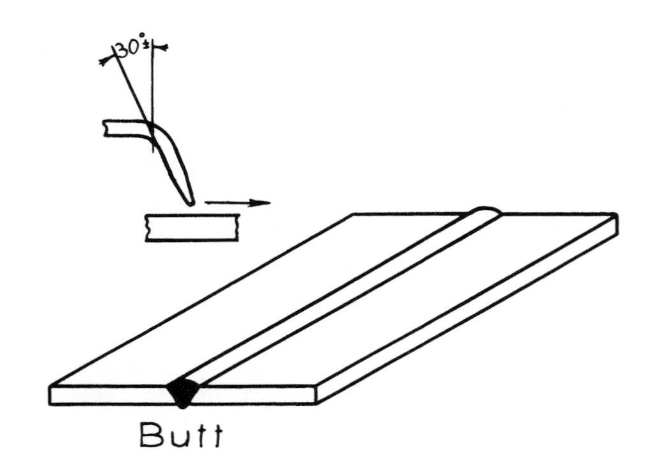

LAB EX-3 BRAZING A FILLET

Material - 1 - 1/8" x 1/2" x 2" mild steel bent to 90° angle.
1 - 1/8" x 1-1/2" x 2" mild steel for the base piece.
1/16 to 1/8 inch brazing rod.
Brazing flux for mild steel.
Tip Size - #6 or #9 tip, neutral flame, equal pressures.

Procedure:

1. Mechanically clean flat base piece and edges of upright piece up at least 1/2" with grinder.
2. Make sure pieces fit tightly together.
3. Preheat total area to 1600° F or cherry red color.
4. Apply brazing flux with rod to area where brazing material is to adhere. Keep temperature at 1600° F.
5. Tin weld with thin coating of brazing filler rod.
6. Build up brazing filler material to 1/16" thick.
7. Allow weld to cool slowly.
8. Clean excess flux from weld by buffing or chisel.
9. Test weld by placing upside down in vise. Metal should shear above brazed weld.
10. Submit completed weld for instructor evaluation.

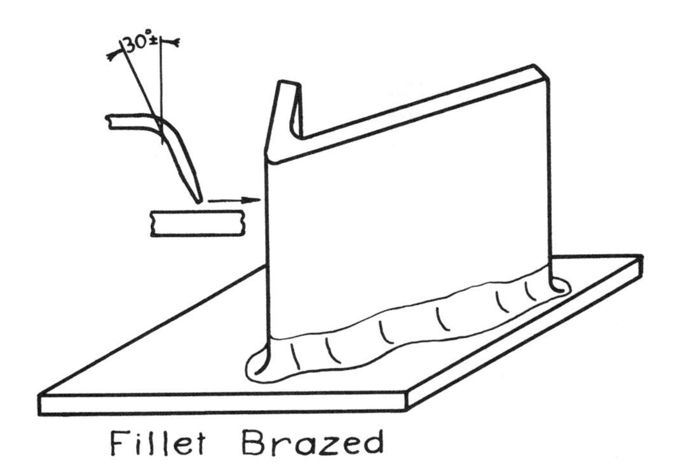

Fillet Brazed

LAB EX-4 CUTTING

Material - Piece of steel plate at least 1/4" thick.
Tip Size - #0 for 1/4", #1 for 1/2", #2 for 1" plate, etc.

Procedure:

1. Use flat bar as straight edge, or draw a straight line with chalk.
2. Adjust gage pressures to approximately 5 psi acetylene and 35 psi oxygen for 3/8" plate, and for heavier material increase oxygen pressure to as high as 75 psi for material 3" in thickness. (See manufacturer's recommendations for correct pressure.)
3. Adjust preheating flame to neutral.
4. Heat edge of metal until molten beads appear.
5. Open the oxygen cutting valve slowly.
6. If cutting stops, close the cutting valve and preheat again. Unless the metal is very dirty or rusty, the heat of combustion is sufficient for preheating the metal ahead of the kerf.
7. Bottom of kerf should be slightly behind the top for straight line cuts.
8. If slag prevents pieces from falling free, hitting one piece with a hammer will separate them.
9. Clean slag from cut with wire brush or buffing wheel, do not grind.
10. Submit straight, bevel and pierced hole cut for instructor evaluation. Follow directions as discussed in manual for bevel cut and pierced hole.

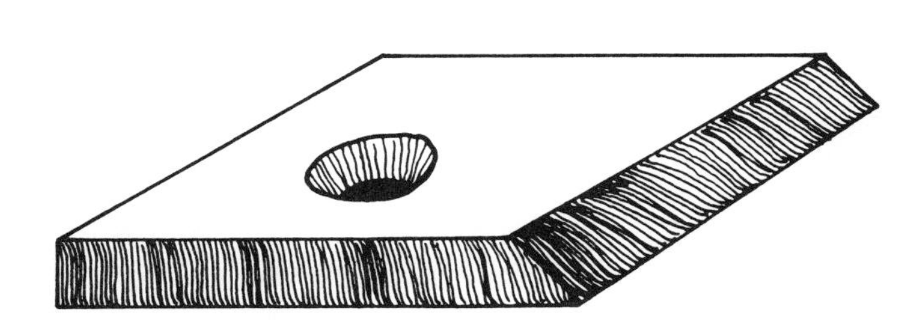

Oxy-Acetylene Cutting
Straight Cut, Bevel
Cut and Pierced Hole

LAB EX-5 FLANGE WELD

Material - Two small pieces of 16 to 26 gage steel.
Tip Size - #2 or #4 for 20 to 26 gage sheet steel.

Procedure:

1. Flange edges to be welded the thickness of the metal.
2. Weld without the use of welding rod by melting down the flanges.
3. Follow procedure outlined in corner weld except that a smaller luminous cone will be needed.
4. Penetration needs to be complete.
5. Test weld by twisting with pliers attempting to pull apart.
6. Submit weld, not tested, to instructor for evaluation.

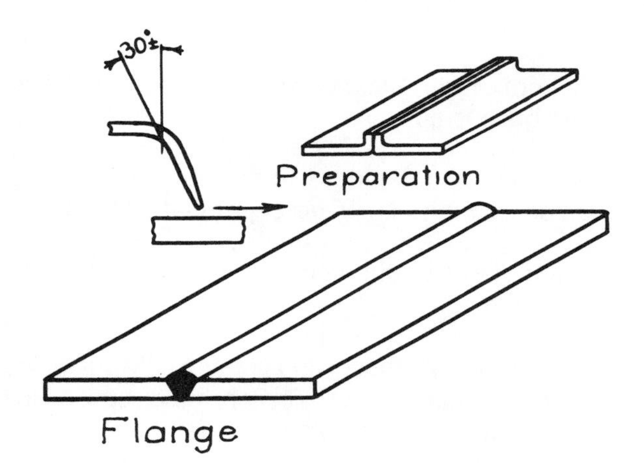

LAB EX-6 BRAZING CAST IRON

Material - Small piece of gray cast scrap.
Brazing rod.
Brazing flux for cast iron.
Tip Size - At least #6 or #9, depending on thickness of metal. Use neutral flame.

Procedure:

1. Fracture or break the piece of cast iron.
2. Vee to within 1/16 inch, single or double vee.
3. Clean thoroughly back at least 1/4" from break or vee.
4. Check flame and gage pressures carefully to avoid blowing.
5. Proceed as in Job #3, 2-1/2 to 3 inches may be tinned at one time. Fill in, then tin another area.
6. Avoid continuous direct contact of flame with brazing filler material. Reheat by applying flame to edge of welded area.
7. Review procedures discussed in oxy-acetylene unit.
8. Allow weld to cool slowly, clean or buff away excess flux.
9. Check for strength by breaking in vise. Cast should fracture along brazed weld.
10. Submit completed weld to instructor for evaluation.

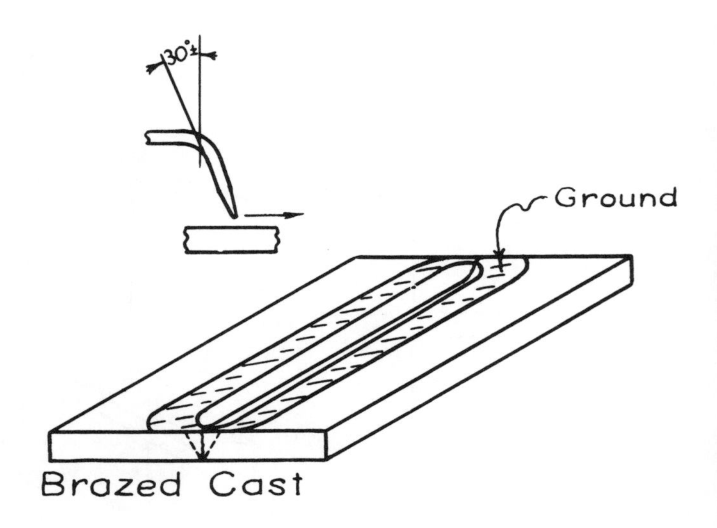

LAB EX-7 FUSION WELDING OF CAST IRON

Material - Small piece of gray cast scrap.
1/8" to 1/4" cast rod.
Flux for fusion welding cast iron.
Tip Size - At least #6, depending on thickness of metal, neutral flame.

Procedure:

1. Fracture or break piece.
2. Vee to within 1/16", single or double vee depending on the thickness of the cast.
3. Clean rust, paint and other foreign material by grinding or buffing.
4. Heat end of rod and dip into cast welding flux.
5. Be sure metal is heated to the melting point, as indicated by a glassy or soapy appearance with star-like luminous particles.
6. Melt cast rod into the puddle formed by the molten cast. Since cast does not flow freely it may be smoothed with the end of the rod or a separate spatula. Cool slowly.
7. Check strength of weld by breaking in vise. Cast should fracture along edge of weld.
8. Submit completed weld to instructor for evaluation.

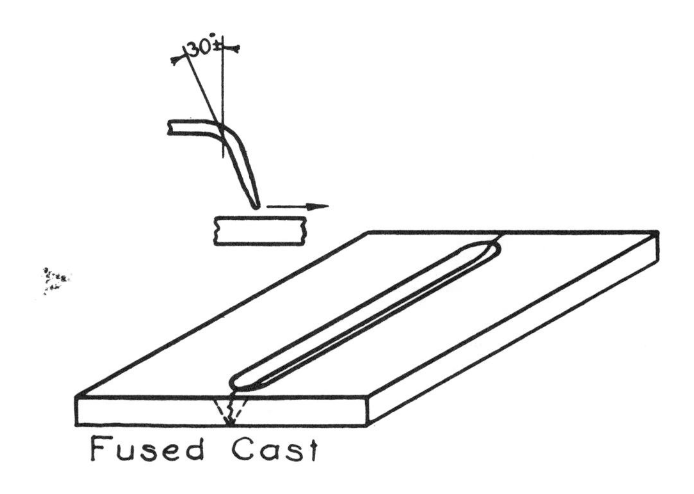

LAB EX-8 SOLDERING RIGID COPPER TUBING

Material - 4" x 1/2" copper tubing. 90° copper elbow.
Propane Torch. Solder and flux.

Procedure:

1. Measure the correct length of copper tubing. Cut tubing to length, either with a hacksaw or tubing cutter. Remove burrs from tubing inside and outside.
2. Mechanically clean outside area of tubing that will be inserted into the fitting. Also clean inside of fitting.
3. Coat inside of coupling and outside of tubing with zinc chloride paste flux.
4. Insert tubing into elbow fitting making sure tubing is in contact with inside shoulder of fitting. Place tubing and fitting in block so joint is in a vertical position.
5. Select 50-50 wire solder.
6. Light propane torch followng all safety precautions. Apply heat evenly to joint. Flux will boil from the joint. Heat joint to melting point of solder, approximately 450°F.
7. Apply the solder to the fitting while maintaining the temperature of the joint. Solder should appear around the edge of the joint around the total diameter of the tubing. Clean the joint, wiping off excess solder and flux residue with a wet cloth or sponge.
8. Submit completed solder to instructor for evaluation.

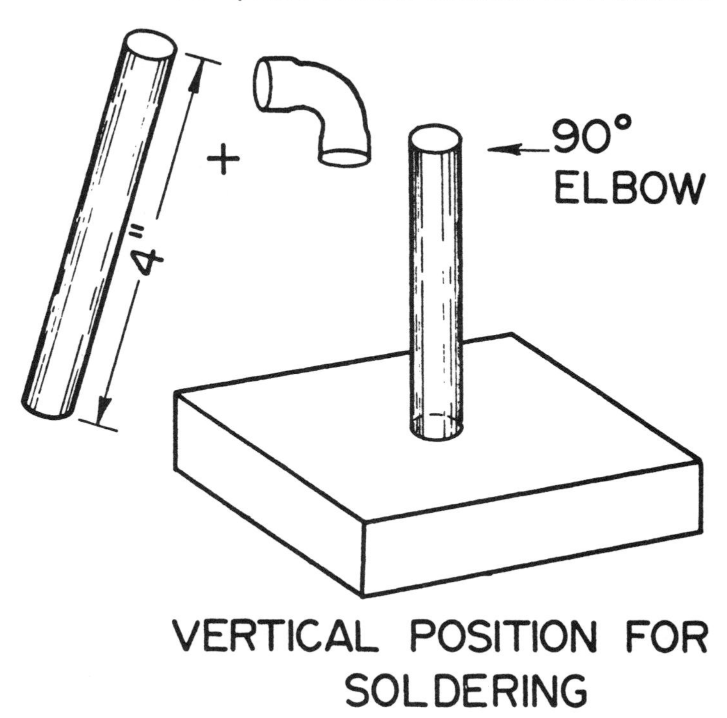

Material - 3 - 26 or 28 gage galvanized iron 2-1/2" x 2-1/2".
1 - 26 or 28 gage blk. iron 2-1/2" x 2-1/2".
1 - bright tin, 2-1/2" x 2-1/2".
Soldering copper or soldering iron.
Solder, 50% lead, 50% tin.
Fluxes according to type of metal.

Procedure:

1. Make the rivet and lock seams with the galvanized iron.

2. Make the lap seams with tin on galvanized iron and black iron or tin plate. These will be sweat joints.

3. Select the correct fluxes for each metal. Apply sparingly.
   a. Galvanized iron - Muriatic acid or commercial HCL.
   b. Black iron - HCL followed by zinc chloride solution.
   c. Tin - zinc chloride solution, sal ammoniac solution, rosin.

4. "Tin" the soldering copper or iron.

5. "Tin" both metal surfaces before sweating seams or patches.

6. Follow all soldering suggestions discussed in this unit.

7. Let solder solidify before moving.

8. **Caution** - the best joints are made with a moderate amount of solder and careful heating of metal by slow movement of the soldering iron.

9. Test practice joints by bending, metal should bend beside soldered seams.

10. Clean soldered joints with wet sponge.

11. This exercise may be done as individual joints or as a total unit.

12. Submit completed exercise without bending to instructor for evaluation.

SOLDERING EXERCISE

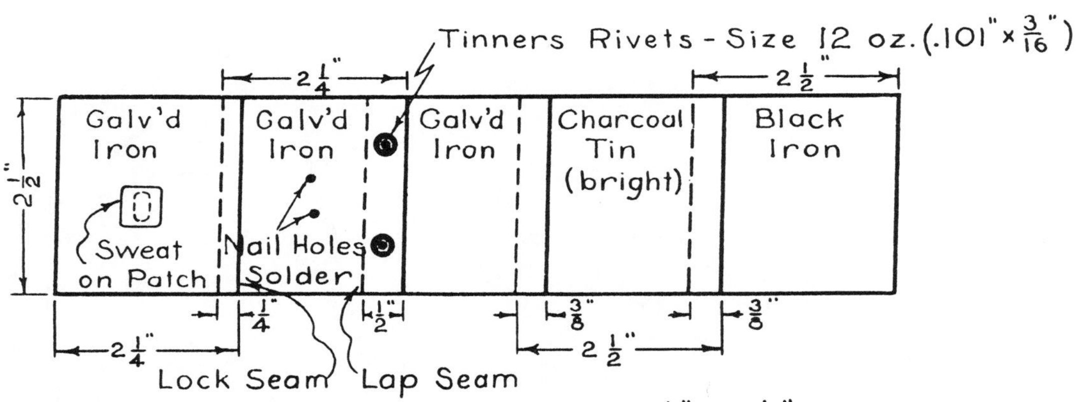

STOCK: 3 Pieces 26 Gage Galv'd Iron 2 1/2" x 2 1/2" — 1 Piece 26 Gage Black Iron 2 1/2" x 2 1/2" — 1 Piece Charcoal Tin 2 1/2" x 2 1/2"

UNIT VI
ARC WELDING

PRECAUTIONS AND SAFE PRACTICES

Although the hazards of arc welding cannot be said to be any greater than those connected with other similar work, it is important that the operator know the safe procedures and practices which will help guard against injury to themselves and to others:

1. **Use a wooden grating** made of 2" x 2" or 2" x 4" sills and pieces 1" x 2" to 3" for the operator to stand on while welding. This is very essential on concrete floors, where there is danger of water being spilled. There is little danger of shock if reasonable care is used.

2. **The danger of electric shock** is increased during periods of high temperture and humidity due to excessive perspiration and the consequent saturation of the workperson's clothing. Never attempt repair of any electrical device without disconnecting from the power source. The operator does not need to turn off welder each time they stop welding.

3. **Clothing** - The body must be completely covered — shoes are preferred to oxfords; leather gloves are desirable; trousers cuffs should be rolled under or cut off; and the head and hair should be protected by a cap. Keep clothing free of oil, grease, or inflammable liquids. Aprons made of asbestos or leather are desirable if much welding is being done. (Clothing suggestions also apply to acetylene welding and cutting.)

4. **Helmet or hand shield** with at least shade No. 10 lens should be used at all times. Shade No.12 should be used when welding heavy material employing 200 to 400 amperes. Approved safety glasses must be worn at all times.

5. The welding equipment must be **completely screened** for arc glare. The machine should be well protected against accidental contact on the part of others.

6. A qualified electrician should be in charge of all maintenance and repairs on the welding equipment - tight connections and good ground connections are important. **Make sure the machine is grounded.**

7. **Special welding cables** with high quality insulation are available and should be used and kept in good condition. They should be kept coiled and free from grease and oil, and not lying in water, oil, ditches, gutters, etc.

8. **Electrode holders and cables should be fully insulated.**

9. **Locate welding jobs in special rooms** or booths to protect other workers or "passers by" from harmful rays and flying chips.

10. Where harmful concentration of gases (nitrous and carbon monoxide), fumes, and dusts are generated by arc welding in confined buildings or areas, **exhaust systems or breathing apparatus should be provided.**

11. Welding operations should not be permitted in or near areas containing flammable materials until all fire and explosion hazards have been eliminated or safe guarded. **Proper fire extinguishing equipment should be stationed at or near welding operations.**

12. When the operator leaves the work area for any length of time, **the main switch should be opened**.

13. After welding operations are completed, **the operator should mark the hot metal** or provide some warning sign to prevent others from coming in contact with the welded pieces.

14. **Good housekeeping** should be maintained at all times. Operators should not discard electrode stubs on the floor or leave tools and other objects that might constitute hazards to others.

15. The usual precautions in **handling electric power** should be observed.

16. The weldor should also realize that **good workmanship** in making sound welds is essential so that others may not be injured because of failure of welded parts.

EYE-SAFETY LENS SHADE SELECTOR

OPERATION	SHADE NUMBER
SOLDERING	2
TORCH BRAZING	3 or 4
OXYGEN CUTTING	
up to 1 inch	3 or 4
1 to 6 inches	4 or 5
6 inches and over	5 or 6
GAS WELDING	
up to 1/8 inch	4 or 5
1/8 to 1/2 inch	5 or 6
1/2 inch and over	6 or 8

SHIELDED METAL-ARC WELDING 1/16, 3/32, 1/8, 5/32 inch electrodes	10
NONFERROUS	
GAS TUNGSTEN-ARC WELDING	
GAS METAL-ARC WELDING .030, .035, & .045 inch wire	11
FERROUS	
GAS TUNGSTEN-ARC WELDING	
GAS METAL-ARC WELDING .030, .035, & .045 inch wire	12
SHIELDED METAL-ARC WELDING	
3/16, 7/32, 1/4, inch electrodes	12
5/16, 3/8 inch electrodes	4
ATOMIC HYDROGEN WELDING	10 to 14
CARBON-ARC WELDING	14

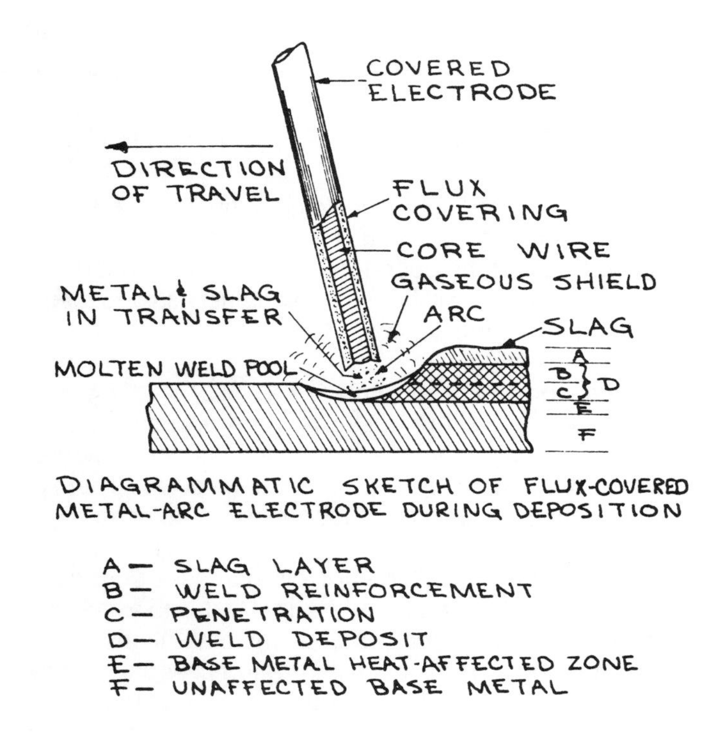

Figure 1. Diagrammatic Sketch of Flux-Covered Metal-Arc Electrode During Deposition.

STICK ELECTRODE WELDING TECHNIQUES

All electric open-arc welding processes depend upon the operation of a miniature electric furnace established between the grounded base metal and the electrode. The arc temperature of about 11,000° F easily melts the base metal and the electrode, mixing and casting a bead. The molten metal must always be protected from the air by a gaseous shield and/or a slag shield. The machine settings and the operator's manipulations determine the shape and size of the bead. See Figure 1.

All successful arc welding depends upon the following factors:

1. **Metal identification** is fundamental because of varying mechanical, physical and chemical factors.

2. **Electrode selection** depends upon metal type, thickness and position of weld.

3. **Amperage** (current) setting depends upon electrode type, size, position and metal thickness. Proper amperage can only be set after initial setting has been tried. Experienced operators freely change amperage settings but **never** during arc operation because this can cause arcing damage in the adjustment mechanism. Amperage settings influence "burn off rate" and, therefore, tend to affect arc length and speed of travel.

4. **Striking the arc** is the act of starting the electric furnace.

5. **Proper arc length** is very important because it influences the amount of heat during welding.

6. **Speed of travel** determines the width of bead and indirectly the strength of the weld.

7. **Angle of electrode** also determines the bead shape as well as controlling slag and gas inclusions.

METAL IDENTIFICATION

A detailed discussion of methods of metal identification is found in the Metals Section of this text. The reader is referred to that section.

ELECTRODE SELECTION

The earliest arc welding was done with DC welders using bare, low carbon steel wire electrodes. In the early 1930's, chemists and physicists cooperated in developing flux-coated electrodes. Modern flux-coated electrodes are the results of exhaustive research of chemical combinations and manufacturing quality control that has made arc welding a highly successful metal joining process under an endless number of welding situations. The American Welding Society has classified electrodes so that specifications allow a wide latitude of choice for many applications. The following explanation of AWS classification and ratings of steel electrodes should be thoroughly understood by all welders.

HOW STEEL ELECTRODES ARE CLASSIFIED

All arc welding electrodes are classified according to definite filler metal specifications. These are prepared by a joint committee of the **AWS (American Welding Society)** and the **ASTM (American Society for Testing Materials)**. Among the more common specifications are those for steel, officially called "Specifications for Steel Arc Welding Electrodes ASTM Designation A233-64T, AWS Designation A5.1-64T."

This steel specification used a classification system based on four factors:

1) Minimum tensile strength of the as-welded deposited weld metal.
2) Type of covering.
3) Welding position of the electrode.
4) Type of welding current.

For identification, each steel electrode is designated by the letter E, followed by a 4- or 5-digit number.

Example 1 AWS E-6010   Chemical Requirement (*C3)

Example 2 AWS E-12018-C3*

* The figure that appears as a suffix on some AWS Classification Numbers indicates the amount of the alloys contained in that eledctrode i.e.: C3 = 0.12 Carbon, 1.00 Manganese, 0.80 Silicon, 0.04 Sulphur, and 0.08 to 1.10 Nickel. This may be omitted in certain more common electrodes.

A. Carbon-Molybdenum
B. Chromium-Molybdenum
C. Nickel
D. Manganese-Molybdenum
E. All other alloys

The first two or three digits represent the first factor (**minimum tensile strength** (TS) of the as-welded deposited weld metal) expressed in thousands of pounds per square inch (1000 psi). For example, E45xx means 45,000 psi TS, E60xx means 60,000 psi TS, E120xx means 120,000 psi TS.

The third or fourth digit refers to the **welding position** in which the electrode can be used, thus:

Exx1x - all positions**
Exx2x - flat and horizontal fillet positions
Exx3x - flat position only

** Note that for all-position welding electrode size is usually limited to a 3/16" maximum. Exceptions are the low-hydrogen and iron-powder types, which are limited to 5/32".

The fourth or fifth and last digit indicates the **type of welding current** with which the welding electrode can be used, and the **type of flux covering**. For full identification, however, the last two digits (sometimes called the usability identification) must be read together, as follows:

Exx10 - DC reverse polarity(electrode positive) only(cellulose sodium).
Exx11 - AC or DC reverse polarity (cellulose potassium).
Exx12 - DC straight polarity (electode negative) or AC (titania sodium).
Exx13 - AC or DC straight polarity (titania potassium).
Exx14 - DC either polarity or AC(iron powder titania).
Exx15 - DC reverse polarity only (low hydrogen sodium).
Exx16 - DC reverse polarity or AC (low hydrogen potassium).
Exx17 - DC straight polarity or AC for horizontal fillet welds with DC either polarity or AC for flat position welding (iron powder iron oxide).
Exx18 - DC reverse polarity or AC (iron powder, low hydrogen).
Exx20 - DC straight polarity or AC, for horizontal fillet welds; and DC either polarity or AC, for flat position welding.
Exx24 - DC either polarity or AC (iron powder).
Exx27 - DC straight polarity or AC, for horizontal fillet welds; and DC either polarity or AC for flat position welding (iron powder, low hydrogen).
Exx30 - DC either polarity or AC.

Thus, for steel, the complete classification number E6010 would signify an electrode that (a) has a minimum tensile strength of 60,000 psi for the as-welded deposited weld metal, (b) is usuable in all welding positions, and (c) can be used with DC reverse polarity only. In the same way, E7024 designates an electrode with 70,000 psi minimum tensile strength, usable in the flat and horizontal positions only, and operating on DC either polarity or AC (iron powder).

The purposes of the **flux coating** are to improve the performance in handling, storage, and operation. Probably the most important single chemical ingredient is the binder that holds the flux securely to the wire under sometimes severe treatment. Other chemicals form slag to protect the molten metal during melting and resolidification. Gas-forming chemicals protect the molten metal during the metal transfer stage by excluding the air. Arc stabilizer chemicals are necessary for an electrode to operate properly with alternating current welders. Earliest flux coated electrodes were made by the dip method; however, the present-day extrusion process makes a more uniform thickness of the flux coating. "**Finger nailing**" is usually the result of nonuniform thickness of the flux coating. This usually contributes to arc blow if DC welding current is used. Finger nailing is the uneven burning of the flux coating causing erratic arc behavior.

Special electrodes for welding cast iron and for hard surfacing will be dealt with elsewhere.

Electrodes are identified either by **type marking** on the flux near the bare end or they may be **color coded** by a system devised by **NEMA (National Electrical Manufacturers Association)**. See Table 1 and Figure 2 for color coding of the more commmon steel electrodes.

The addition of iron powder to the flux coating of electrodes has greatly increased the rate of metal deposition as well as improved arc striking ability. The so-called drag rod AWS E7024 and others have as much as 50% iron in the flux coating. Figure 3 shows a comparison between powdered metal coated electrodes and conventional electrodes emphasizing operational characteristics of the two types.

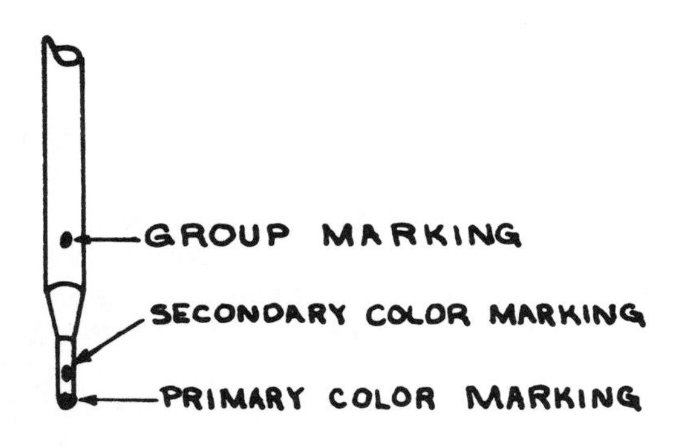

	Primary	Secondary	Group
6010	none	none	none
6011	none	blue	none
6012	none	white	none
6013	none	brown	none
7014	black	brown	none
7018	black	orange	green

Figure 2. Electrode Color Markings.

CONVENTIONAL COATING

POWDERED METAL COATING

SIDE VIEW

Larger drop transfer, thin coating, shallow cone allows rod to stick or freeze unless proper arc gap is maintained by operator.

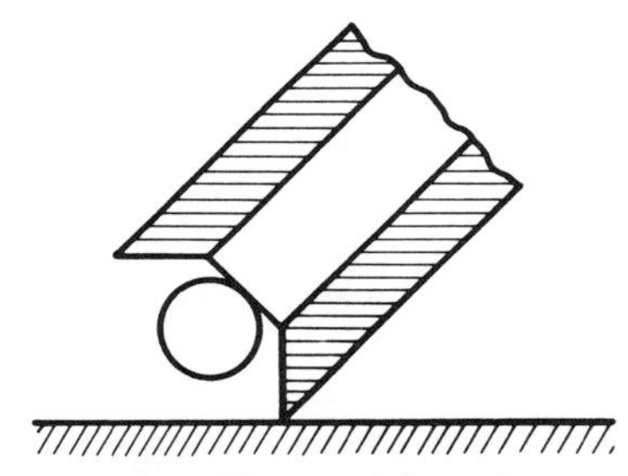

Electrode will not stick or freeze to work. Small molten globule cannot bridge arc gap due to depth of cone formed by coating.

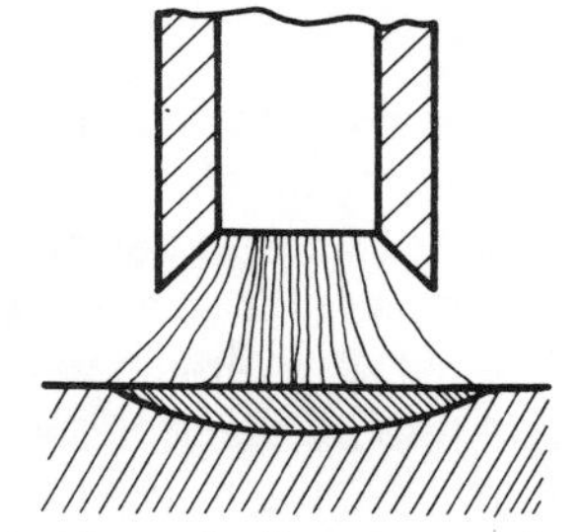

Thin coating, shallow cone allow arc to flare... spatter. Operator must accurately control arc length for maximum results.

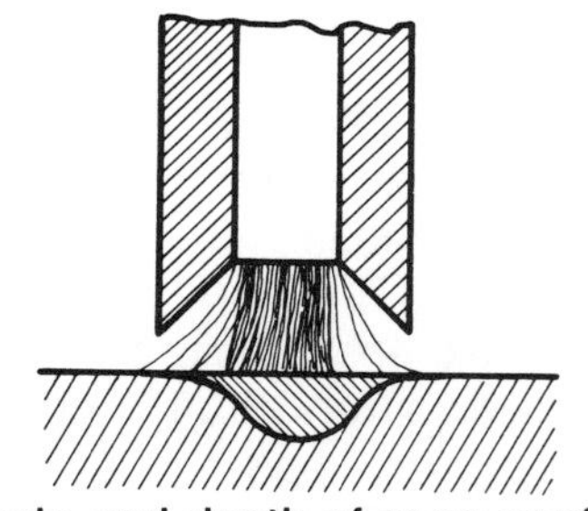

Angle and depth of cone confines arc... directs heat... increases arc density... minimizes spatter... insures adequate penetration without undercutting.

Figure 3. Comparison of Flux Coatings.

TABLE 1. ELECTRODE COLOR CODING AND MANUFACTURER'S CROSS-REFERENCE

AWS Classification	Color Markings End	Spot	Group	Polarity	Manufacturer Numbers AIRCO	Lincoln	A.O. Smith	Hobart	Marquette
E6010	—	—	—	DC RP	Easy-arc 10 78	5	SW 10	#10	#105
E6011	—	Blue	—	AC DC RP	230	180	SW 14	335	#130
E6012	—	White	—	AC DC SP	387 323	Fleet-weld 7	SW 11,12 7, 29	#12, 77,& 212	#120
E6013	—	Brown	—	AC DC SP	90	37	SW 16 or 15	447 313 13	140 151
E6014	Black	Brown	—	AC DC SP	Easy-arc 14	47	SW 15	Rocket 14	—
E7018 (low hydrogen)	Black	Orange	Green	DC RP AC	Easy-arc 328	LH70 LH71	SW 47	LH 718	LH 7016
E6024 (Iron powder)	Black	Yellow	—	AC DC RP	Easy-arc 12	Jet-weld-1	SW 44	Rocket 24	#12
E6027 (Iron powder)	—	Silver	—	AC DC RP	Easy-arc 27	Jet-weld 2	—	Rocket 27	—
E-Nickel for cast	Orange	Blue	White	AC DC RP	375	Soft-weld	SW Nickel Nickel cast	SW Nickel Nickel cast	#99
E-Steel for cast	Orange	—	—	AC DC RP	77	Ferro-weld	SW 5	Strong cast	#40

Electrode Selection According to Job Conditions:

A. Fast-Freeze-electrode that deposits a weld that solidifies or freezes rapidly-applicable to vertical or overhead welds. E6010 and E6011.
B. Fast-Fill-ability to deposit metal rapidly-opposite of fast freeze electrode. E7018, E6024 and E6027.
C. Fill-Freeze-characteristics between fast-freeze and fast-fili electrodes. E6012, E6013 and E6014.

TABLE 2. PHYSICAL PROPERTIES OF STEEL ARC WELDING ELECTRODES.

A.W.S. Classifica tion	As welded				Stress Relieved[5]		
	Tensile[1] Strength	Yield[2] Point	Elongation[3]	Impact[4] Resistance	Tensile Strength	Yield Point	Elongation
E6010	62-72,000	50-60,000	22-30	75	60-70,000	46-56,000	28-36
E6011	62-77,000	50-65,000	22-28				
E6012	67-80,000	55-69,000	17-22	55	67-80,000	55-69,000	22-27
E6013	67-80,000	55-65,000	17-29	55	67-78,000	55-63,000	24-30
E6014	72-80,000	60-70,000	17-29	55	67-77,000	55-70,000	24-30
E6018 E7018	79-84,000	62-72,000	25-27	80-100	78-80,000	67-69,000	27-29
E6018 E7018	83-93,000	72-85,000	25-29	80-100	80-90,000	68-80,000	25-29
E6024	72-90,000	60-86,000	17-25	45	72-86,000	60-82,000	22-32
E6027	69-73,000	57-61,000	25-30	65	62-70,000	50-59,000	25-32

1. **Tensile Strength** - given in (psi) pounds per square inch - the resistance of a material to a force that is acting to pull it apart. Maximum load in pounds divided by cross-sectional area in square inches.

2. **Yield Point**- a point at which the metal continues to stretch without an increase in load pounds. Point at which the stress strain curve ceases to be a straight line. Not point of Failure.

3. **Elongation** - given in percentage based upon amount of stretch over 2 inches in a 2 inch specimen. Dividing the excess over 2 inches by initial length of 2 inches gives the percentage of elongation. A good measure of ductility.

4. **Impact Resistance** - given in ft. - lbs. - charpy V-notch at room temperature - a 2 inch specimen, .394" square supported at both ends 1.575" between supports and broken by blow opposite the notch. Normal arc electrodes show impact resistance of 50-80 ft. lbs.

5. **Stress Relieved** - the weldment is heated to some temperature below transformation but sufficient to allow the metal to stretch slightly.

TABLE 3. MILD STEEL ELECTRODE RATINGS BY JOB APPLICATION AND PHYSICAL PROPERTIES

	E6010	E6011	E6012	E6013	E7014	E7018	E7024	E7028
Groove butt welds, flat (1/4")	4(b)	5	3	8	9	9	9	10
Groove butt welds, all positions (1/4")	10	9	5	8	6	6	(a)	(a)
Fillet welds, flat or horizontal	2	3	8	7	9	9	10	9
Fillet welds, all positions	10	9	6	7	7	6	(a)	(a)
Current	DCR	AC DCR	DCS AC	AC DC	DC AC	DCR AC	DC AC	AC DCR
Thin material (1/4")	5	7	8	9	8	2	7	(a)
Heavy plate or highly restrained joint	8	8	6	8	8	9	7	9
High-Sulfur or off-analysis steel	(a)	(a)	5	3	3	9	5	9
Deposition rate	4	4	5	5	6	6	10	8
Depth of penetration	10	9	6	5	6	7	4	7
Appearance, undercutting	6	6	8	9	9	10	10	10
Soundness	6	6	3	5	7	9	8	9
Ductility	6	7	4	5	6	10	5	10
Low-temperature impact strength	8	8	4	5	8	10	9	10
Low spatter loss	1	2	6	7	9	8	10	9
Poor fit-up	6	7	10	8	9	4	8	4
Welder appeal	7	6	8	9	10	8	10	9
Slag removal	9	8	6	8	8	7	9	8

(a) Not recommended

(b) Ratings on a 1-10 basis with higher number indicating electrode is outstanding for particular application or physical property.

AMPERAGE

Proper **amperage setting** is a very critical part of arc welding. When starting to weld on a new machine, it is important to learn first how to set the amperage, then the effect of changing the settings. Seldom can an operator select the right setting on the first try. Amperage affects the burn-off rate of the electrode and therefore, influences the rate of metal deposition and the speed of welding. The type of electrode being used also influences amperage setting, for example, the 7024 requires much higher setting than does the 6011 due to the higher iron content in the 7024. Amperage determines heat, according to Joule's Law - I^2 R= heat. Actual amperage is greatly influenced by arc length.

STRIKING THE ARC

The ability of the weldor to **strike the arc** successfully is where arc welding begins. Too often the beginner minimizes its importance, apparently thinking that the welding arc is incidental to the heat required for proper fusion. Therefore, a great volume of light from the arc site means that welding is not being done. Proper arc striking is attained by the so-called "**peck**" method or by the "**scratch**" method. "Peck" ("A") is used with DC welders whereas "scratch"("B") is used on both AC and DC. See Figure 4. In either case, the objective is to cause the electrons to flow through an air gap. The heat from the higher amperage of a short arc melts the base metal and electrode metal. Metal transfer from the electrode to the base metal is influenced by arc length and chemical flux coatings. "**Freezing**" of the electrode is a beginner's first frustrating experience. This can be remedied by raising amperage and using a more distinct scratching motion at moment of contact of electrode with work. Usually a slightly long arc is held momentarily at the start of a bead in order to heat the cold metal. This promotes a "**hot start**" or admixture of base and electrode metal.

Manual arc welding involves unavoidable changes of arc length (short to long). Increasing arc length accidentally or deliberately increases arc voltage and decreases amperage (Heat = I^2R). As illustrated in Figure 5, in slope A there is a change of 15 amps when the desired short arc is lengthened. In slope B (a flatter slope), there shows a 40 amp drop from short to long arc. Practically, the flat slope is most useful for position welding where heat control is necessary by arc manipulation. The steeper slope is most useful for high production flat welding. Dual Control DC welders are more expensive but more versatile.

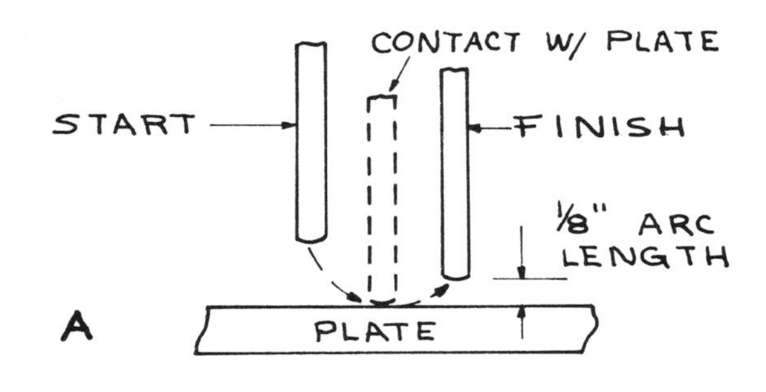

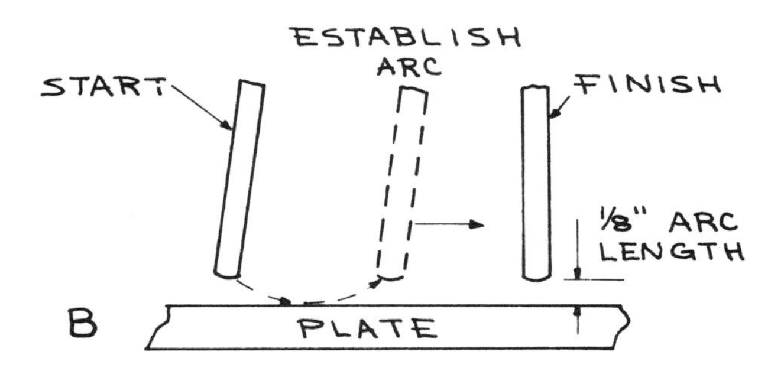

Figure 4. There are Two Methods of Starting, or Striking, the Arc.

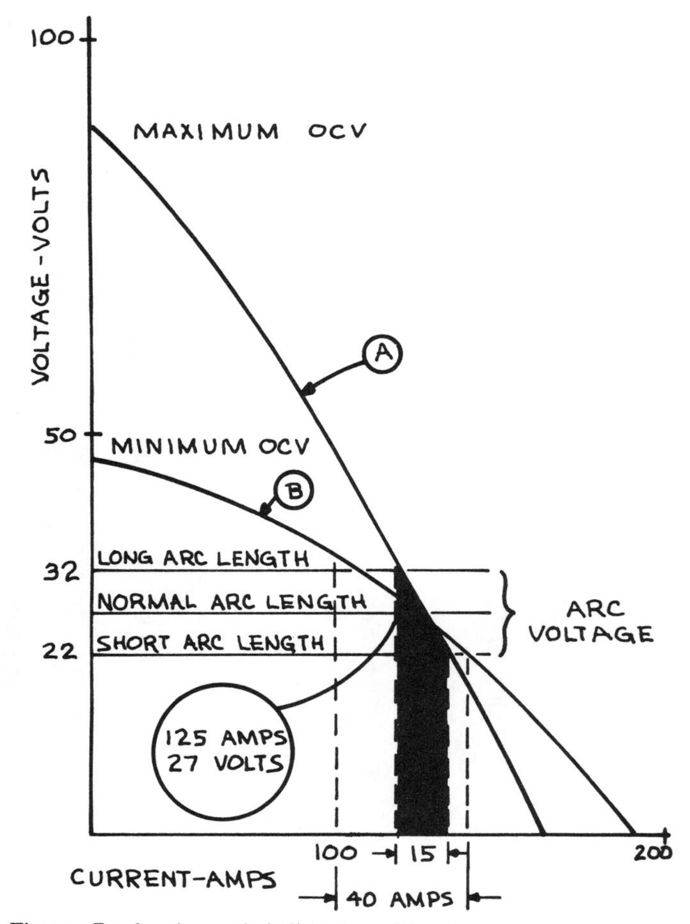

Figure 5. Arc Length Influences Heat.

ANGLE OF ELECTRODE

The arc has a definite directional force; therefore, a skilled operator uses this force to an advantage. In flat position welding, the electrode is perpendicular from side to side and tilted in the direction of travel about 15°. See Figure 6.

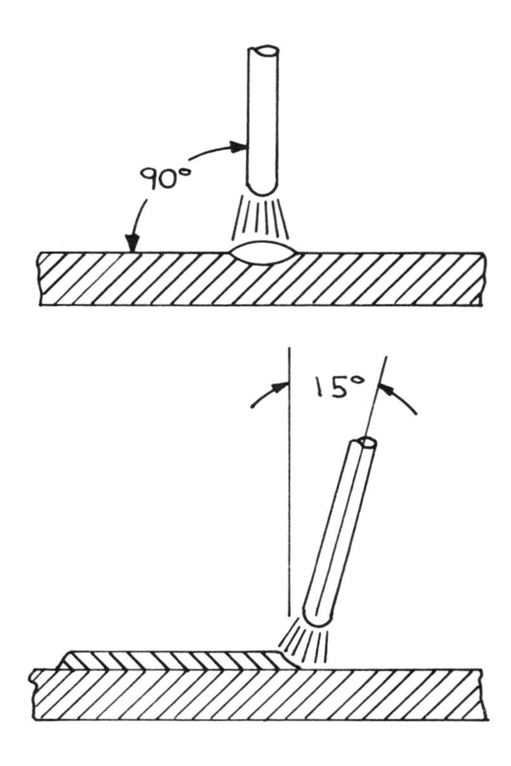

Figure 6. Proper Angle of Electrode

PROPER ARC LENGTH.

In manual stick-electrode welding, the ability to maintain proper **arc length** is a variable that must be under complete control by the operator. Figure 5 explains the influence of the arc length on heat. Arc length influences the appearance of a bead as shown in Figure 7. Weld spatter shows electrode metal loss, a form of waste. Therefore, in order to get maximum heat and to make best use of electode metal, a short arc is always used. However, there are times when the weldor deliberately lengthens the arc to control heat. A welding arc is measured from the bottom of the molten pool. As a rule of thumb, arc length should be **equal to the diameter of the wire in the electrode** being used. A useful idea to remember is that if you have difficulty seeing because of the small amount of arc light given off, you are probably near the right arc length. Arc length should be held uniformly short throughout the length of the bead to get even penetration and metal deposition if a strong bead is wanted.

SPEED OF TRAVEL

The compound movement of the electode forward and downward (for short arc) is probably the most frustrating experience in learning to weld. Travel speed influences **bead width**, **penetration**, and **general shape** of the bead. Speed must be uniform not erratic. Most people can weld with a more steady movement if both hands are on the electrode holder. Operator comfort certainly influences speed of travel; therefore, best quality welding can always be done in a flat or down-hand position. A common rule of thumb is that the **bead should be about twice the diameter of the electrode wire**. See Figure 7.

RUNNING THE BEAD

Having "hot started" the bead, the arc is then shortened as the electrode is moved forward at a steady rate. See Figure 7. If speed is erratic, poor fusion and strength are the results of jerking forward motion. It is evident that the electric cupola furnace that casts the bead is fed iron from the top by pushing the electrode into the furnace. The compound movement of the electrode forward and downward simultaneously can be a learning experience for a beginner. The bead width is thus determined by forward speed. A rule of thumb calls for a bead width twice the diameter of the electrode wire. The end of a bead is called a crater which can be a stress point if not filled properly. Crater filling techniques vary with electrode types and individual skill. Every bead should have a uniforn width middle with individual ripples evenly spaced. Finally, the good bead should have a filled crater. Beginners should develop techniques by running beads about three inches long because muscular coordination tends to decrease the longer the beads become.

"**Read the Bead**" is the best way to evaluate amperage settings as well as other factors that determine successful arc welding. The weld profiles shown in Figure 7 illustrate the importance of evaluating every bead to determine its quality. Any weldor who fails to "read the bead" will never attain high success in arc welding.

Further success in "bead reading" can be attained by studying Figure 8, a chart showing common welding troubles and how to correct them.

Finally, beginners should follow suggestions given by manufacturers of electrodes as to amperage setting and applications.

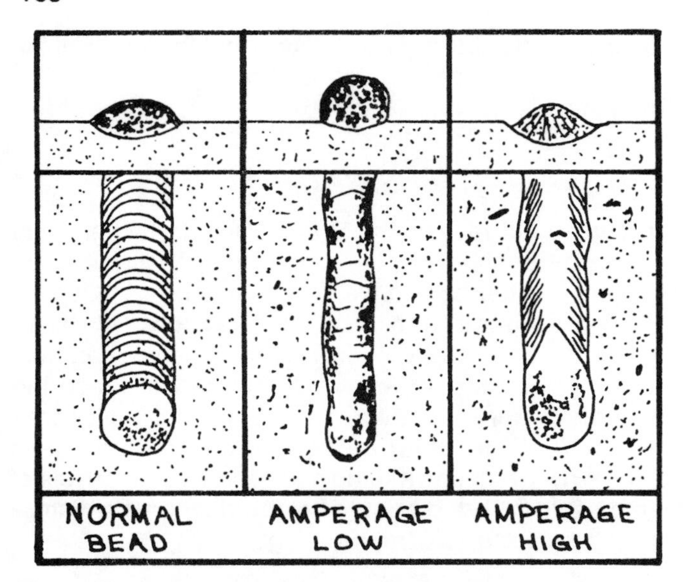

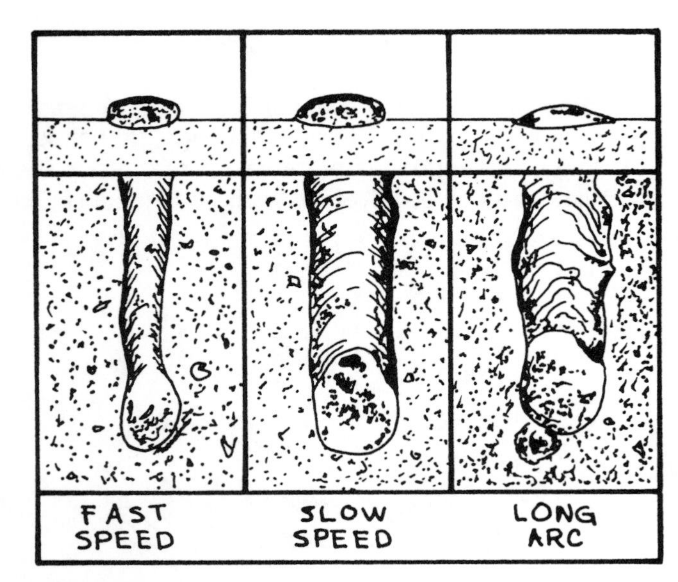

Figure 7. Amperage Setting, Speed of Travel, Length of Arc and Angle of Electrode Influence the Quality of the Weld.

CONDITION	CAUSE	CORRECT BY
Undercutting	• Excessive welding current	• Reduce current
	• Arc too long	• Shorten arc
	• Electrode being held at 90 degree angle	• Tilt electrode in direction of travel
	• Travel too fast	• Reduce speed of travel
Convex bead (bulging)	• Arc too short • Travel too slow	• Lengthen arc • Increase speed of travel
Slag included in bead	• Current too low • Arc too short	• Increase current • Lengthen arc
Bead is porous	• Excessive current	• Reduce current
	• Travel too fast	• Reduce speed, weave electrode to allow time for gas to escape
Cracks	• Small throat (light penetration)	• Check valve of current • Reduce arc length • Reduce travel speed
	• Craters	• Improve welding technique, back step, fill craters
	• High quenching rate	• Preheat for heavy plate and low - temperature plate
	• Contaminated deposit	• Remove slag, use light penetration electrode, improve welding technique to prevent pick-up from base metal

Figure 8 Common Welding Troubles and How To Correct Them.

CONTROLLING DISTORTION

Welding involves chemical, metallurgical and mechanical changes determined by the magnitude of the heating and cooling cycles of the joining process. Metallurgical and chemical changes are dealt with elsewhere. Atomic vibrations due to heat cause volume changes of expansion followed by contraction during cooling. The ductility of welded parts determines to a great extent, the mechanical changes that occur. Rigid metals tend to show evidence of stress cracks or residual stress. Ductile metals tend to distort or warp. Restrained low carbon steel heated lowers its elastic limit or yield point thereby tending to develop permanent or plastic deformation. The average change in linear dimensions for low carbon steel is 0.8 x 10^{-8} in. per in. for each degree F. change in temperature. The result of weld shrinkage is shown in Figure 9.

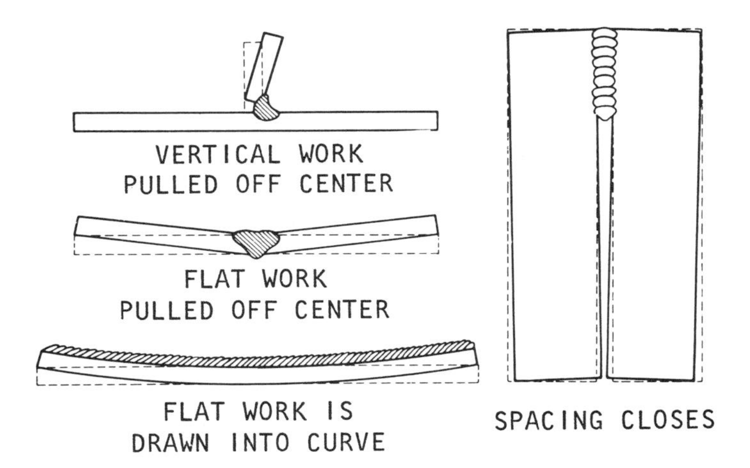

Figure 9. The Result of Weld Shrinkage.

Avoid distortion by:

1. **Do not overweld** - know the requirements of the weld and meet those needs for strength or water-tightness suggestions. Since lap welds are essentially fillet welds, Figure 10 would also apply. The size and reinforcement of groove welds are shown in Figure 11. Additional suggestions appear in Table 4.

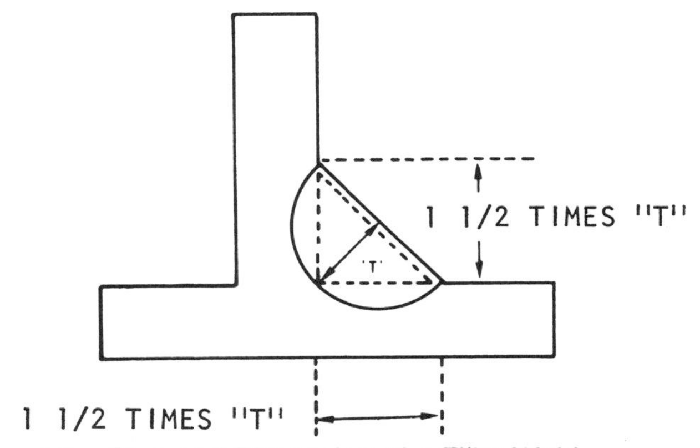

Figure 10. Size and Dimension of a Fillet Weld.

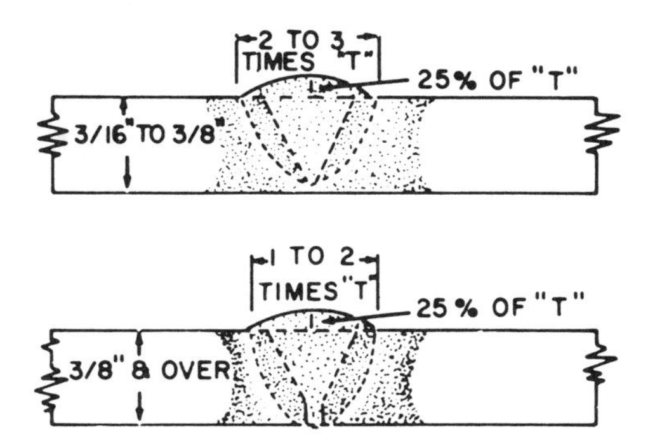

Figure 11. Specifications for Thick Metals.

TABLE 4. REINFORCEMENT AND WIDTH OF WELD BASED ON METAL THICKNESS.

Thickness	Reinforcement
up to 1/16 in.	thickness (T)
1/16 to 1/8 in.	75 per cent T
3/16 to 1/4 in.	25 per cent T
over 1/4 in.	1/8 in. max.

Metal Thickness	Width of Weld
up to 1/16 in.	6 to 8 T
1/16 to 3/32 in.	4 to 6 T
3/32 to 1/8 in.	3 to 4 T
1/8 to 1/4 in.	2 to 3 T
over 1/4 in.	varies

2. **Avoid continuous welds** except for fluid-tight requirements. Examples of intermittent and staggered intermittent welds are shown in Figure 12.

3. **Use fewer beads (passes)** with larger electrodes. Lateral distortion is about one degree per pass.

4. **Use 60 ° included angle on edge prepared joints.**

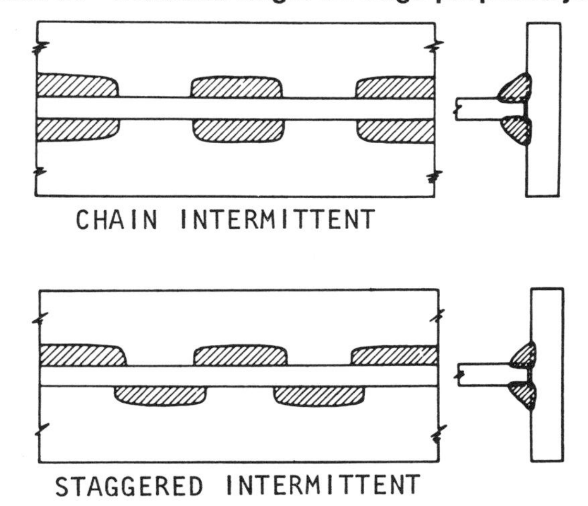

Figure 12. Intermittent Fillet Welds.

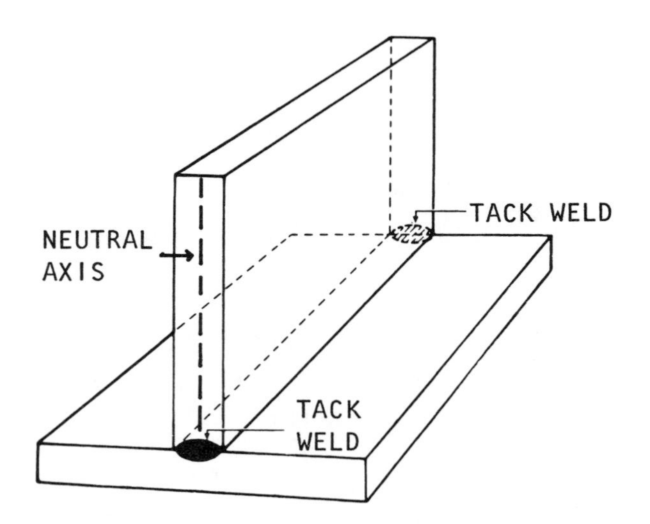

Figure 13. Metal Preparation and Neutral Axis.

5. **Weld near the neutral axis**. See Figure 13.

6. **Use back-step welding**. See Figure 14, part A.

7. **Use wedging** shown in Figure 14, Part B.

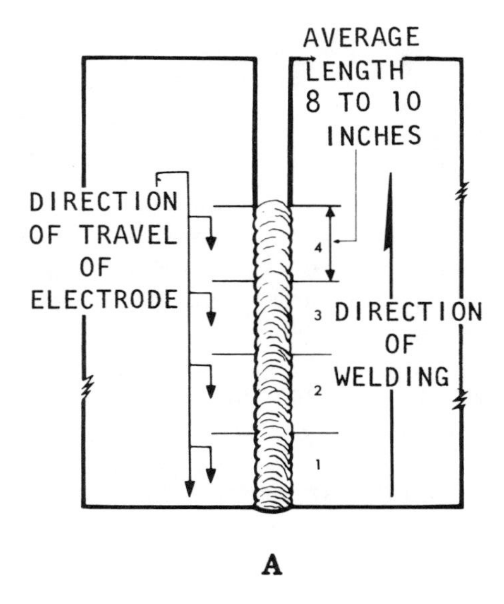

A

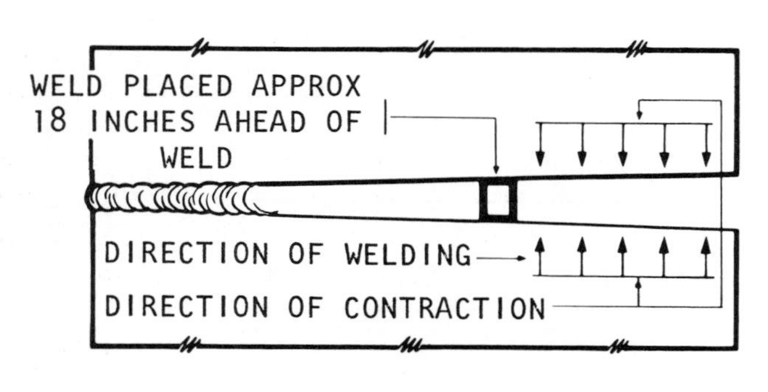

B

Figure 14. Methods of Counteracting Contraction.

Another principle for controlling distortion is to use planned shrinkage to bring welded parts into proper position. Planned shrinkage involves prebending with clamps. Presetting parts out of position will allow shrinkage to return parts to desired position on welding.

A final principle for distortion control is the use of restraints thereby causing the weld metal to stretch . **Tack-welds** are the most common restraining device, "C" clamps, vise grips, fixtures and jigs are frequently used. Figure 15 shows wedged and bolted fit-up fixtures. Limited peening also controls distortion.

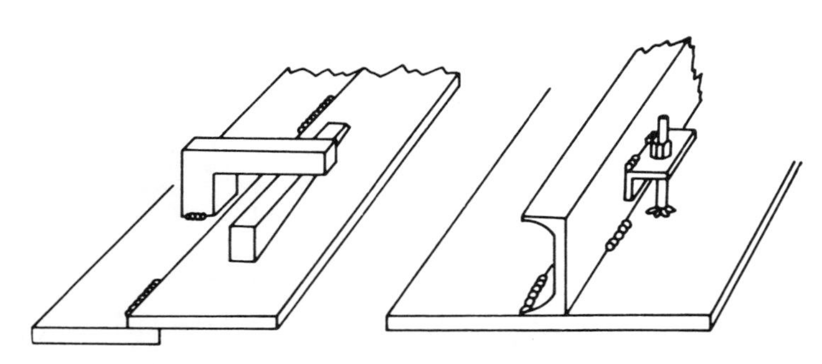

Figure 15. Using Fit-Up Fixtures.

HOW TO CHOOSE AN ARC WELDER FOR SHIELDED METAL ARC WELDING (SMAW)

The common metal-joining process of stick electrode arc welding has become complicated. Today the purchaser of a welder has a choice of four entirely different types. Each has its particular merits and some drawbacks.

The four types of manual arc welders - (1) alternating current,(2) direct current motor-generator, (3) direct current rectifier and (4) the combination. AC-DC are compared in welding performance, benefits and limitations, and their best applications.

The conventional stick electrode welder is sometimes called a "constant current machine". It is also called a "drooper" because its voltage drops as welding current increases, thus its volt-ampere output curve "droops".

There are many different ways of classifying welding machines; one of the basic classifications is the type of control. There are two basic types; **single-control machine** and the **dual-control machine**. Single-control machines have just one dial which changes the current output of the machine. Dual-control machines have two controls, one for fine and one for coarse current adjustment.

The single-control welding machine is designed to produce a number of characteristic output curves similar to Figure 16. The characteristic curve is obtained by plotting the terminal voltage when loading the welding machine with varying amounts of pure resistance load. Note that as the amperage output of the machine increases, the voltage of the machine decreases. The "open-circuit" voltage is a fixed figure based on the design of the machine and is usually in the neighborhood of 75 V. The single-control machine changes the current output and is adjustable from a minimum figure to a maximum figure which is usually greater than the rated output of the machine. The shaded area on this particular curve is the output voltage or "arc voltage" of the machine during welding. By adjusting the current control output a great number of curves can be obtained. The dotted line shows the machine adjusted to 150 amps. On tap or plug-in machines, there will be a number of curves corresponding to the number of taps that are on the machine.

Dual-control machines are normally direct-current generator-type welding machines. Dual-control machines offer the operator much more flexibility in selecting conditions for making different types of welds. These machines, in effect, have what is know as "slope control".

A welding machine produces two types of voltage. One, known as "**open-circuit**" voltage , has the voltage produced when the arc is not in operation. The other, known as

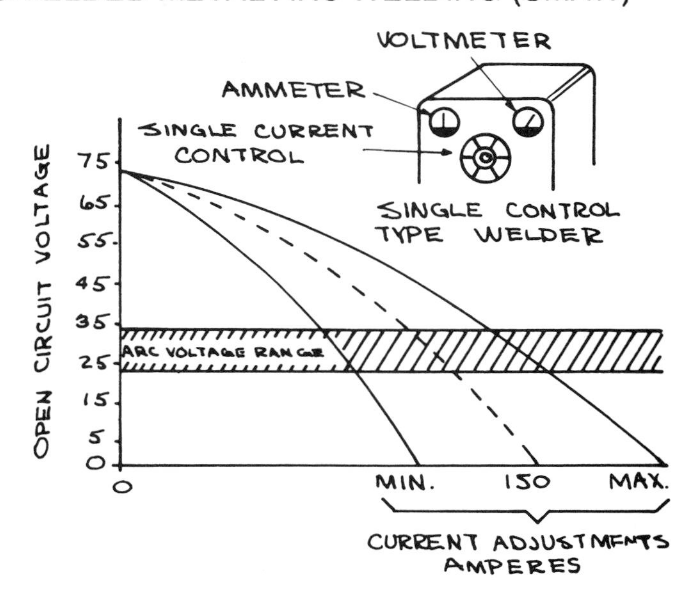

Figure 16. The Single-Control Machine Produces Various Volt-Ampere Curves Based on a Single Open-Circuit Voltage.

"**arc voltage**", has the voltage output of the machine while the arc is in operation. A dual-control machine has two control adjustments; a coarse adjustment and a fine adjustment. A coarse adjustment sets the current output of the machine and varies the output from the minimum to the maximum current. The fine control actually changes the current setting between the coarse control setting points.

On motor generators, however, this fine control has an additional function. Under nonwelding conditions, the fine control also changes the "open-circuit" voltage of the machine. This relationship is shown in Figure 17. Note the open-circuit voltage can be changed from approximately 60 V to 100 V. This is accomplished by the fine adjustment dial. While welding, this same dial has no effect on the welding-arc voltage. It is only controlled by the length of the welding arc.

The higher no-load, or "open-circuit", voltage provides for easier arc starting with all types of electrodes. It also provides a steeper curve through the arc-voltage range. The change of slope, which is controlled by the open-circuit voltage setting, does have an effect on arc characteristics.

A study of Figure 18 will help to illustrate this situation. A **short arc is a lower-voltage arc**; and with the same settings of the machine, is a higher-current arc. This is shown by the meters and the diagram of the short arc.

On the other hand, the **long arc is a higher-voltage arc,** but the current is lower, at the same machine setting. The slope of the characteristic curve, which is changed by changing the open-circuit voltage, actually changes the characteristics of the arc.

It can be seen that with the flatter slope an equal change in arc voltage will produce a greater change in welding current. This produces the harsh or digging arc. This type of arc is popular for pipe welding. With the steeper curve, produced by the higher, open-circuit voltage, an equal change in arc length will produce less of a change in current output. This is a softer or quieter arc, and is more useful for sheetmetal welding. Thus, the dual-control generator welding machine allows the most flexibility to the weldor. Note: Transformer welders with coarse and fine adjustment knobs are not considered dual-control machines because the open-circuit voltage is not changed.

Constant current machines can also be classified as motor generator, AC transformer, AC-DC rectifier welders, engine-driven welders, and many, many more. The direct-current rotating generator was the earliest type of welding machine and still enjoys a very favorable reputation. Generators can be driven by electric motors, gasoline engines, diesel engines, or even from power take-off shafts of any type of rotating device.

Another very popular welding machine is the **transformer welder**. It is usually the least expensive, lightest, and smallest of any of the various types of welding machines. The transformer welder takes power directly from the utility lines and transforms it to the voltage required for the arc. Then by means of various magnetic circuits, inductors, etc., it provides the voltage and ampere characteristics necessary for welding. In alternating-current welding, the polarity of the welding current changes 120 times each second. This cannot be seen, of course, and is not notice able by the weldor.

The welding current output of a transformer welder can be adjusted in many different ways. Perhaps the simplest method of adjusting output current is to use a **tapped secondary winding**. Plugs or a tap switch can be used. This is a popular method employed by many of the limited input welding transformers. Exact current adjustment, however, is not possible. This type of machine is normally used on farms and light industry.

The advantages of the transformer and rectifier have been combined into machines known as AC-DC welders. By means of a special switch, the output terminals are connected to the transformer, or the rectifier, so the operator can select either AC current or DC straight or DC reverse-polarity welding current to satisfy his welding requirements. In some types of AC-DC welders, high-frequency oscillators, plus water and gas control valves, are installed. This then makes the machine ideally suited for tungsten inert-gas welding as well as for manual, coated-electrode welding.

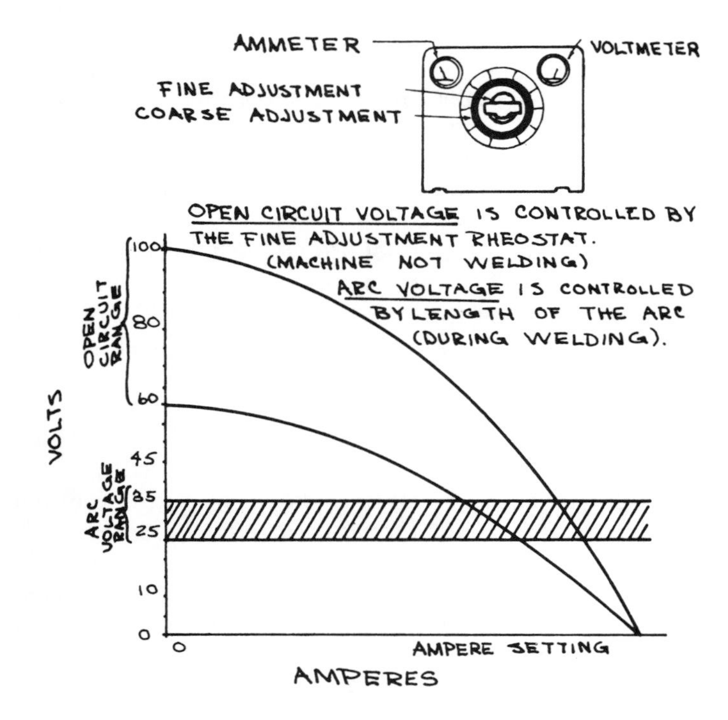

Figure 17. Dual-Control MGDC Welders Allow the Operator to Vary the OCV.

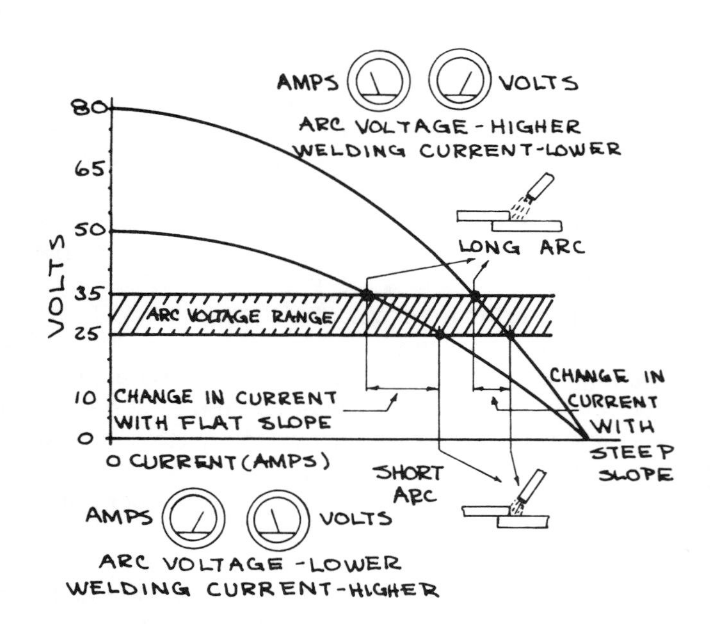

Figure 18. The Relationship Between Possible Current Changes Due to Arc Length as Related to Two Volt-Ampere Curves.

TRANSFORMER TYPE ARC WELDERS

In order to understand transformer arc welders, one must first understand the construction of a simple transformer and its function. A transformer is an electrical device having a laminated silicon iron cone and two electrical circuits called **primary and secondary windings**. These windings are not connected directly but are installed to prevent connection. The transformed electricity is moved from the primary winding to the secondary winding by **induction**. Induced current flows in the secondary winding because it is within the electro magnetic field of the primary winding. Current is inversely proportional to the number of turns. Voltage is directly proportional to number of turns in the two windings. Figure 19 shows a step-down transformer where the secondary winding has half the number of turns of the primary. The result shown on the volt meter is that the output is equal to half the input. Figure 20 shows a step-up transformer. Note that each has the appropriate number of windings for the job it is to do.

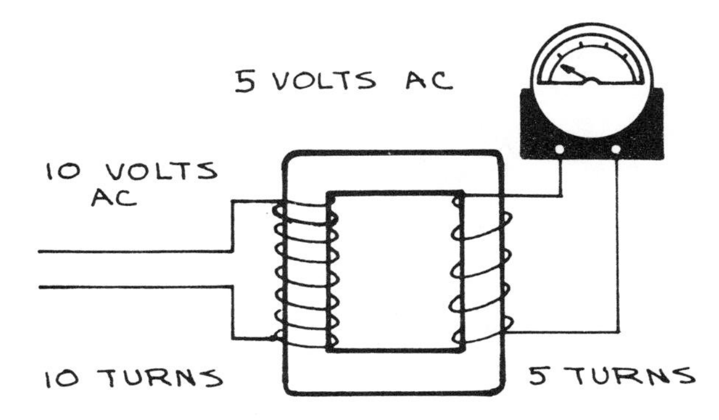

Figure 19. Step-Down Transformer.

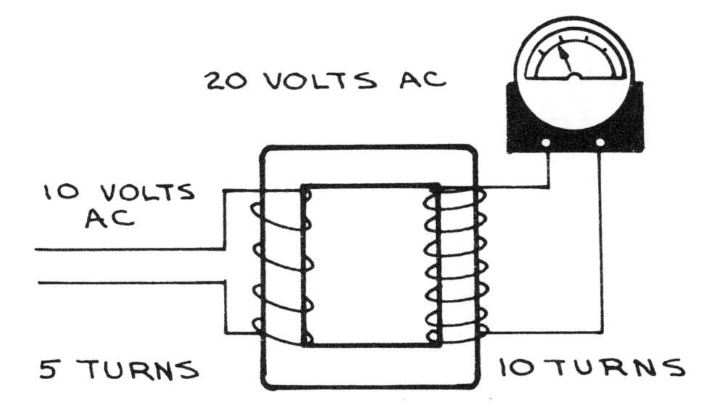

Figure 20. Step-Up Transformer.

THE EQUIPMENT

AC: An AC welder is basically a single-phase transformer. Practically all AC welders in service are of the single-operator static transformer type. They are standardized by the National Electrical Manufacturers' Association in ratings of 180, 200, 300, 400, 500, 750 and 1000 amps. The maximum current of each welder is 25% more than its rated output which is a safey factor avoiding damage to the machine. Single-phase welding transformers are made for 230 to 575 volt power supply, and 25, 50 or 60 cycle electricity.

Methods of Controlling Output Current: Single operator, AC arc welding machines may be divided into two general classes with respect to means employed to control the output current. The first class uses a **constant voltage transformer**, and controls the output current by means of an adjustable resistor or an adjustable reactor, the latter being the more common. The second class includes those welding machines in which the **internal reactance of the transformer is adjustable**.

I. ADJUSTABLE REACTOR IN OUTPUT CIRCUIT.

A. A constant current transformer with **tapped reactor coil** as illustrated in Figure 21 has separate reactors making contact to taps on the reactor winding. The more windings of the reactor coil included in the output the lower will be the output or welding amperage.

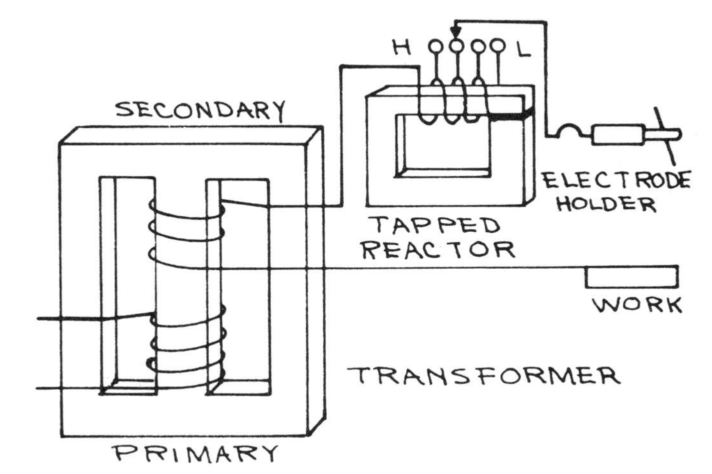

Figure 21. Constant Current Transformer with Tapped Reactor.

B. Shown in Figure 22 is a constant current transformer with **magnetically adjustable reactor**. The adjustable air gap in the magnetic circuit of the reactor adjusts the reactance. An iron block is moved by a screw to adjust the air gap which adjusts the reactance of the coil. As the bar is moved down the reactance in the reactor coil is increased which lowers the welding current going to the electrode.

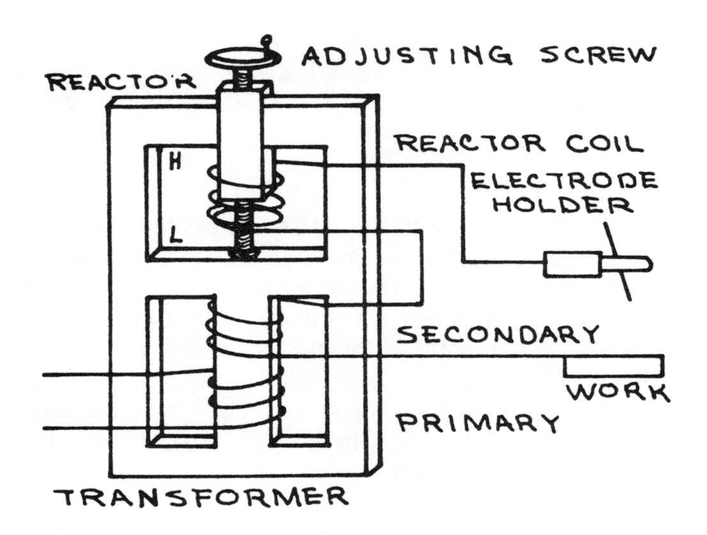

Figure 22. Constant Current Transformer with Magnetically Adjusted Reactor.

C. Constant current transformer with **electrically adjustable reactor** - the magnetic circuit is saturated by means of a direct current winding on the center leg of the reactor as shown in Figure 23. The direct current is provided by a rectifier energized by connections to the main transformer. The adjustment of the rheostat will vary the direct current, the saturation of the iron core of the reactor, thus controlling the output. As the rheostat is moved to the right in the diagram, the welding amperage will be decreased. The rheostat sets up resistance to the DC current flow in the center of the reactor. This results in lower welding amperage.

II. ADJUSTMENT OF INTERNAL REACTANCE.

A. Figure 24 shows a diagram of a constant current transformer with **adjustable flux leakage air cap**- a movable iron member in the magnetic circuit is used to adjust the reactance between the primary and secondary coils. Separating the coils further or shortening the air gap increases the leakage flux between the coils which reduces the output current. As the bar is moved down between the coils, the welding amperage is decreased because the bar tends to break up or weaken the magnetic field that is formed around the two coils.

B. The diagram in Figure 25 shows a constant current transformer with **adjustable coil spacing** - adjustment is accomplished by changing the position of the primary and secondary coils in respect to each other. As the coils are moved closer together, the amperage is increased because the induced current has a narrower air gap to pass through.

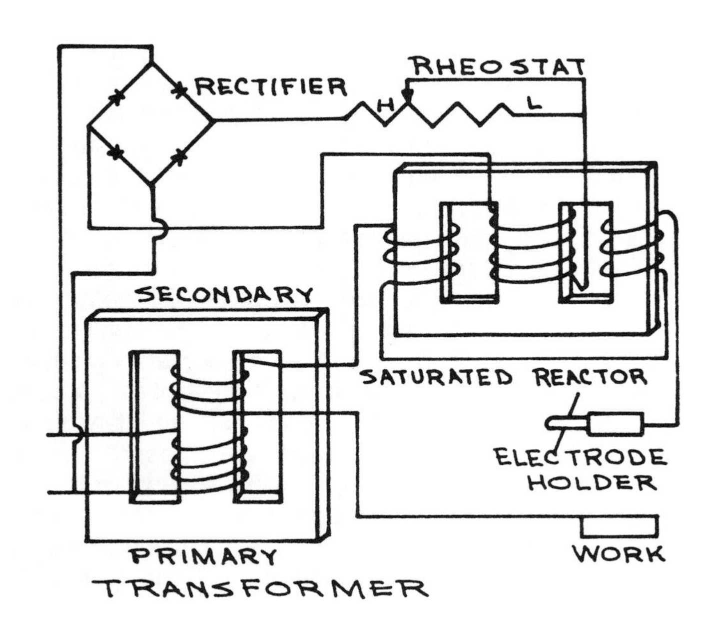

Figure 23. Constant Current Transformer with Electrically Adjustable Reactor.

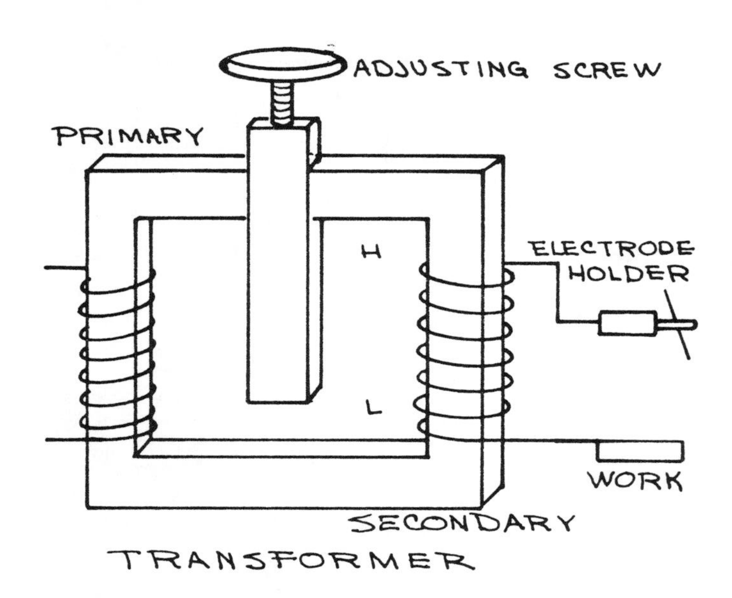

Figure 24. Constant Current Transformer with Adjustable Flux Leakage.

DC Motor Generator: This type of welder consists of a generator and driving electric motor (or gasoline engine for field service) plus generator and motor controls. Illustrated in Figure 26 are the two types of DC welder generator circuits.

Single-operator DC welding generators are made in ratings from 150 to 600 amperes. They are standardized by the National Electrical Manufacturers' Association in three classes: (1) 150 and 200 amp, 30 volts (suitable for light industrial arc welding and gas-shielded arc welding); (2) 200, 300 and 400 amp, 40 volts (manual and machine

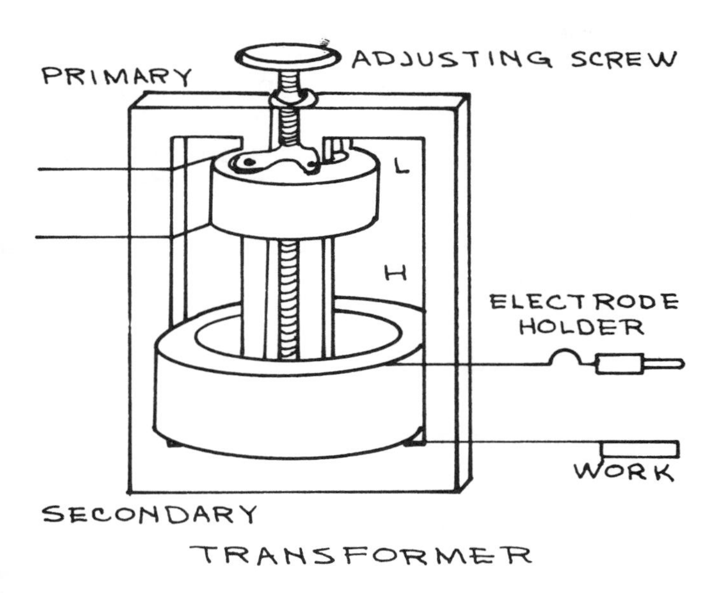

Figure 25. Constant Current Transformer with Adjustable Coil Spacing.

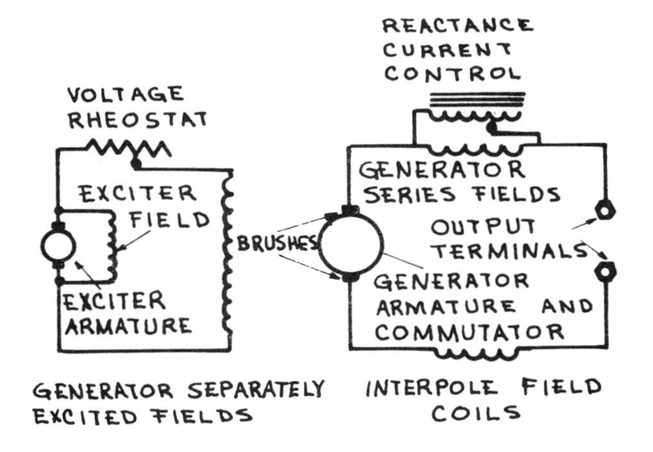

SEPARATELY EXCITED, DIFFERENTIALLY COMPOUNDED WELDING GENERATOR CIRCUIT

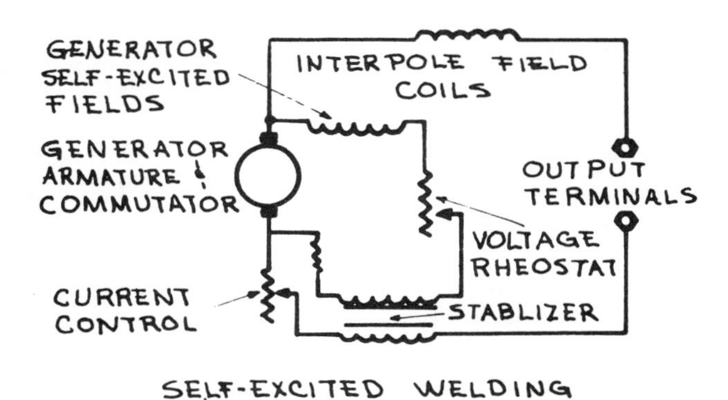

SELF-EXCITED WELDING GENERATOR CIRCUIT

Figure 26. Two Types of DC Welder Generator Circuits.

welding in industrial plants, field erection) and (3) 600 amp, 40 volts (submerged-arc welding and carbon-arc cutting).

Induction motors that drive welding generators are mostly three-phase. There are not many single-phase motors used, since AC welders meet the need for single-phase equipment. Motor-driven machines are available for 220, 440, and 550-volt, and 25, 50, or 60 cycle power motors. Note Figure 27 showing the pictorial wiring diagram of DC motor-generator welding machine wih an electric motor drive. Internal combustion engines or power take-off devices provide the generator drive for portable or field service machines.

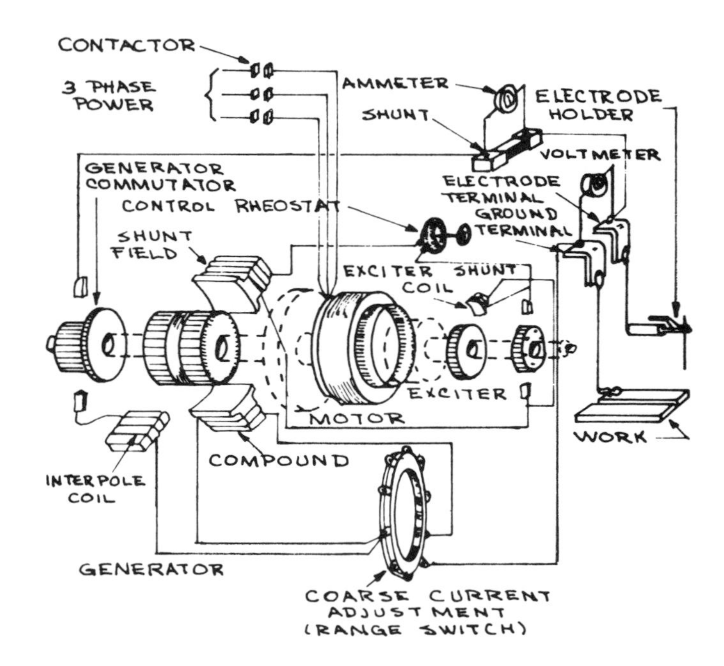

Figure 27. DC Motor-Generator Welding Machine with Electric Motor Drive.

DC Rectifier: The rectifier type welder, schematic diagram in Figure 28, uses a three-phase transformer to change the power supply voltage, and a dry-plate rectifier stack of selenium plates. In a selenium cell, the thin layer of selenium lies between front and back electrodes. Though neither electrode, nor the selenium itself, is an asymmetric conductor, the "barrier layer" between the selenium and front electrode acts as a one-way street for electric current.

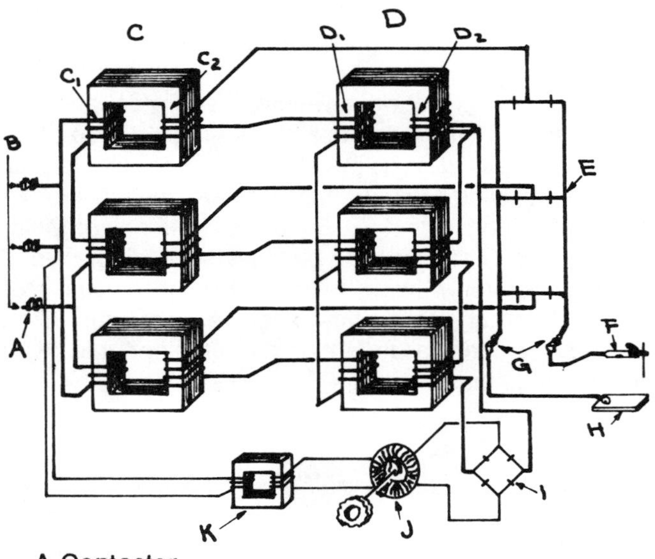

A-Contactor
B-3 Phase power
C-Transformers
  C_1-Primary coil
  C_2-Secondary coil
D-Saturable reactors
  D_1-Load coil
  D_2-Control coil
E-Main rectifier (converts AC to DC)
F-Electrode holder
G-DC output terminals
H-Work
I-Control rectifier (converts control current from AC to DC)
J-Variable transformer (current adjustment)
K-Control transformer (supplies power for control circuit)

Figure 28. Schematic Diagram of 3-phase Rectifier Arc Welder.

Silicon rectifiers AC-DC and DC arc welders are becoming more common. Aluminum heat sinks provide cooling for these smaller devices.

The rectifier welders combine the single or three-phase transformer, a single or three-phase adjustable reactor and a single or three-phase, full-wave, high-voltage rectifier. Current adjustment is by movable coil or by a saturable reactor, as with AC equipment as shown in Figure 25. A wide range of usable current is provided. Rated amperage output of the new NEMA design standards for rectifier welders approximately parallels the standards for motor-generator welders. At present, the most popular ratings are 300 and 400 amps.

AC-DC Transformer Rectifier: The sine wave diagram in Figure 29 shows the polarity reversals of AC rectified to DC welding power sources. Single phase DC was single phase AC. Three-phase DC was three-phase AC. The close spaced ripples of three-phase DC makes a smoother arc performance. In recent years more efficient and economical capacitors have been installed to smooth the ripple configuration. As a result single phase AC-DC welding machines have become popular.

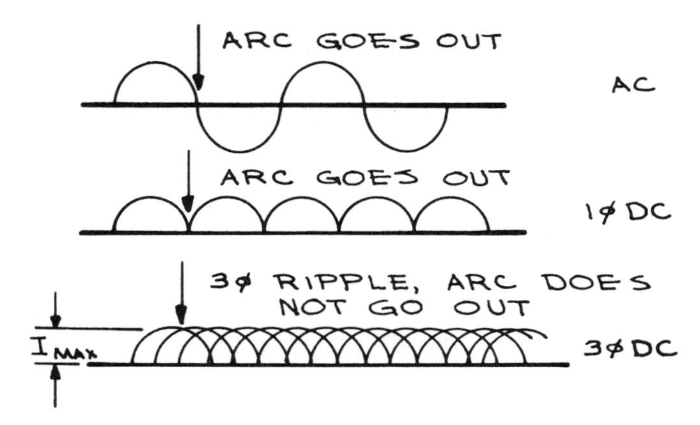

Figure 29. Sine Wave Comparison of Arc Welder Power Sources.

WELDING PERFORMANCE

By far the greatest single item in welding costs is labor. Hence, a welding performance that will save on labor is far more important than equipment cost or power consumption. Power cost and amortization of equipment are almost insignificant in comparison with the weldor's time. The cheapest welding machine in the long run is the one that allows the most work to be done with the least number of defective welds.

What makes good welding performance? To answer this question adequately, it is necessary to explain the nature of the welding arc.

Examination of oscillograms of a welding arc, or of the arc itself as shown in high-speed motion pictures, reveals that it is by no means continuous. Large, medium and small globules of molten metal are constantly short circuiting the arc by building temporary metallic bridges from electrode to work piece. Since this is an indispensable condition to welding, any welder must be designed to cope with it.

The transient currents caused by these frequent short circuits across the welding arc become particularly important when the welding operator tries to "crowd" or shorten the arc, which must be done when welding with low-hydrogen electrodes or in welding vertically. A surge of extra current is needed at such time to remove each molten metal bridge quickly before it can chill and freeze the electrode to the work, the unhappy condition that weldors call "sticking".

If the momentary current at short circuit is only slightly higher than during arcing, the welding machine is known as a "**soft-arc**" machine. In contrast, a forceful or "**peppy**" arc is created when the current builds up rapidly to a high peak during metal transfer. No factor of welder performance is more importnat than arc force.

Closely related to current transients caused by short circuiting are the voltage transients. A large drop of molten metal will leave a long, relatively cool gap in the hot gases surrounding the arc. The current will drop momentarily until the gap becomes ionized. If ionization is delayed too long, the arc may "pop out" and have to be restarted.

A high open-circuit voltage in the welder indicates a high degree of arc stabilization. A low open-circuit voltage does not permit smooth and popout-free operation, especially in the low-current range. Very high OCV can be a shock hazard. 80 OCV is relatively safe and therefore quite common. It should be remembered that shock hazard threshold limits are ill-defined. Electrical shock tolerance seems to vary with individuals up to a certain point. Another factor influencing electrical shock is the relative humidity and temperature relationships.

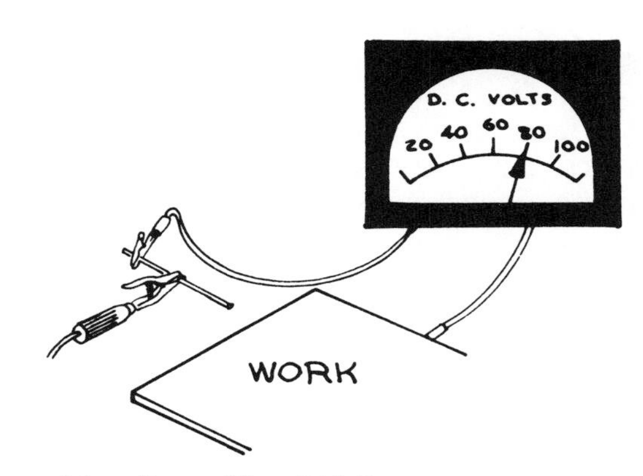

Figure 30. Open Circuit Voltage.

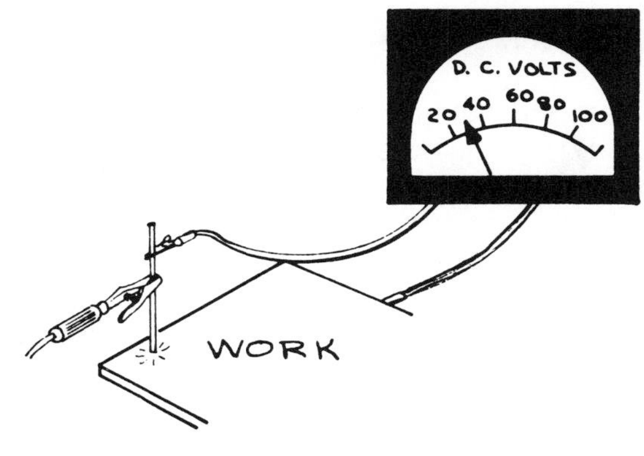

Figure 31. Arc Voltage

Open circuit voltage (OCV) is shown in Figure 30. Emf is the electrical pressure available that influences arc striking ability. Arc voltage shown in Figure 31 is the emf (electromotive force) maintained during arc operation. Arc voltage varies with arc length but is always lower than OCV.

AC: An AC arc is inherently forceful. Very little trouble because of electrode sticking is ever experienced with a well designed AC welder. An early difficulty with AC welding was trouble in starting. This is overcome on some makes of welders by a "hot start" circuit that provides a surge of extra current at the instant the electrode touches the work. One AC welder also has two arc-stabilizing capacitors in the secondary circuit. These have the effect of providing at least five additional volts of open-circuit voltage without the increased shock hazard. To do a satisfactory job of welding with low-hydrogen electrodes at low currents, it is necessary to have at least 75, preferably 80 open circuit volts. This make of welder has been designed with a 75 open-circuit voltage, and the arc-stabilizing capacitors will furnish the extra five volts needed for smooth starting.

DC Motor-Generator: Most welding generators will furnish a substantial peak current during metal transfer and hence have a forceful, "peppy" arc. Popouts are counteracted on one make of motor-generator by an "interactor" that stores magnetism and transforms it into volts as needed. The instantaneous high recovery voltage reduces current sag, thus insuring a steady arc. Even on tough jobs like all-position pipe welding, the arc is easily maintained. When the current dial is set for the desired amperage, voltage is automatically adjusted to give best welding performance at the particular setting. Freedom from the influence of variations in power line voltage fluctuations favors the rotating DC welding power source. Figure 32 graphically shows the effect of power line voltage fluctuations on DC welders.

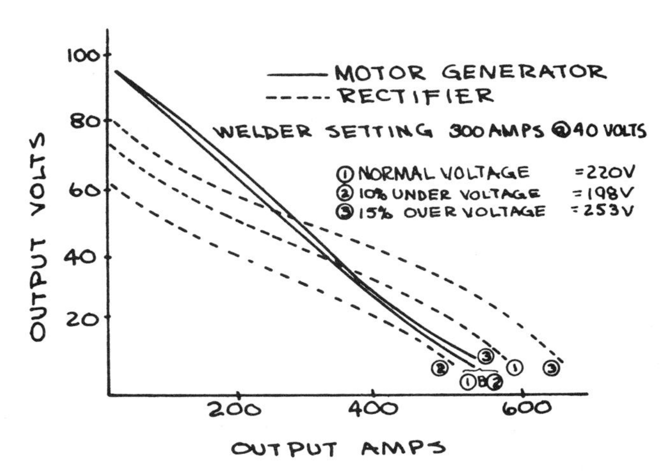

Figure 32. Motor-Generator Welder is Relatively Free From Power Line Variations Resulting in Less Interference to Welding.

Another factor of performance in favor of the motor-generator welder is that of extra reserve power. A comparison of typical volt ampere curves, Figure 33, for rectifiers and motor-generators clearly shows an extra useable power range for motor-generator sets not to be found in rectifiers of the same rating. It is true that most welding applications do not require this added performance. However, on some applications such as arc-air cutting and cutting with special electrodes, superior performance can be gained with a motor-generator. Should a rectifier source be selected for this type of work, it could require the purchase of the next larger size to do the work of a motor-generator.

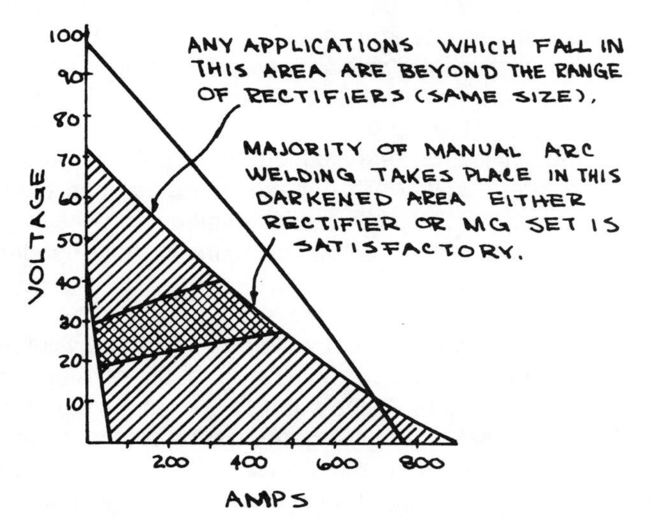

Figure 33. Volt-Amp Curve for 300 Amp Rectifier, Cross Hatched, Superimposed on Volt-Amp Curve for 300 Amp Motor-generator.

DC Rectifier: Early recifier welders were of the soft-arc type and provided small current at short circuit. Their usefulness was limited to sheet metal and to welding by the tungsten-arc gas-shielded process. This is no longer universally true. The best modern rectifier welders will deliver an arc comparable in force to that of the motor-generator welder.

The needed improvement in the basic rectifier welder is effected by incorporating auxiliary devices to increase the transient short-circuit currents. One method is to use **parallel rectifiers**, each of which is supplied at less than normal arc voltage. A second method is to use a **saturable core** reactor, also used for current control in some makes of rectifier welders. A rectifier having a saturable core reactor can be made to have varying degrees of arc forcefulness.

A third method, without the problems entailed by continuously operating rectifiers or saturable core reactors, is to provide a **capacitor discharge** circuit. A bank of high-capacitance, low-voltage DC capacitors is connected to the welding circuit and will discharge through the weld whenever a transient short circuit occurs. The added boost of current quickly clears away the molten drops of metal that are interrupting the arc. Arc force is provided for short-arc welding jobs with low-hydrogen electrodes, and for welding in the difficult vertical-up and overhead positions against the force of gravity.

Within the limits of safe operation, the open-circuit voltage of a rectifier welder should be high to insure arc stability. When a high open-circuit voltage is obtained at the expense of a reduction on power factor and/or efficiency, there is no cause for alarm. The increased power charge matters little in comparison with the saving in welding time.

AC-DC Transformer Rectifier: The advantages of a quiet welding power source, capable of operating all electrodes, and economically available to single phase power users tends to cause prospective buyers to seriously consider this type of machine. However, arc blow may be a problem when operating on DC current.

ARC BLOW

When steel or another ferromagnetic material is welded, the magnetic field of the welding current may act upon the current passing through the arc column to move and lengthen the arc. This is called "**arc blow**", a very apt name since the effect is almost exactly as if a strong wind were blowing against the arc. The cause of arc blow, however, is not wind but the magnetism of the work piece produced by the welding current.

AC: Arc blow is essentially absent with the AC welder. Alternating current does not permanently magnetize steel. The temporary magnetism present is by no means as great as DC magnetism. The changing cycles of alternating current produce both a magnetizing and a demagnetizing effect. As a result, the plate can carry AC magnetic flux in only a thin film on the surface.

DC Motor Generator: A DC magnetized plate is magnetized permanently and magnetized through its entire thickness. Hence, arc blow is a condition often encountered with direct current welding. It is particularly bad in corners or near the edges of the plate, where the current is able to flow to the arc from only one direction.

DC Rectifier: It is not true, as has been claimed, that a rectifier welder will minimize or eliminate arc blow. Direct current is direct current, regardless of the source. The same arc blow will be encountered with a rectifier welder as with a motor-generator type.

A condition of electrode burn-off characteristics called "**finger-nailing**" is especially troublesome with heavy flux coated electrodes. Poorly manufactured (flux unevenly extruded over the wire), improper storage (moisture) and unnecessary rough treatment can be causes of "finger-nailing". Arc blow is likely to begin as a result.

Controlling arc blow may be a difficult problem especially for a novice. The following suggestions may help counteract arc blow problems:

1. Relocate the ground and weld toward it.
2. Make multiple tack welds.
3. Reduce the current setting.
4. Use back-stepping.
5. Hold a very short arc.
6. Switch to AC.

BENEFITS OF EACH TYPE

AC: Freedom from arc blow is probably the major advantage of AC welding. The weldor does not need the skill and experience required in fighting arc blow. Welding is generally faster, both because of the absence of arc blow and because AC welders can use higher currents and hence larger-diameter electrodes.

The AC welder, having a minimum of movable parts and no wearing moving parts, requires little maintenance. A comparatively uncomplicated electrical machine, its initial cost is low compared to other types of welders. Furthermore, it has a high electrical efficiency and is economical in power consumption.

DC Motor-Generator: Flexibility is the greatest single advantage of the DC motor-generator set, which over long years of tested service has become the general-purpose welder of the industry. It may be used to weld all metals and alloys that are weldable by the arc process and can weld in all positions anywhere.

A forceful, penetrating stable arc is characteristic of motor-generator sets.

DC Rectifier: Perhaps the outstanding special advantage of the rectifier welder is its good performance where low welding currents must be used. Its lack of rotating parts gives it quiet operation and reduced maintenance.

The versatility of polarity allows a ratio of about 2/3 - 1/3 concentration of the arc heat on the electrode or the work. Electrode positive - DCRP (**Reverse Polarity**) puts 2/3 of the heat on the electrode - 1/3 on the work. Electrode negative -DCSP (**Straight polarity**) 1/3 of the heat on the electrode - 2/3 on the work.

AC-DC Transformer Rectifier: The main benefits of the AC-DC combination welder is it combines the benefits of both the AC transformer and DC rectifier machine as previously discussed.

LIMITATIONS OF EACH TYPE

AC: With present electrodes, AC welding must be confined largely to arc stabilized electrodes. AC welding is not well suited to thin-gage materials on which low currents are required.

DC Rectifier Welder: Besides difficulties caused by arc blow, the natural soft arc limits the usefulness of the rectifier welder, unless it is equipped with some means of boosting arc force during short circuits. Care must be taken with even short-time jobs that require more than 100% of rated current because of the damaging effects of overload on the rectifiers.

AC-DC Transformer Rectifier: The limitations of AC and DC rectifier welders applies as previously discussed.

POWER FACTOR

Power factor is important to the welder owner if the utility that serves the customer bills for the unused portions of their incoming power. A high power factor will save money; a low power factor is always to be avoided. The efficiency of the welder is the ratio of consumed power to total power employed in actual welding. It is by no means the same thing as power factor.

AC: An AC welder has a high load efficiency, normally between 80 and 90%. At no-load, the power intake is very low, another way of saying that current is being used only when there is an actual welding arc. The combination of high full-load efficiency and low no-load input naturally makes for favorable power consumption. An AC welder consumes less power than a DC motor-generator set of the same rating.

Power factor is not good (normally about 40%) unless the AC welder has been equipped with power - factor - correcting capacitors. However, practically all AC welders on the market are equipped with such capacitors, which are generally arranged to give a power factor of about 80% lagging at rated load. As the load decreases, the power factor will increase until unity power factor is reached at about half load. Below half load, the welder will have a leading power factor. At light loads or at no load, the welder capacitors therefore tend to correct the power factor of other equipment on the line.

DC Motor-Generator: Overall efficiency of an average motor-generator welder is about 60%. Its no-load power input is high. Power factor is usually good and may be improved by static capacitors (similar to the power-factor-correcting capacitors on AC welders), although there is a limit to the amount that can safely be added. Power factor may be as high as 90% when the welder is under full load. When idling, the direct-current welder has a low power factor.

The undesirable combination of high no-load input and low

no-load power factor may be compensated for by an automatic shut-down device. This will disconnect the motor from the power line at a specified interval after welding has stopped, and the welder will then draw no power until the arc is restarted.

DC Rectifier: Overall efficiency of the rectifier welder is rated about 66% under full load, and improves at reduced loads until the welder is about 73% efficient when operating at about 20% of the peak load. In a 400-amp welder, the full-load losses were found to be 8 kw as compared to 9 kw for a motor-generator welder of the same rating. At no load, however, the rectifier consumed 3 kw less than the motor-generator, an important consideration in view of the fact that a welder is ordinarily idle about 50% of its time.

The efficiency of a rectifier welder declines with age, owing to deterioration of the selenium rectifier stack. The aging of the rectifier stack usually results in an increase in forward resistance, a decrease in reverse resistance. Performance is not markedly affected for some time, since the plate losses can be overcome by turning up the current adjuster. When the welder has completed 10,000 to 15,000 hours of rated load service, however, the plate losses are apt to cause such a high operating temperature that failure will result.

The "forward drop", the voltage across the rectifier in the current-carrying direction, provides an excellent indication of the condition of the rectifier. It is generally agreed that when the forward drop was doubled, the rectifier has reached the end of its life. The forward drop can be measured by short circuiting the DC terminals, adjusting for rated current through the short circuit and measuring the voltage across the AC terminals of the rectifier.

Power factor of the rectifier welder is high. Test of several competitive makes revealed power factors ranging betwen 78 and 80.5%.

AC-DC Transformer Rectifier: Power factor for the AC-DC combination machine is as previously discussed for the AC and DC rectifier welders. In most cases this machine is power factor corrected or can be purchased as an option.

The details of power factor are described in the following discussion and illustrations. Welding current is the result of induced current. Circuits with induced current such as electric motors and arc welders produce a power factor product of less than one or unity. Power factor is the ratio of true watts to apparent watts or it is the factor by which the volt-ampere product must be multiplied in A.C. current in order to obtain the actual power or wattage. Circuits with resistance only such as light bulbs, heater coils, or electric irons have a power factor of 1.0 or unity. If diagrammed, the current (amperage) and voltage are in phase, have the same frequency and pass through the zero point together. The power curve is always positive as shown in Figure 34.

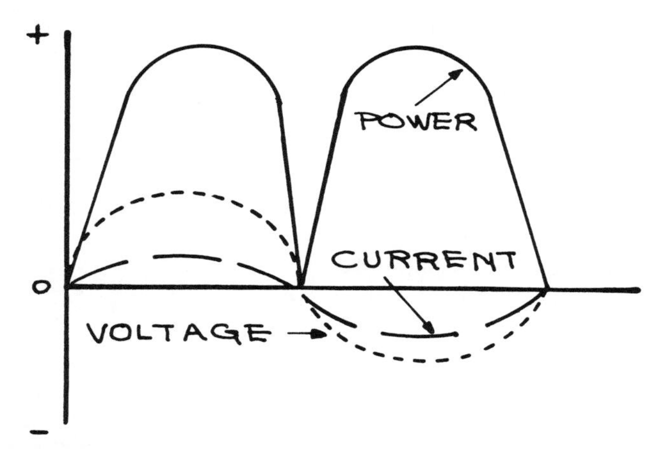

Figure 34. Unity Power Factor of a Resistance Load.

The induced circuit produces a curve in which the current (amperage) lags the voltage which results in an uneven power curve, in that part of the time the power curve is below the line. A negative product results when the voltage is multiplied by amperage. This produces a lagging power factor or less than 1.0. The amount less than one is related to the degrees that the current lags the voltage. Induced current is diagrammed in Figure 35.

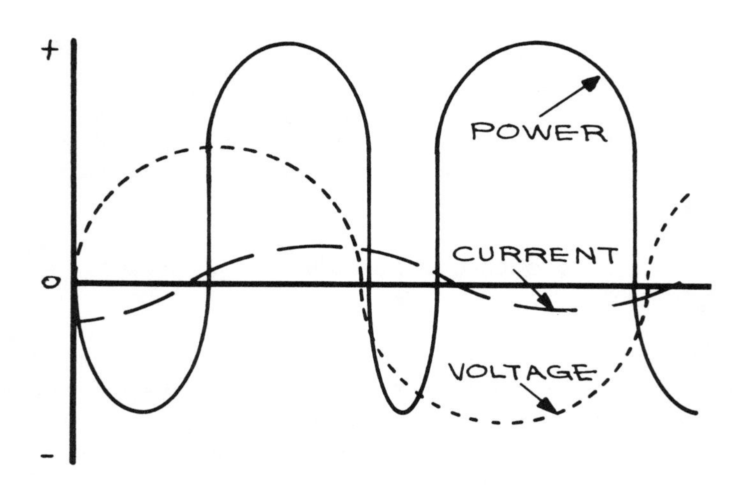

Figure 35. Lagging Power Factor Caused by Induction Loads.

Capacitance causes the current to lead the voltage and results in a leading power factor, a power factor greater than 1.0. Capacitance is illustrated in Figure 36.

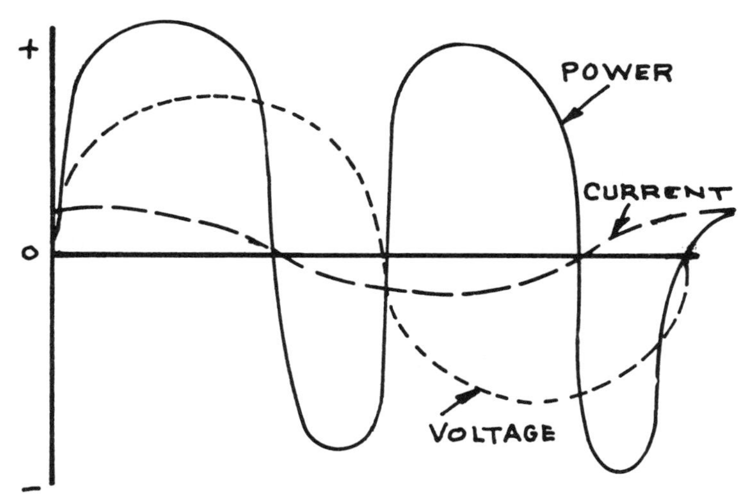

Figure 36. Leading Power, a Result of Capacitance.

Thus, a capacitor of the correct size across the primary or input circuit in the arc welder corrects the lagging power factors to a predetermined percentage. Normal power factor in arc welding machines not corrected is .55 to .65, while corrected machines will usually have a power factor of .75 to .80 depending upon the size of the capacitor connected across the input circuit.

DUTY CYCLE

What is this "Duty Cycle"? "Duty Cycle" is an engineering term used to express the same idea as the production man's "Work Factor". It is a measure of a welder's ability to perform at its "ampere-volts" rating under practical welding conditions. It indicates the percentage of working time and idling time for which the machine is safely designed.

For example, a machine rated 200 amps on 50% duty cycle allows for 50% working time to be actually welding. The other 50% is supposed to be running idle while work is being fitted up, electrodes changed, welds cleaned, etc. A machine rated 200 amps on 60% duty cycle provides for 60% actual welding, with only 40% running idle. Hence, the 60% duty cycle provides for 20% more actual welding time. **Duty cycle tests are completed at full amperage rating**.

It is obvious that a welder designed for a higher duty cycle will heat up less, and have longer life with freedom from trouble, than a lower duty cycle machine. That applies even when used under lighter loads — or under overload. A welder designed for the higher duty cycle must necessarily be more liberally designed and represents a greater value to the user than one designed for lower duty cycle.

How duty cycle is tested? Machines designed for various duty cycles must stand tests prescribed in industry specifications for conventional ratings. They are tested in **ten minute cycles** for four hours, involving a total of 24 test cycles. During that period, temperatures are checked constantly and no part of the machine may reach a temperature higher than the maximum allowed in the specifications. Different temperatures are allowed for different parts, according to the different classes of insulating materials used for the various parts.

For example, in testing a machine rated 200 amperes at 40 volts on 60% duty cycle a resistance load of 200 amps at 40 volts (8 kw) is applied for 6 minutes, at the end of which period the load is automatically cut off and the machine allowed to run idle for 4 minutes. This 10 minute on and off test cycle is repeated continuously for four hours without stopping the machine. The same procedure is involved in testing machines of other duty cycle ratings, the test cycle for a 50 cycle machine being 5 minutes loaded with 5 minutes idle; for a 30 cycle machine 3 minutes loaded with 7 minutes idle, and so on. Remember, duty cycle ratings are at full-rated amperage therefore at lower amperages the duty cycle percent would go up and in fact all machines have a point where duty cycle is 100%.

TEMPERATURE RISE

Temperature rise and duty cycle tests are related. Temperature rise is commonly from 90 to 110° C. Temperature rise is the **degrees in C above ambient or room temperature that the secondary coil can withstand without causing damage to this coil**. Temperature rise is based upon the quality of the insulation used on the coils. A machine having 110° rise has better quality insulation than does a machine having 90° rise. The better quality insulation has the ability to dissipate away the heat produced in the coil at a higher rate. Some welder manufacturers place small fans in their machines to aid in cooling of the coils or maintaining a temperature lower than the rise for which the machine is designed. This air movement tends to draw and deposit dust internally making periodic cleaning with air necessary.

AC and Motor-Generator: With either an AC welder or a DC motor-generator welder, over-heating occurs when the welder is operated for longer intervals or at higher currents than its specified duty cycle. Most AC or DC motor-generator welders have rated duty cycles of either 50% or 60%. The duty cycle may be increased at lesser current than the rated output, or, conversely, the welding current may be increased at a lower than rated duty cycle. A 400-amp AC welder, to give a specific instance, may be operated at its maximum current of 500 amp on a 38% duty cycle.

AC-DC and DC Rectifier: With rectifier welders, the above concept of duty cycle applies only to the transformer components. The rectifier reaches its ultimate constant temperature in two or three minutes, as compared to about 15 minutes for a motor-generator system. Further operation after "steady state" temperature has been attained will have little effect upon the rectifier, provided - and this is of the utmost importance - that it is not overloaded.

When a rectifier welder is overloaded, the heat changes the thin selenium barrier layer into an allotropic form of selenium that will not rectify. The only form of selenium that will work has a low transition temperature. When this temperature is exceeded, the rectifying power is gone.

If many short-time jobs are to be done in excess of 100% of rated current, the owner of a rectifier welder is probably headed for trouble. Reducing the duty cycle when the rectifier welder is being run at near-maximum current will not help the situation. The operator should either purchase a rectifier of larger amperage or shift to another type of welder.

Why is arc voltage important? Remember that an electrical "load" is measured in watts, which is the product of amperes times volts. A 200 ampere welder rated at 30 volts is designed for a load of 6,000 watts (6 kw), while a welder rated 200 amperes at 40 volts is designed for a load of 8,000 (8 kw). Hence, the latter has 1/3 more capacity.

Some welding machines, electric motor driven and gasoline engine driven as well as AC transformers are rated at 25 or 30 volts in the smaller sizes, and on the basis of lower duty cycles, they are included in the "Safe Load" table. Information in Table 5, indicates the loads under which

TABLE 5. SAFE LOADS FOR AC OR DC WELDING MACHINES.

Rating			Welding	SAFE LOAD CURRENT (AMPERES) AT VARIOUS DUTY CYCLES								
Amps.	Volts	Duty Cycle	Range (Amps)	20% Duty	30% Duty	40% Duty	50% Duty	60% Duty	70% Duty	80% Duty	90% Duty	100% Duty
100	25	20%	45-100	100	83	71	63	58	53	50	47	45
180	25	20%	20-180	180	148	126	113	103	95	89	84	80
200	30	50%	30-250		*250	239	200	182	168	157	148	141
200	40	60%	15-300	*300	283	245	219	200	185	173	163	155
200	25	100%	30-225							225	211	200
250	30	30%	30-250		*250	215	192	175	162	152	143	136
250	40	60%	30-325		325	307	274	250	232	217	204	194
250	30	100%	30-275							*275	263	250
295	30	20%	35-295	295	244	208	186	170	158	147	134	132
300	40	60%	35-450	*450	424	367	328	300	289	259	244	232
300	40	60%	30-400		*400	367	328	300	289	259	244	232
300	40	100%	30-325							*325	316	300
400	40	60%	50-500		*500	490	438	400	370	347	327	310
400	40	60%	35-640	*640	566	490	438	400	370	347	327	310
500	40	60%	100-625		*625	612	547	500	462	432	408	387

*NOTE:

These figures are arbitrarily limited to the maximum advertised welding range although higher values are indicated from the standpoint of temperature rise only, which must be modified to consider the entire load (watts) when making "Safe Load" recommendations.

properly rated arc welders may be used safely in practical work involving various duty cycles (work factor). For example, note that the 60% duty cycle 200 ampere welder is good for 283 amps. when actually welding 30% of the time, and for 155 amps. when welding continuously (100% duty cycle). The 50% duty cycle 200 ampere welder however, is good for only 260 amps. and 140 amps. respectively under the same conditions.

NOTE: Such work as automatic welding and pipe thawing are 100% duty cycle operations, therefore the welding machine must be operated at less than 100% of the rated amperage for these operations.

Figure 38. Open circuit voltage, arc voltage and amperage can be demonstrated with this laboratory device. Variations in arc voltage and amperage can be shown by varying arc length using AC or DC current.

TYPES OF ARC WELDING MACHINES

Figure 37. This machine combines standby power with welding power for stick electrode arc welding. It also provides 115 volt electrical tool power outlets. As a standby generator it will deliver 5 kw 230 volt 60 cycle AC power.

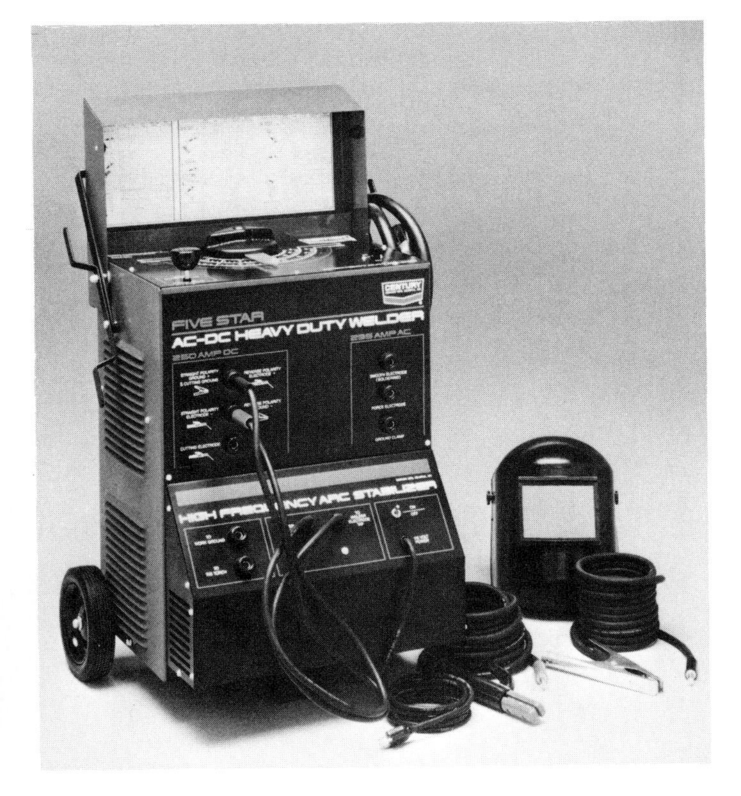

Figure 39. This versatile AC-DC arc welding machine provides quiet welding power with continuous amperage adjustment up to 250 amperes. It operates on 230 volts single phase thus bringing DC to rural power users.
Courtesy Century Mfg. Company.

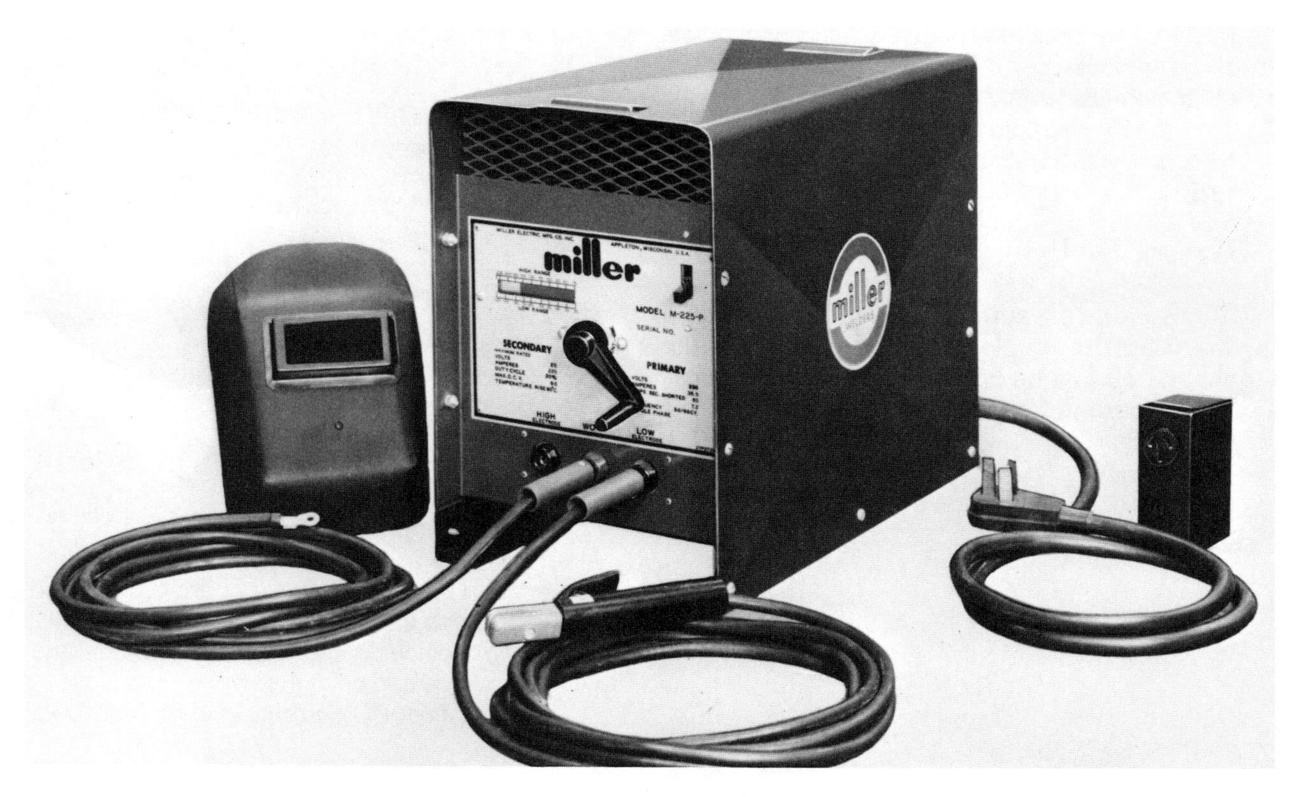

Figure 40. This 20% duty cycle, limited input type AC arc welder operates on 230 volts AC 60 cycle single phase power. It delivers 225 amps at 25 arc volts. This machine has the additional feature of two possible volt-ampere curves as illustrated in Figure 41. Courtesy Miller Electric Mfg. Co.

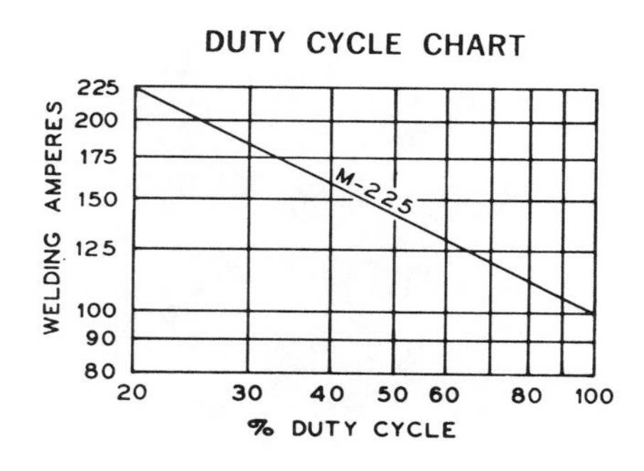

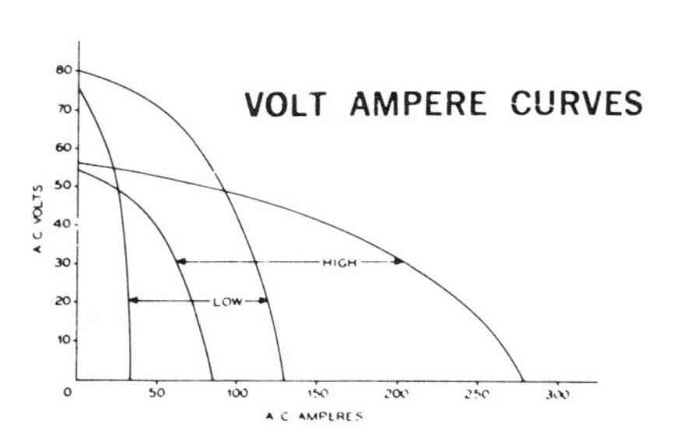

Figure 41. Duty Cycle Chart and Volt-Ampere Curves for High and Low Settings Miller M-225 Welding Machine.

INDUSTRIAL WELDING PROCESSES

GAS TUNGSTEN-ARC WELDING (GTAW)

GTAW began during WW II as a special welding process for welding light aircraft metals. It was then called **TIG (Tungsten Inert Gas)**. Trade names like Heliarc, Heliwelding and Ion-Arc have also been used since the beginning. GTAW will be used in this discussion. The process uses a non-consumable tungsten electrode that forms a heat source to melt the base metal. An inert gas either argon or helium is provided at the arc to protect the molten metal and the electrode. The welding power source is a constant current type known as "drooper". It may be DC or AC or AC-DC. The latter type is the most common. Figure 42 shows a block diagram of a GTAW system.

In application the GTAW torch is held much like an oxy-acetylene torch in that a filler rod may or may not be applied by hand as needed. The torch consists of a handle or body, a collet which grips the tungsten electrode and a collet cap that protects the unused end of the electrode. Finally a gas nozzle directs the flow of inert gas.

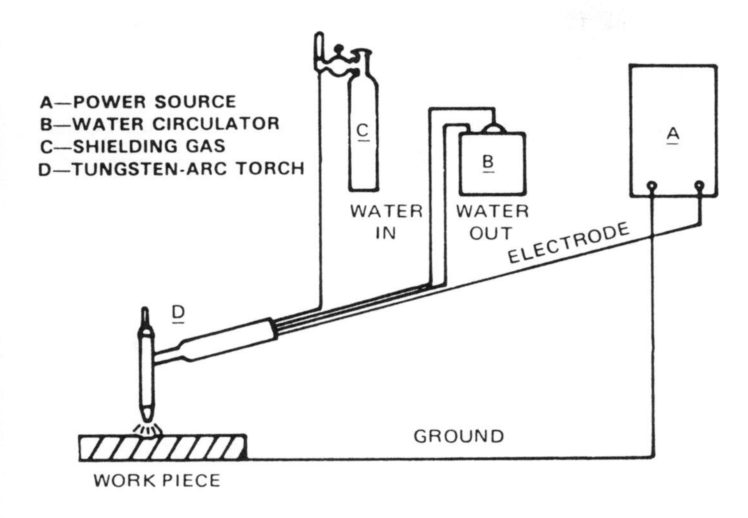

Figure 42. Gas Tungsten Arc Welding Equipment. Courtesy Miller Electric Mfg. Co.

A flow meter is required to supply an even amount of gas measured in cubic feet per hour (cfh).

An AC-DC, single-phase transformer-rectifier with high frequency (HF) is the most desirable GTAW power source. The machine should have a water-cooled torch and a remote control mechanism to control current. This control will allow the operator to supply more heat at the beginning of the weld and then apply the heat as needed for completion.

Shielding Gases for GTAW: Argon and helium are the two gases used either singly or mixed usually the former. Argon is a heavier than air absolutely inert gas obtained from the liquefaction of air. For large users it is supplied at 99.99% pure in liquid form therefore dry (-300°F). See Figure 43.

Helium is a lighter than air inert gas obtained from certain natural gas deposits in the Texas-Oklahoma region. The federal government controls the supply of helium which is expected to be exhausted by 2000 A.D. See Figure 44.

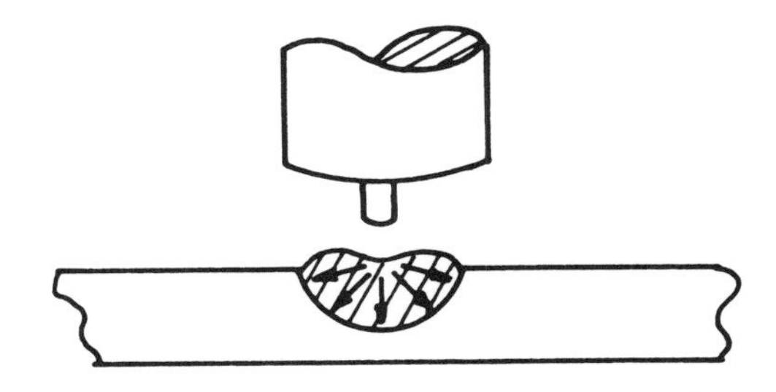

Figure 43. Argon Shielded Gas Tungsten Arc Weld.

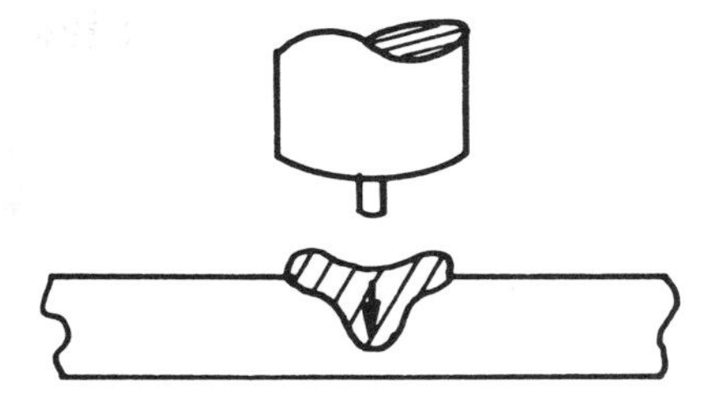

Figure 44. Helium Shielded Gas Tungsten Arc Weld.

Types of Electrodes: The common types of electrodes used today for GTAW are as follows: Note Table 6.

1. Pure tungsten.
2. 1 + 2% thoriated tungsten.
3. 1/2% of zirconium tungsten.

The addition of thorium and/or zirconium increases current carrying capacity, improves electron emission for easier starts and increases contamination resistance. Pure tungsten electrodes are cheaper and easily form a ball which is useful for welding aluminum. When welding metals other than aluminum and magnesium, the electrode end is pointed to increase current density. This also helps control arc blow since DCSP is usually used. Note Figure 48.

GTAW electrodes are generally 7 inches long and range from 0.020" to 1/4" in diameter. Tungsten electrodes are made by sintering (powder metallurgy) and are centerless ground. They are extremely brittle.

TABLE 6. COMMON ELECTRODES FOR GTAW

CHEMICAL REQUIREMENTS

AWS-ASTM Class. No.	Tungsten Min. %	Thorium %	Zirconium %	Total Other Elements, Max. %
EWP	99.5			0.5
EWTH-1	98.5	0.8 to 1.2		0.5
EWTH-2	97.5	1.7 to 2.2		0.5
EWZr	99.1		0.3 to 0.5	0.5

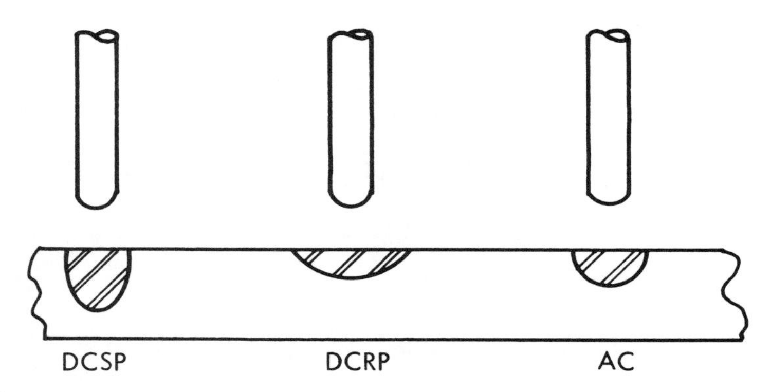

Figure 45. Typical Deposition Characteristics.

The choice of welding current (AC or DC) is determined by the welding conditions and the type of metal. Figure 45 shows penetration profiles of the three possibilities. In other words the polarity aspect of DC is apparent. Two-thirds of the heat of the arc is on the positive side of the arc. AC being a combination of DC+ and DC- has adavantages. The instability of the sine wave characteristic of AC requires electrical stabilization in the form of a superimposed HF current as shown in Figure 46. This high voltage, high frequency, low current has certain advantages as follows:

1. Easy arc starting without touching electrodes to the work.
2. Arc stability when AC is used.
3. A longer arc can be maintained for certain operations.
4. Expensive tungsten electrodes have longer life.
5. Allows for wider current ranges on a given electrode.

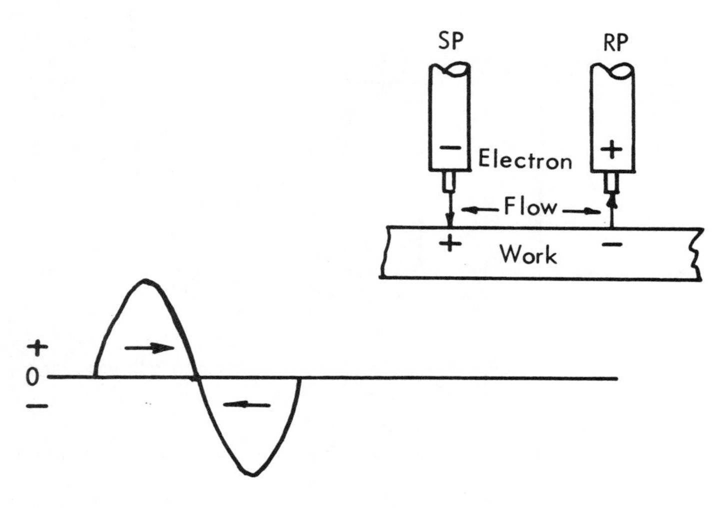

Figure 46. AC Sine Wave Trace and Polarized Electrodes.

The inert gas ionizes in the arc plasma providing a very useful metal cleaning aspect of GTAW. As shown in Figure 47, DCSP electrons hitting the plate causes greater heat as compared with the electrode end. In DCRP gas ions hit the plate and electrons leave the plate surface automatically removing oxide. This is the reason AC is preferred for welding aluminum, magnesium and their alloys.

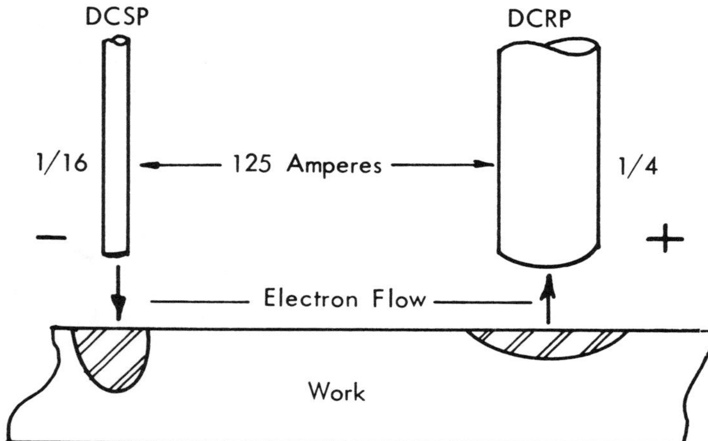

Figure 47. Tungsten Electrode Diameters, Heat Distribution and Penetration Patterns.

Electrode end preparation as illustrated in Figure 48 is important for successful GTAW applications. Aluminum and Magnesium welding requires a half sphere (ball) end ACHF is the proper electrical mode. The ball end is obtained by gradually increasing amperage (foot control) with DC+ mode carefully watching the electrode end. This should be done on a piece of copper.

The welding of other metals usually requires DC negative. Grind the electrode to a sharp point with all grinding marks ending at the point.

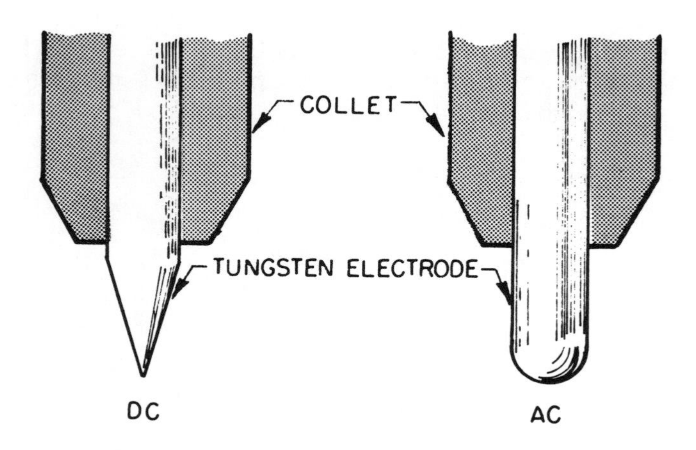

Figure 48. Electrode End Preparation.

SUBMERGED ARC WELDING (SAW)

About 15% of the total carbon steel weld metal in the U.S. is deposited by SAW. The process is especially useful for very high rate of metal deposition with one pass fill on edge prepared, thick plate involving one or more filler wires fed into the arc covered by a granular flux which become electrically conductive at arc heats. Resultant welds have unusually good ductility, high impact strength, uniform density, low nitrogen content, and high corrosion resistance. Direct current welding power sources are most frequently used because of better control and easier arc starting. AC is usually used on multiple wire jobs. Figure 49 shows a diagrammatic sketch of SAW illustrating the fact that the usual arc welding helmet is not necessary, the arc being covered by flux.

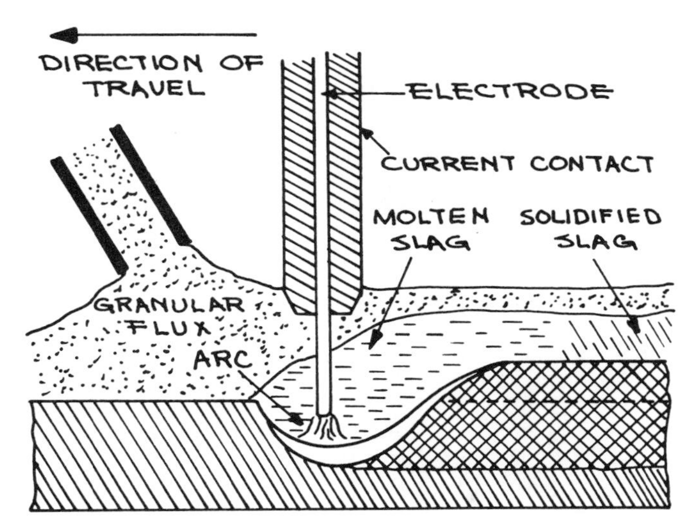

Figure 49. Diagrammatic Sketch of Submerged Arc Welding Operation and Heavy Plate.

Figure 50. This Motor-Generator DC Welder is Equipped for Submerged Arc Welding. It Will Also Run Stick Electrodes, Solid Wire and Covered Wire.

Many SAW machines are the stationary, automatic production type. However, Figure 50 shows a manually operated SAW machine with the flux hopper on the torch.

GAS METAL ARC WELDING (GMAW)

Gas metal arc welding GMAW was developed from gas tungsten arc welding (GTAW) in the early 1950s. Figure 60 shows that the filler wire (electrode) in coiled form, is mechanically driven and energized by the welding power by a contactor in the welding torch at the welding zone. As shown in Figure 53 inert gas or inert gas mixtures protect the site of the weld from the atmosphere. CO_2 is frequently used in welding steel. High current densities on small diameter wire results in metal transfer by the spray or short circuit method. Short circuits occur about 100-200 times per second. Metal deposition rates are about four times faster than GTAW. Power sources must be constant voltage type. Direct current electrode positive (DCRP) is used on 95% of the applications. An example of this type machine is shown in Figure 51. DCSP can be used on surfacing operations. The short-circuiting type welding is best for thin steel, aluminum and stainless steel. Spray GMAW is best used for larger diameter wire and higher currents on thick metal. Short-circuiting GMAW is considered an all-position industrial process, although maintenance welding shops are also using these machines. Flux cored wire applications as illustrated in Figure 52 tend to obscure the site of the weld and leave a slag cover. Solid wire filler gives a clear view of the weld and no slag. OCV is set on the machine control. Arc voltage influences arc length and the shape of the bead. Amperage is controlled by the rate of wire feed (inches per minute). Optimum setting of the short-circuiting machine is frequently made by sound of the arc and by the bead shape that results.

Figure 51. This Gas Metal Arc Welder is Equipped With CO_2 for Short- Circuiting Welding With .045" Steel Wire.

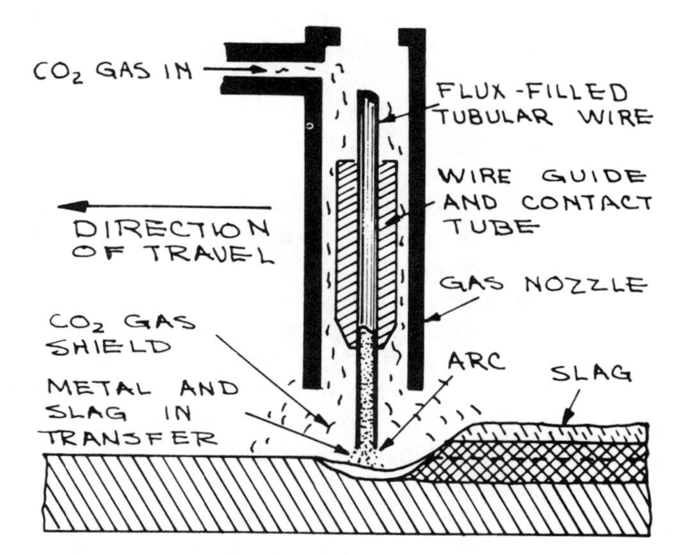

Figure 52. Flux Cored Wire With or Without Shielding Gas May be Used in GMAW.

GAS METAL-ARC WELDING BASICS

In gas metal-arc welding (GMAW), an open electric arc is established between the machine-grounded work and a consumable, bare, solid wire electrode. The heat of the arc melts the base metal and admixes the electrode wire constantly fed into the puddle which is protected by gas. The process began in 1948, a development from GTAW, a non-consumable electrode arc welding process. See Figure 53.

The following are advantages of GMAW in both manual and automatic production applications:

1. Welding can be done in all positions.
2. High rates of metal deposition.
3. Excellent filling ability for poor-fit joints.
4. High weld quality.
5. No electrode stub loss.
6. No slag removal necessary.
7. Less distortion due to narrow, deep weld profile.
8. High visability because of no smoke.
9. Less operator training.
10. Metals weldable include aluminum, stainless steel, low alloy and mild steel.

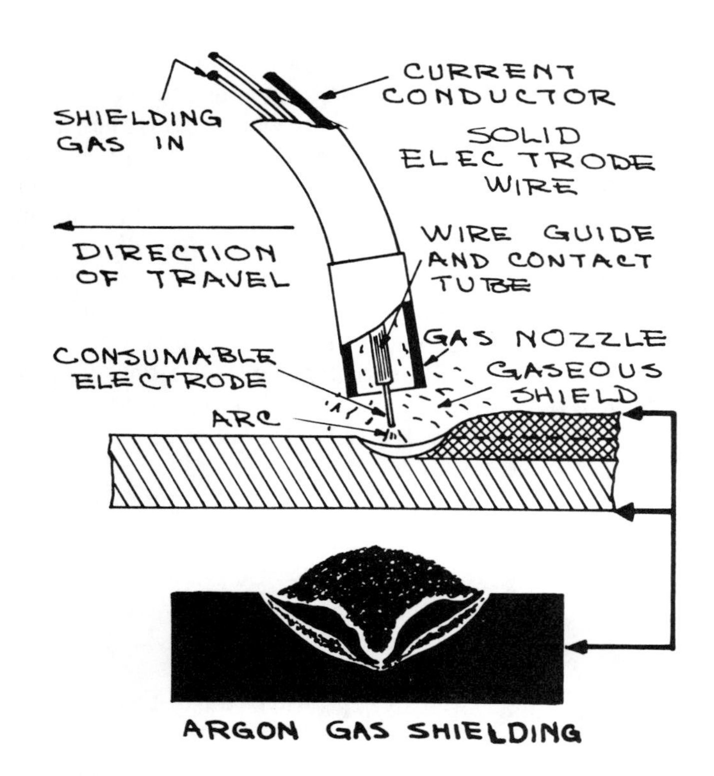

Figure 53. Gas Metal-Arc Welding Process Sometimes Referred to as "MIG" Because of the Deposition of a Consumable Metal Electrode Under an Inert Gas Shield.

Some disavantages are:

1. Welding power source expensive.
2. Shielding gas expensive.
3. Higher electrode wire cost.
4. Most machines require three-phase input power.
5. Not as versatile as SMAW for maintenance, for example, welding cast iron, cutting and carbon arc torch applications.

Metal Transfer: Metal transfer from the electrode wire is accomplished by one of three ways; namely, **short circuit, spray** and **globular**. Short circuit and spray are most commonly used.

Short circuit transfer starts when the end of the mechanically driven electrode touches the molten base metal as shown in Figure 54. "A" shows contact being made. "B" shows a necking out of the heated wire that separates at "C". At "D", a full-length arc forms to heat the base metal and flatten the bead. At "E" the sequence repeats.

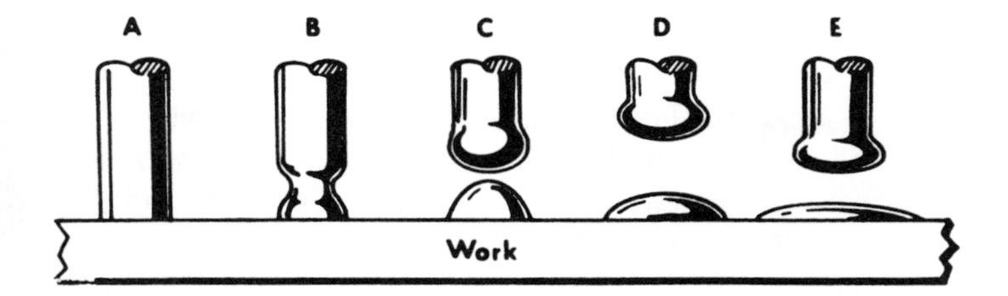

Figure 54. Short Circuit Transfer.
Courtesy Miller Electric Mfg. Co.

Short circuits followed by arcing occur about 100 times a second making a sound somewhat like a rapid firing small caliber gun.

As shown in Figure 56 type of metal transfer is affected by amperage, voltage, type of shielding gas and wire size.

Shown in Figure 55 is the current-voltage relationship versus time in a typical short circuit cycle.

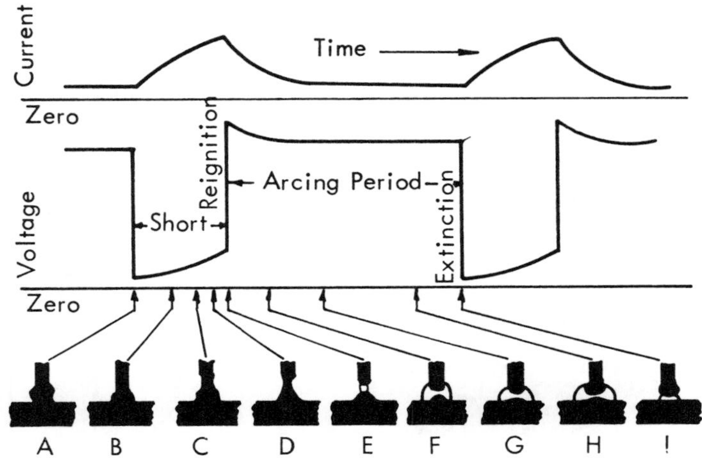

Figure 55. Current-Voltage vs. Time for Typical Short Circuit Cycle

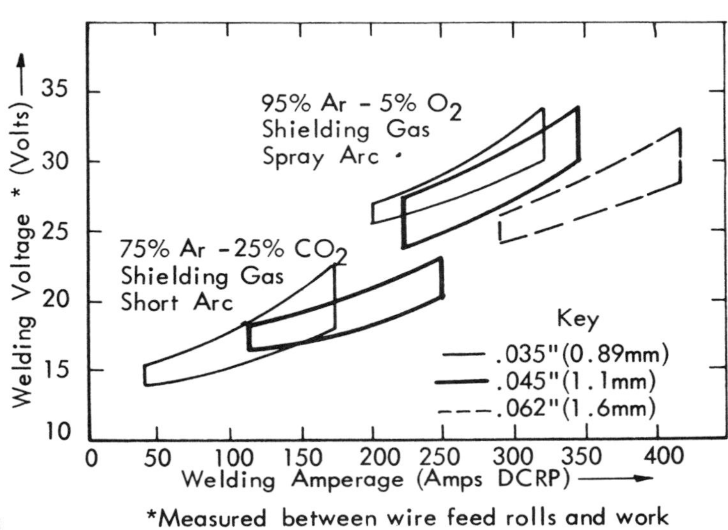

Figure 56. Arc Voltage-Welding Current Relationship to Achieve Good Arc Stabiltiy For Short Circuit and Spray Arc Welding Carbon Steels.

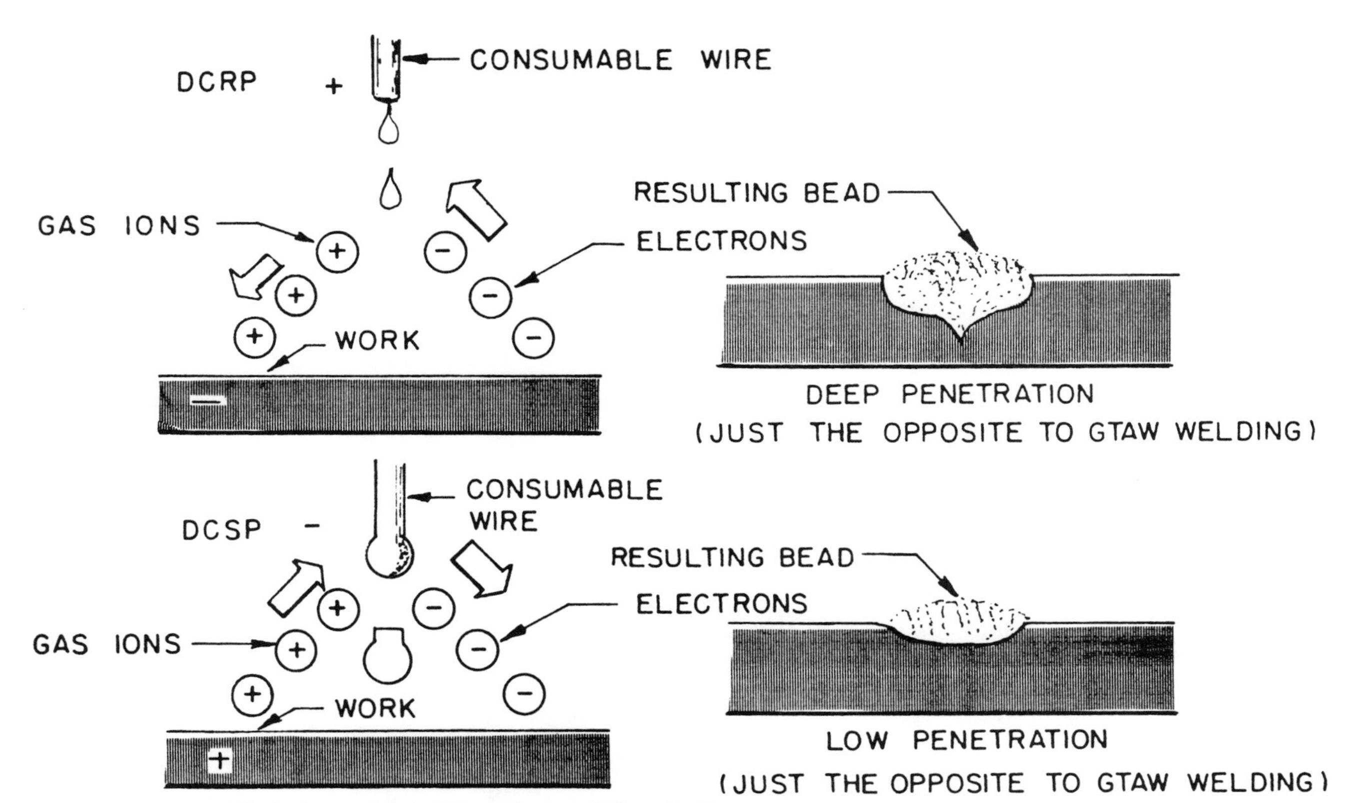

Figure 57. Effect of Polarity on Metal Transfer and Penetration.

From Figure 56, it can be seen that short circuit metal transfer takes place in an amperge range of 50 to 250 with the two most common wire sizes .035" and .045" using 75% argon and 25% carbon dioxide. To attain spray metal transfer, the shield gas is changes to 95% argon and 5% oxygen. The volt-ampere relationship from short circuit to spray takes place at a transition range which varies with the wire size used.

As shown in Figure 57 polarity affects metal transfer and penetration. Electrode positive (DCRP) is used in steel welding where greater strength is desired. Note the deep penetration when using reverse polarity. For application such as hard surfacing alloys where less penetration is

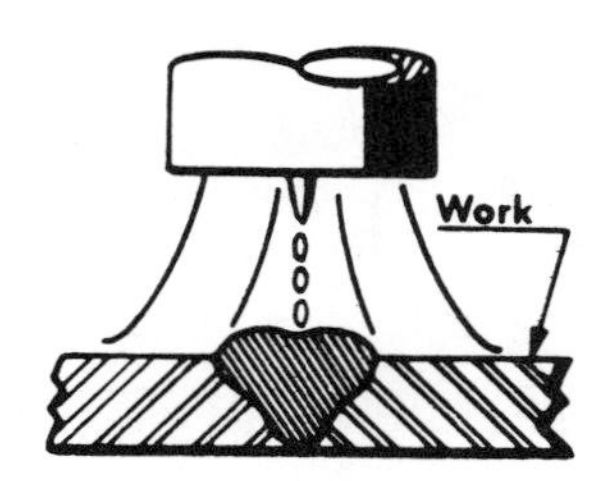

Figure 58. Spray
Courtesy Miller Electric Mfg. Co.

desired, the operator would select electrode negative (DCSP). Note the shallow penetration when using straight polarity illustrated in Figure 57.

Figure 58 shows metal transfer by spray. Spray transfer SMAW uses larger wire sizes than for short circuit welding. This makes a larger molten pool thus restricting the applications to flat and horizontal fillets. Very high rates of metal deposition are obtainable.

Figure 59 shows globular metal transfer. It is the least desirable metal transfer because of shallow penetration and excessive spatter. Globular transfer takes place at high amperages if more than 15% of the shielding gas CO_2 is used. It can occur if too low amperages are used.

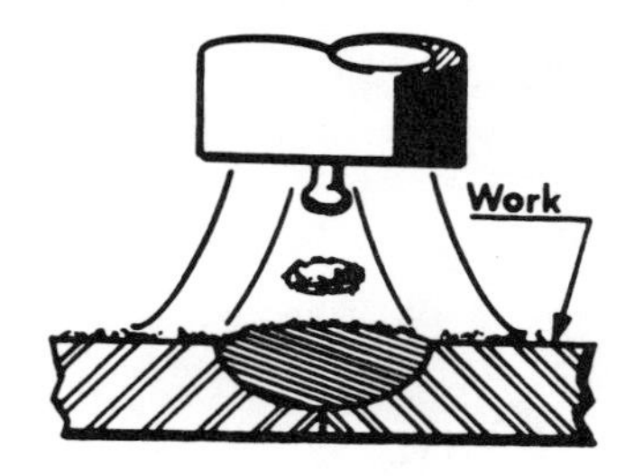

Figure 59. Globular
Courtesy Miller Electric Mfg. Co.

Equipment and Supplies: Equipment and consumable supplies for GMAW are as follows: See Figure 60.

Power Source: The welding power source **must** be DC, **electrode positive**. The nature of AC makes it impossible to control. In order to control current surges the welding power source must be constant voltage as contrasted with constant current of SMAW. See Figure 61. The most common machine is a transformer rectifier using three-phase power input. This type of a machine is very versatile in that there is amperage control, voltage control and slope (V/A curve) control. Low-priced, single phase units are available but are of limited usefulness.

As shown in Figure 62, rectified three phase power is smoother than single phase.

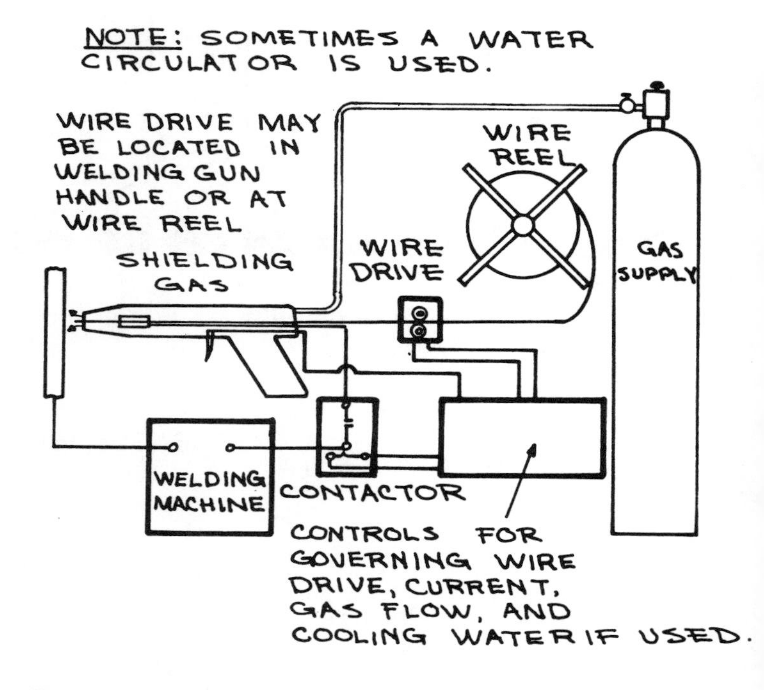

Figure 60. Schematic Diagram of SMAW Equipment.

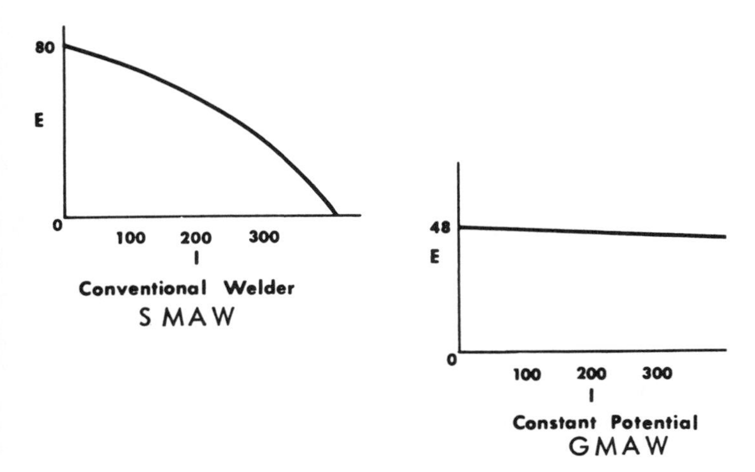

Figure 61. Comparable Volt-Ampere Curve.
Courtesy Miller Electric Mfg. Co.

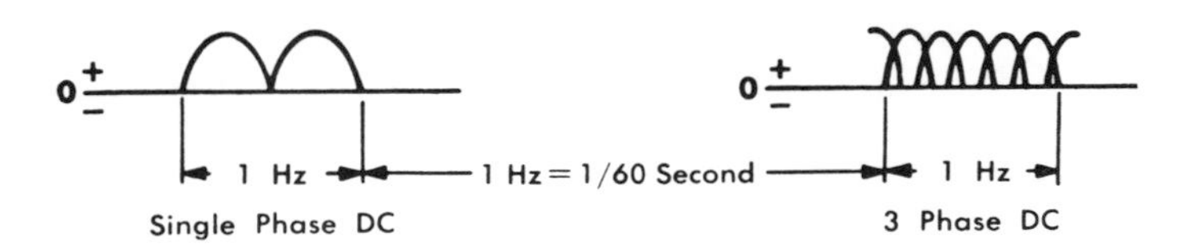

Figure 62. Single Phase & Three Phase Current
Courtesy Miller Electric Mfg. Co.

At present, single phase units have fixed slope and are therefore useful only for thin steel such as auto bodies.

Amperage is controlled by the speed of wire feeder. Slow speed gives low amperage whereas high amperage is obtained by faster speed (inches per minute). It is apparent that a constant even feeder mechanism is absolutely essential. Therefore, any restriction to the free movement of the wire must be removed. **Cleanliness is important.** Sharp kinks or bends in the cable between the feeder and the gun should be avoided. Alignment of incoming and outgoing guides in relation to driver rolls is important. Spring tension on the drive rollers should be enough to move the wire without deforming or scaling. Welding current is applied to the electrode by a copper contractor within the gun. The contractor hole should match the wire size being used. Contractors wear out and should be replaced as necessary. They also should be kept clean from spatter build-up which restricts gas flow and may short circuit with the gas cup. As shown in Figure 63, amperage, current density and wire size have relevance to the weld penetration profile.

A common rule-of-thumb useful for initial amperage is as follows: One ampere for each .001% plate thickness. Example: 1/8" thick steel is .125" so initial amperage would be 125 amperes. Fit of parts, position of weld and type of

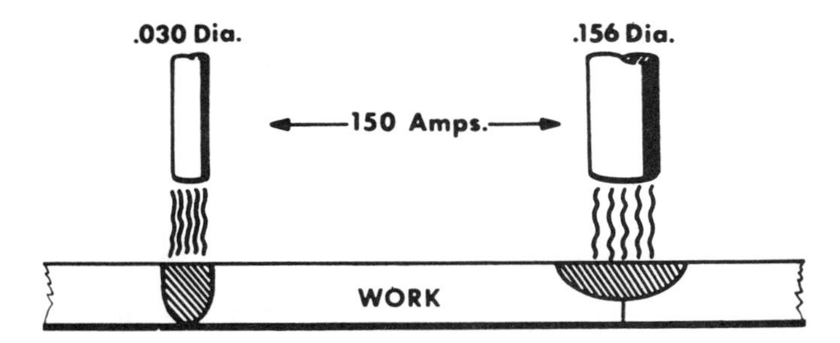

Figure 63. Current Density Comparison.
Courtesy Miller Electric Mfg. Co.

joint must be related to operator skill and technique in final amperage setting (wire feed rate).

Voltage and amperage can be read on the respective gages on the machine. It must be remembered that these **valves are accurate at the machine not at the bead.** The shape of the bead is determined to a great extent by arc voltage. A high, convex or ropey bead is the result of low voltage. A wide, flat bead means high voltage.

Change only one variable at a time when adjusting for optimum welding conditions. Visual inspection of the bead as well as listening for a "healthy arc" will assist the operator in the final adjustment.

Slope: Figure 64 shows the volt-ampere characteristics for a SMAW power supply. Theoretically, a constant voltage or potential power source would have a straight line with no drop in voltage for increased amperage. Slope in a SMAW power source is used to limit the short circuit current so that spatter is reduced when short circuits between the wire electrode and work are cleared. This is shown in Figure 55. Optimum slope for short circuit metal transfer should give a smooth bead with little or no spatter and clean bead solidification pattern. Slope values are expressed as volts drop per 100 amps. Slope setting on a machine obviously does not account for resistance between the machine and the bead such as power cable size and condition, connec-

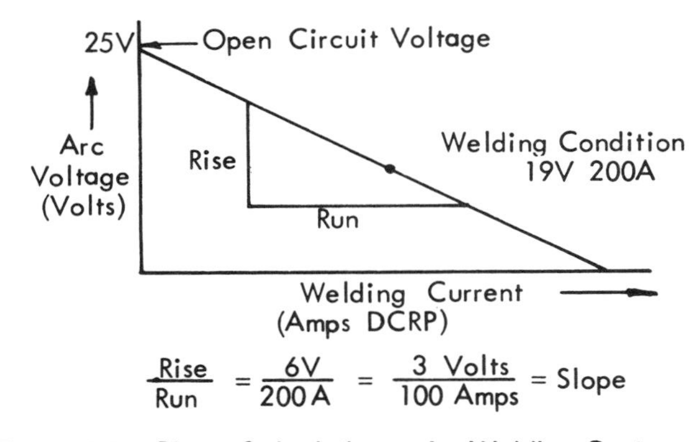

Figure 64. Slope Calculations of a Welding System.

tions, loose terminals, worn contactors, etc. Trial and error adjustment of one variable at a time is the only successful way to make the power source perform satisfactorily.

Welding Gun or Torch: GMAW manual electrode holders are usually called guns probably because they have triggers that control welding current (on-off), gas flow and water flow (if gun is water-cooled). The gun is aimed at the joint; the trigger is pulled and welding begins if the wire touches the grounded work. Operator skill is apparent because angle of electrode, speed of travel and distance from the bead determine the shape of the bead. The gun must be held close enough to give good gas coverage **not for holding a short arc.** Arc length is automatic. The operator may weld forehand (lead the bead) or backhand (drag the bead). Leading is usually used for fast welding whereas dragging allows manipulation for larger bead size.

Wire Feeder and Control Box: The wire feeder pulls the wire off the spool and pushes it to the gun at a very constant uninterrupted rate. Care and maintenance is extremely important. The solenoid controls for gas, water and welding power are found in the feeder box. They are activated by the trigger on the gun.

Electrode stick-out should also be considered as electrical stick-out due to the resistance in the welding wire. in GMAW the gas nozzle directs the flow of protective gas. Therefore both electrical performance and gas protection must be proper. Note figure 65.

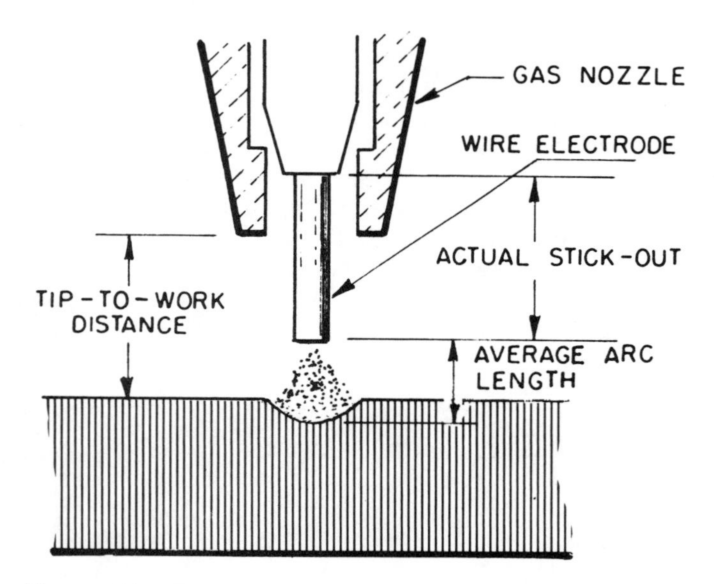

Figure 65. Electrode Stick-Out.

Inside corners might require greater stick-out than outside corners. Spatter accumulations in the gas nozzle can obstruct gas coverage. Flux-cored (FCAW) wire stick-out is usually specified by wire supplier instructions. It may vary from three fourths inch to one and one-half inch. GMAW stick-out usually is three-eighths to one half-inch.

GMAW Wire Selection: The selection of a welding wire depends upon:

1. The shielding gas used.
2. The type of joint design and position.
3. The mechanical and service requirements.

This discussion will deal with the selection of carbon steel wire only. One of the greatest causes of weld discontinuity or defects is porosity in the form of oxides. Therefore, GMAW wires are rich in deoxidizers such as silicon, manganese, aluminum, titanium and zirconium.

Silicon content ranges from .40% to 1.00%. Over 1.00% tends to cause crack sensitivity in the weld.

Manganese content ranges from 1-2% acting as a strengthener as well as a deoxidizer.

Aluminum, Titanium & Zirconium are found in very small quantities of about .20% combined.

Carbon content of GMAW wire ranges from .05% to .12%. The low range used where CO_2 is the shielding gas.

The following are the five most used GMAW wires listed by AWS designation:

1. Type E60S-2- This wire works best with 75% AR - 25% CO_2 shielding gas giving excellent mechanical deposits. However, the bead tends to peak in the middle due to lack of fluidity of the puddle.

2. E60S-3- This is an economical wire using Mn and Si as dioxidizers. The arc tends to be harsh leading to more spatter. However, the bead contour is flat. Glassy islands of silicon are found on the surface but easily removed. This is not a good wire for "rimmed" or non-deoxidized steel. It works best with Ar 75-CO_2 25 mixture.

3. E70S-4- This is a medium-priced wire for killed or semi-killed steel (deoxidized). It forms a concave fillet. It is extremely fluid and may be best used flat.

4. E70S-5- This is a wire recommended for low alloy, high-strength steels. It works best in 75% Ar - 25% CO_2 shield. It makes a good, flat bead with little spatter. It is a medium-priced wire.

5. E70S-6- This is made especially for CO_2 gas covering in welding mild killed steel. It uses powerful Mn and Si deoxidizers.

Note Table 7 for a listing of other GMAW wires for welding aluminum and stainless steel.

Shielding Gas Selection: The most common shielding gases for steel welding are:

1. **CO_2** - economical welding on sheet and plate but may produce fine spatter.

2. **75% Ar + 25% CO_2** - can be mixed on the site or bought mixed. This gas is excellent for thin sheet metal. It shows easy puddle control and minimum burn-though.

3. **50% Ar + 50% CO_2** - is a shielding gas that is excellent for heavy plate in all positions. It has better wetting characteristics than the 75 Ar - 25 CO_2.

4. **95% Ar + 5% Oxygen** - is an excellent shielding gas for spray metal transfer. GMAW is done at high amperages with minimum slope.

Gas flow from the cylinders must be controlled by flow meters rated in cubic feet per hour (cfh).

Table 8 reveals a complete listing of GMAW gases and specific applications.

HELPFUL HINTS ON USE AND CARE OF SEMI-AUTOMATIC WELDERS

Efficient operation of semi-automatic welding equipment will depend on proper application and the weldor's care of the equipment. Although it has a similarity to the stick electrode, there are certain features in the application of semi-automatic equipment which should be carefully noted.

Don't push the gun into the arc like an electrode. If you do, you'll melt the nozzle tip off. The wire driver pushes the wire into the arc - sit back and let the machine do the work. All you have to do is lead it down the joint. You can proceed straight down the seam or you can weave the arc back and forth as you would do with a stick. Fill craters at the end of the weld just as you always have.

Either a forehand or backhand technique can be used. The forehand, or leading arc technique, lends itself to accurately tracking joints at fast travel speeds. Under a given set of conditions, a forehand technique will produce a flatter bead shape than a backhand technique.

A backhand technique lends itself to producing large wide beads or fillets. The arc force holds the molten metal back from the center. This technique is especially effective when using slow-cooling wires such as flux-cored wires.

Semi-automatic equipment has to be kept clean, in proper adjustment, and good mechanical condition. Erratic wire feeding will result in porosity in the weld and is a signal that the wire feeding system is getting ready to stop functioning.

Care of nozzles - keep the gun nozzle contact tube, and wire feeding system clean to eliminate wire feeding stoppages. The nozzle itself is a natural spatter collector. It surrounds the contact tube and provides a good target area as the spatter flies out of the arc stream.

TABLE 7. GMAW WIRE SELECTION.

ALUMINUM WIRES	
ER-1100 ER-4043 ER-5183 ER-5554, 5556 ER-5654	To weld aluminum or similar composition
STAINLESS STEEL WIRES	
ER-308L	For welding types 304, 308, 321, 347
ER-308L-Si	For welding types 301, 304
ER-309	For welding types 309 and straight chromium grades when heat treatment is not possible. Also for 304-clad
ER-310	For welding types 310, 304-clad and hardenable steels
ER-316	For welding 316
ER-347	For welding types 321, and 347 where maximum corrosion resistance is required

TABLE 8. SHIELDING GASES FOR GAS METAL-ARC WELDING

MATERIAL	PREFERRED GAS	REMARKS
Aluminum alloys	Argon	With DC reverse polarity removes oxide surface on work piece
Magnesium aluminum alloys	75% He 25% Ar	Greater heat input reduces porosity tendencies. Also cleans oxide surface
Stainless steels	Argon + 1% 0_2	Oxygen eliminates under-cutting when DC reverse polarity is used
	(Argon + 5% 0_2)	When DC straight polarity is used 5% 0_2 improves arc stability
Magnesium	Argon	With DC straight polarity removes oxide surface on work piece
Copper (deoxidized)	75% He, 25% Ar (Argon)	Good wetting and increased heat input to counteract high thermal conductivity Light gauges
Low-carbon steel	Argon + 2% 0_2	Oxygen eliminates under-cutting tendencies also removes oxidation
Low-carbon steel	Carbon dioxide (spray transfer)	High quality low current out of position welding low spatter
	Carbon dioxide (buried arc)	High speed low cost welding accompanied by spatter loss
Nickel	Argon	Good wetting decreases fluidity of weld metal
Monel	Argon	Good wetting decreases fluidity of weld metal
Inconel	Argon	Good wetting decreases fluidity of weld metal
Titanium	Argon	Reduces heat-affected zone, improves metal transfer
Silicon bronze	Argon	Reduces crack sensitivity of this hot short material
Aluminum bronze	Argon	Less penetration of base metal Commonly used as a surfacing material

Note () = Second choice

TABLE 9. GMAW ADJUSTMENT SUGGESTIONS.

FAULT OR DEFECT	CAUSE AND / OR CORRECTIVE ACTION
Porosity	A. Oil, heavy rust, scale etc. on plate B. Wire-may need wire higher in Mn and Si C. Shielding problem; wind, clogged or small nozzle, damaged gas hose, excessive gas flow, etc. D. Failure to remove glass between weld passes (Silicon Oxide) E. Welding over slag from covered electrode
Lack of Penetration	A. Weld joint too marrow B. Welding current too low; too much electrode stick-out C. Weld puddle rolling in front of the arc
Lack of Fusion	A. Welding voltage and/or current too low B. Wrong polarity, should be DCRP C. Travel speed too low D. Welding over convex bead E. Gun oscillation too wide to too narrow F. Excessive oxide on plate
Undercutting	A. Travel speed too high B. Welding voltage too high C. Excessive welding currents D. Insufficient slow-up at edge of weld bead
Cracking	A. Incorrect wire chemistry B. Weld bead too small C. Poor quality of material being welded
Unstable Arc	A. Check gas shielding B. Check wire feed system
Poor Weld Starts or Wire Stubbing	A. Welding voltage too low B. Inductance or slope too high C. Wire extension too long D. Clean glass or oxide from plate
Excessive Spatter	A. Ar-CO_2 or Ar-0_2 instead of CO_2 B. Decrease percentage of H_e C. Arc voltage too low D. Increase inductance and/or slope
Burnthrough	A. Welding current too high B. Travel speed too low C. Decrease width of root opening D. Use Ar-CO_2 instead of CO_2
Convex Bead	A. Welding voltage and/or current too low B. Excessive electrode extension C. Decrease inductance D. Wrong polarity, should be DCRP E. Weld joint too narrow

If the spatter builds up thick enough, it can actually bridge the gap and electrically connect the insulated nozzle to the contact tube. If you accidentally touch the nozzle to a grounded surface when this happens, there will be a blinding flash. It is quite likely the nozzle will be ruined. If the face of the nozzle is burned back, the shielding gas pattern over the arc will be changed to the extent that the nozzle will have to be replaced.

To remove the spatter, use a soft blunt tool, such as an ice cream stick, for a prying tool. If the nozzle is kept clean, shiny, and smooth, the spatter will almost fall out by itself. Just as a matter of interest, many materials have been tested for nozzles - copper, wood, asbestos, ceramics, glass and glasscoated metals. It has been observed that cool, clean, smooth shiny copper does the best job of all.

Do not clean guns or torches by tapping or pounding them on a solid object. Gun nozzles bend - distorting the gas pattern. Threads get damaged, making it difficult to replace nozzle tips and contact tubes. High-temperature nozzle insulation breaks, resulting in arcing in the handles, trigger switches get broken, and - in the case of water-cooled guns - leaks are sure to develop.

Contact tubes - Here is an area of compromise. The contact tube transfers the welding current to the electrode wire. New contact tubes have smooth round holes of the proper minimum diameter. The hole has the proper minimum diameter. The hole has to be big enough to allow a wire with a slight cast to pass through easily.

With use, the wire wears the hole to an oval shape, especially on the end closest to the arc. The wire slides easier, but the current transfer is not as good. Arcing in the tube results, and the wire can actually be stopped from moving completely. When this happens, the contact tube has to be replaced. It is not a good idea to salvage these tubes.

When the hole in the contact tube wears too big, trouble can also be expected with spatter flying up into the bore and wedging against the wire. This will result in an increased amount of friction on the wire, slowing it down. It also produces a long arc that burns back to the contact tube and fuses the wire to the tube. Worn contact tubes also allow the wire to flop around as it emerges from the gun or torch, making it difficult to track a seam.

Bird nesting - Occasionally trouble is experienced with the wire sneaking out sideways between the wire feed cable and the drive rolls. This happens with greater frequency when small wires are pushed because the column strength or stiffness of the small wire isn't great enough to withstand the drive motor push. Worn contact tubes and other sources of friction previously discussed signal a "bird nest."

Bird nesting can be minimized by making sure the wire feed cable inlet guide is close to the rollers - very close, in fact just so it does not touch the rollers. Alignment of the inlet guide to the rollers should also be perfect so the wire does not have to make a reverse bend.

Wear industrial eye protection, proper face, hands and body protection at all times when welding or working around industrial welding equipment.

IDEAS FOR BETTER WELD DESIGN

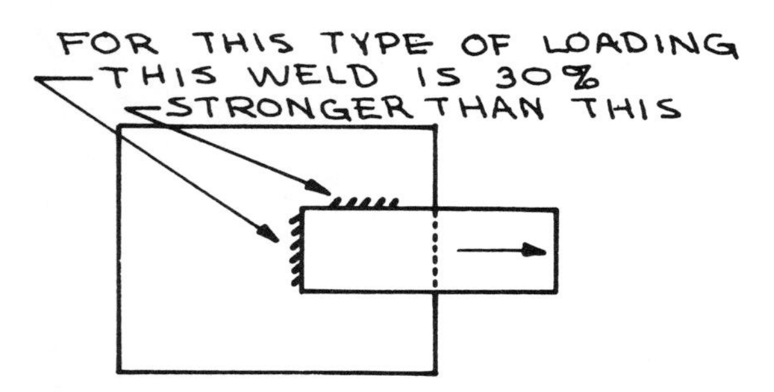

1. Welds in Transverse Shear are Stronger than Welds in Longitudinal Shear.

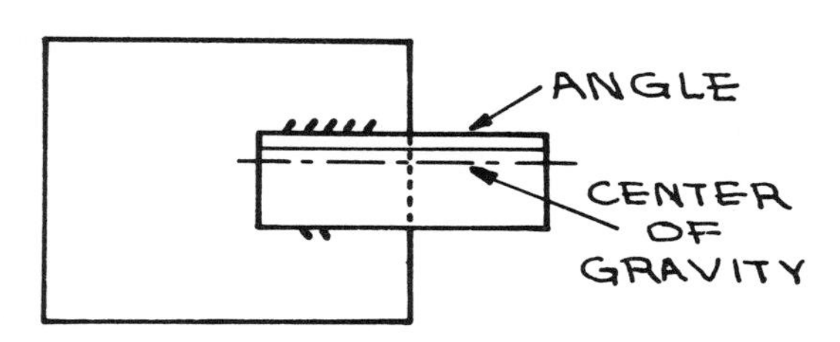

2. Lengths of Welds Proportioned to Loading Provide Better Load Distribution Results.

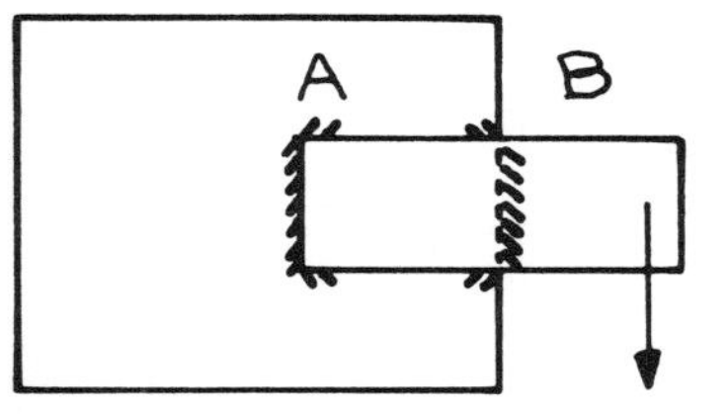

3. Placement of Welds to Resist Turning Effect of One Member of the Joint.

4. Examples or Lap Weld Having Poor Distribution of Stress Through Weld. Satisfactory for Many Jobs, Though 5 and 6 are Better.

5. Example of Lap Weld Having a More Even Distribution of Stess Through Weld. Better than 4 and More Commonly Used.

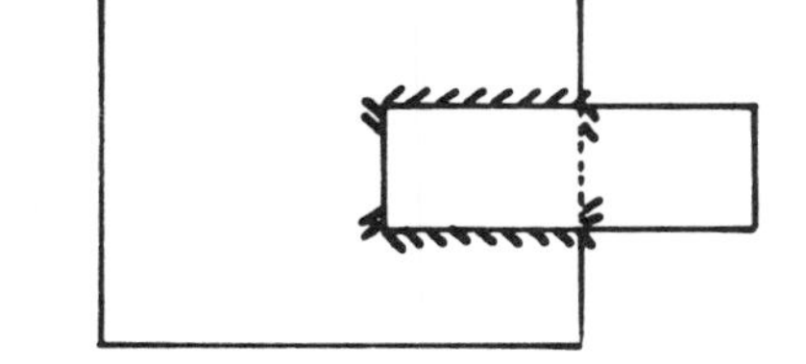

6. Example of Lap Weld in Which There is Fairly Uniform Transfer of Stress Through the Weld. Better than 4 and 5 for Severe Fatigue and Impact But More Expensive.

7. Example of Welds Hooked Around the Corners to Obtain Resistance to Tearing Action on Welds when Subjected to Eccentric Loads.

PREFERRED TYPE OF JOINTS FOR FRAMING STEEL STRUCTURE

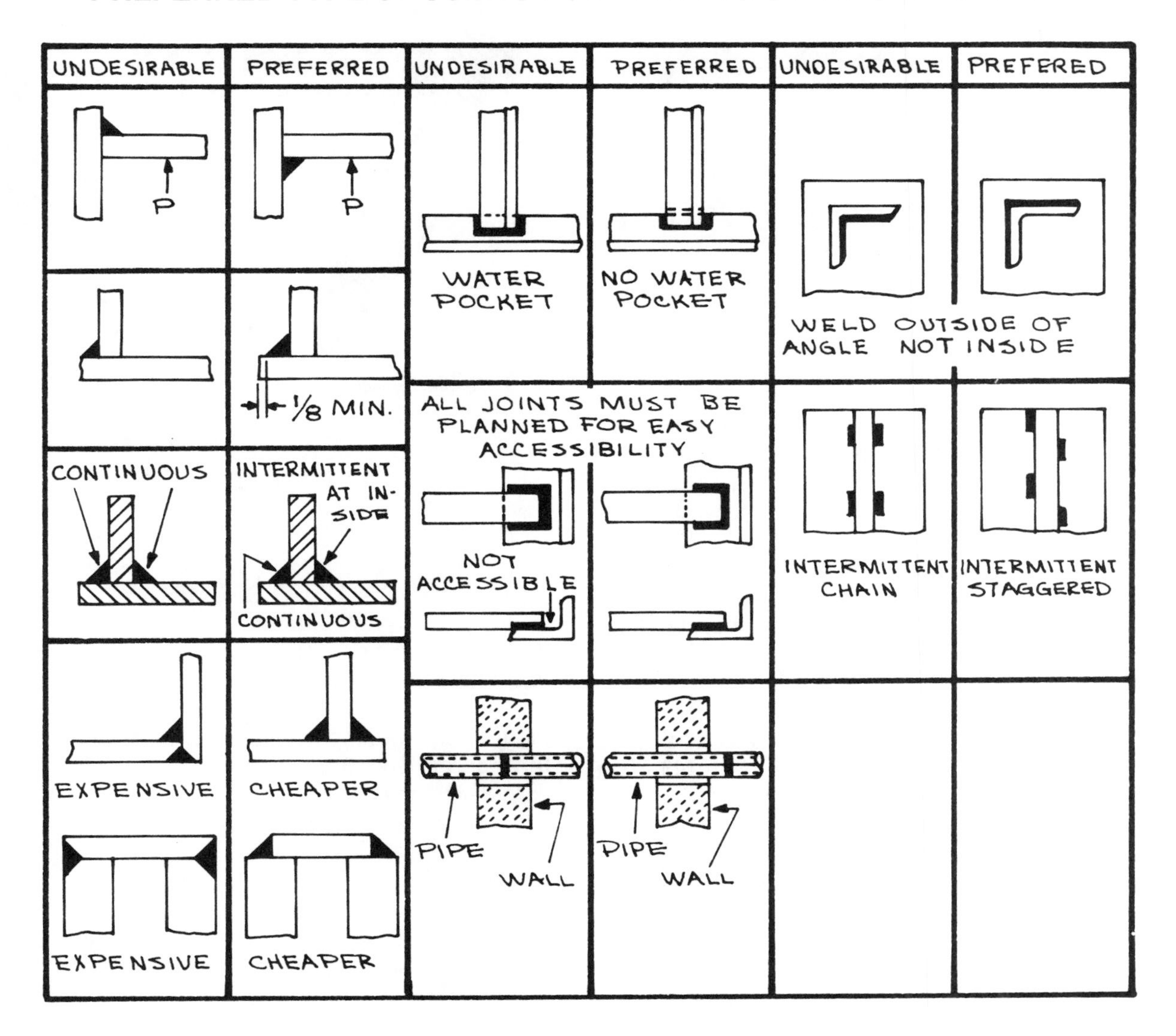

STRENGTH OF FILLET WELDS

Since it is possible to overweld or underweld a fillet weld, it is necessary to determine the size and length of fillets for all connections using them.

The strength of a fillet weld is based on the effective **throat thickness** defined as the shortest distance from the root to the face of the diagrammatic weld. Therefore, for an equal leg (45 degrees) fillet weld the throat is 0.707 times the normal leg size of the weld, Figure 66.

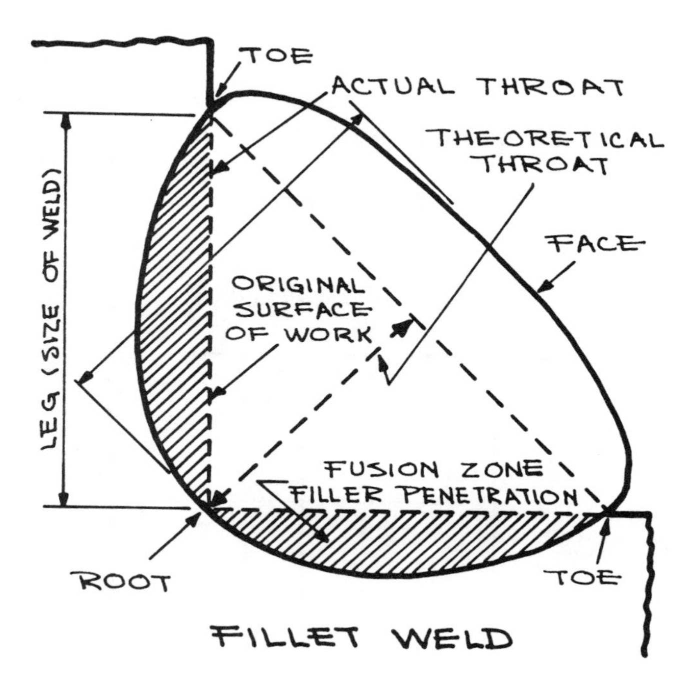

Figure 66. Fillet Weld Nomenclature.

A fillet weld may be concave, flat or convex. However, the ideal fillet weld is assumed to have a flat surface or only slightly convex and at 45 degrees to the plates being connected, Figure 66.

A concave fillet weld produces a smooth change in section at the joint but is more susceptible to shrinkage cracks, especially in the higher carbon steels. The critical dimension in this shape of fillet weld which must be measured is the throat size.

A convex fillet weld has less tendency to crack as a result of shrinking while cooling. It is relatively free from undercut but may result in excess weld metal if excessively convex, which will not add to the allowable strength of the weld but will increase the cost. The critical dimension in this shape of fillet weld which must be measured is the leg size.

The strength of a fillet weld depends upon the direction of the applied load which may be parallel or transverse to the weld. In both cases, the weld fails in shear, but the plane of rupture is not the same. The weld will fail on the throat plane which has the maximum shear stress.

STRENGTH OF BUTT WELDS

A butt weld is used in connecting two members to transmit the full capacity of the smaller one. To develop the full force in the member, the butt weld must be made the same thichness as the plates. This is called a full strength weld. In a connection of this type, there is no need to calculate the unit stresses in the weld, nor attempt to determine its size, since a butt weld has equal or greater strength than the mild steel plates being joined. This is also true with low alloy high tensile plates when the weld metal matches the physical properties of the plate. Study Figure 67 for the nomenclature of the groove (butt) weld.

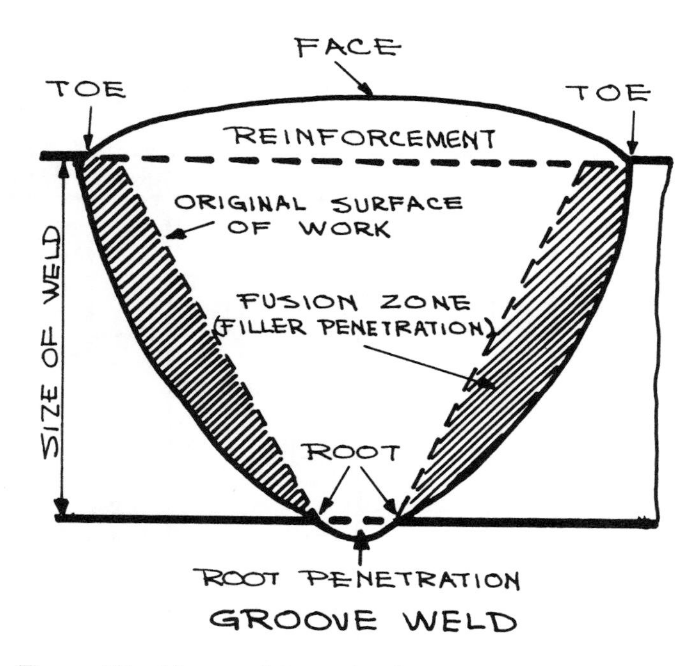

Figure 67. Nomenclature of a Groove (Butt) Weld.

WELD GEOMETRY

Welding being a metal joining process, a certain amount of basic nomenclature is in order.

Figure 68 shows the five basic types of joints, namely: (1) **butt**, end to end on one plane; (2) **T or tee**, a corner 90° corner; (3) the **lap** and (4) **corner** are basically 90° corner variations; and (5) the **edge** is a variation of a butt weld.

Figure 69 shows the four positions in which welding is done using the butt and fillet type joints. The forces of gravity acting on fluid weld metal make flat position welding the most successful and productive.

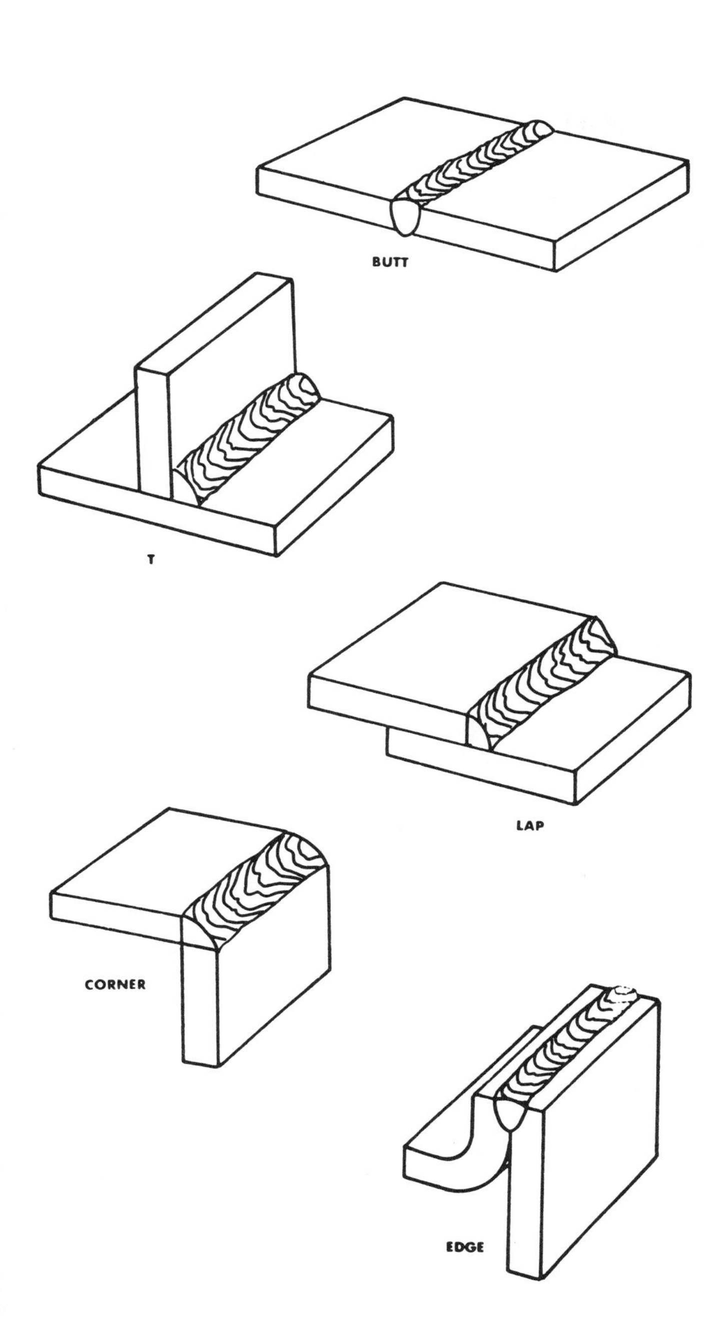

Figure 68. Basic Weld Joints.

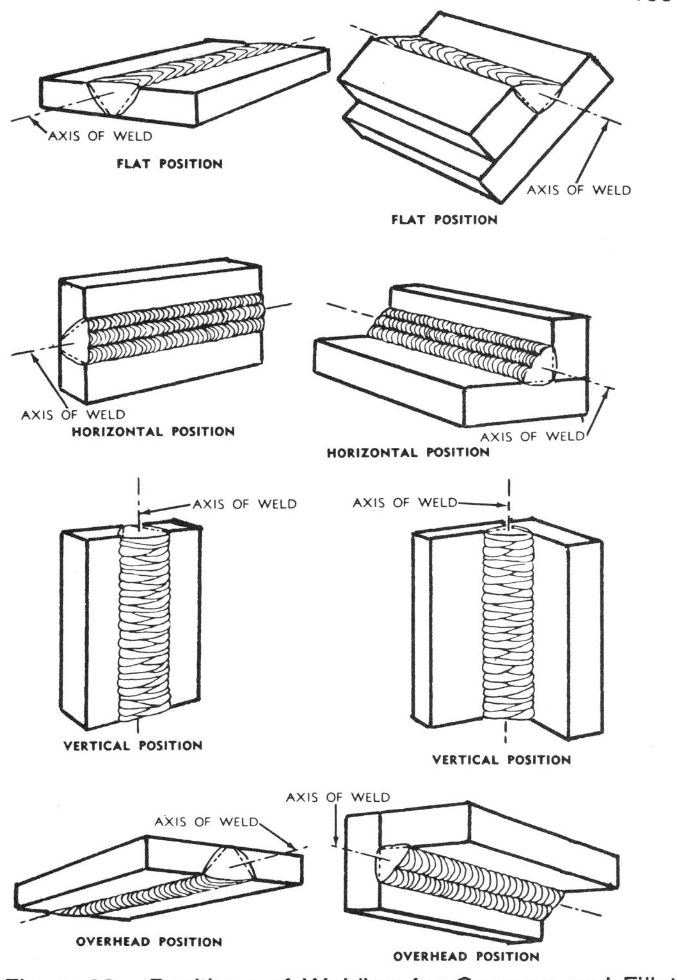

Figure 69. Positions of Welding for Grooves and Fillet Welds.

PLASMA ARC CUTTING (PAC)

Plasma is a gas (compressed air) which has been heated to an extremely high temperature, about 30,000°F, and ionized so that the gas becomes electrically conductive. The plasma arc cutting unit as shown in Figure 70 uses this plasma to transfer an electric arc to the workpiece. The metal to be cut is melted by the heat of the arc and then blown away by compressed air.

Compressed air is used as both the plasma and secondary gas. Incoming gas is internally channeled (Zone A, Figure 71) to the proper exiting ports. A high frequency arc between the electrode and the tip (Zone B) initiates the pilot

Figure 70. A Plasma Arc Cutting Unit.

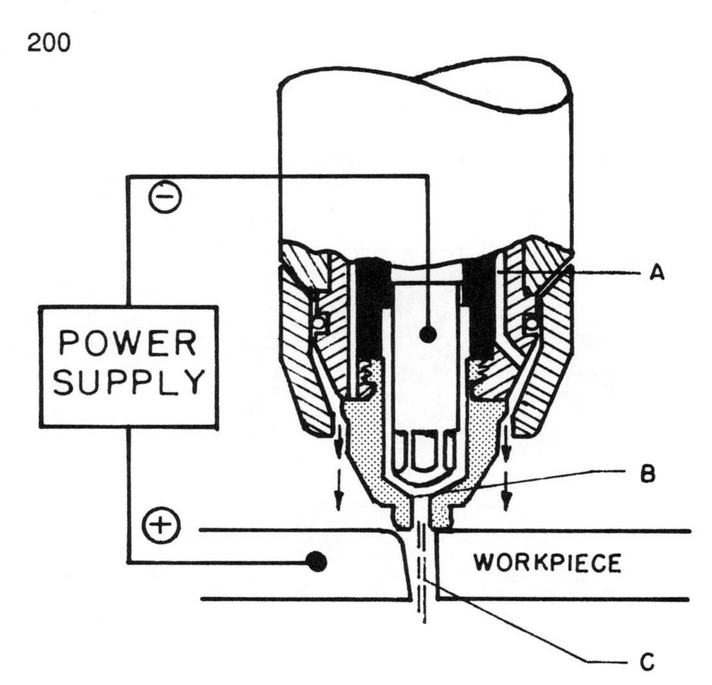

Figure 71. The Plasma Arc Cutting Unit.

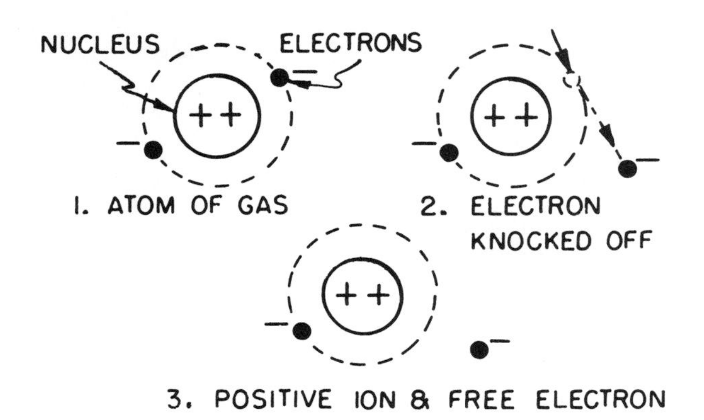

Figure 72. What Plasma Is.

arc by heating and ionizing the plasma gas. When the tip of the torch is brought close to or in contact with the workpiece, this ionized gas produces a stiff, constricted cutting arc (Zone C) that delivers a high concentration of heat to a very small area.

The design of PAC cutting torches utilizes a secondary gas, shown by the small arrows in Figure 71. The secondary compressed air is channeled out around the tip, where it cools torch parts and protects the torch from spatter.

THEORY OF OPERATION

Plasma and secondary gases - The single source of compressed air for the plasma and secondary gas flows through the PAC unit to the torch at the pressure set on the external regulator. A solenoid valve turns the air on and off. A gas pressure interlock shuts the system down if input air pressure falls below approximately 40 psi (2.8 kg/cm^2).

High frequency - High frequency is used to initiate and maintain the pilot arc. Both high frequency and pilot arc shut off when cutting arc is established.

The accelerated motion of particles in a gas, caused by heat from an electric arc, results in collisions which can knock off electrons and leave positive ions as illustrated in Figure 72. This electrically ionized gas is called plasma. It is often considered the fourth state of matter, after solid, liquid and gas.

Pilot arc - When the torch sequence is started, the high frequency relay closes and a pilot arc is established between the electrode and cutting tip. Note Figure 73. This pilot arc creates a path for transferring the cutting arc to the work.

Cutting arc - The bridge rectifier converts the AC power to DC power for the cutting arc. The negative output connects to the torch electrode through the torch lead. The positive output is connected to the workpiece through the work cable.

Interlocks - A pressure switch (PS1) acts as an interlock for the air supply. Note Figure 73. If the air supply pressure is below approximately 40 psi, the pressure switch will open, shutting off power to the main contactor, and the gas pressure L.E.D. will go out.

Two thermal overload switches in the PAC power source (TP1 and TP2) are located in the main transformer coil and above the pilot resistor, respectively. If one of these components is overheated the appropriate switch will open, causing the temperature light to go out and shutting off power to the main contactor. When the overheated component cools down, the switch closes and allows operation of the system.

Plasma arc cutting equipment can be used for cutting ferrous and nonferrous metals. The operator in Figure 74 is using the Plasma arc cutting unit to cut stainless steel. Plasma arc cutting (PAC) is most efficient for long, continuous cuts in carbon steel, aluminum and stainless steel. Cuts made with the plasma arc on metals up to 3" thick are generally faster than those made with oxyacetylene cutting. Carbon steels 1" thick or less can be cut up to five times faster with PAC than with oxyacetylene unit.

Cutting speed is important for productivity, but speed also has an effect on the metallurgical properties of the cut metal. Stainless steels, for example, form hard, brittle chromium carbides if heated and held at high temperatures. Carbides reduce the metal's corrosion resistance. With plasma cutting, the cut made in stainless remains corrosion resistant.

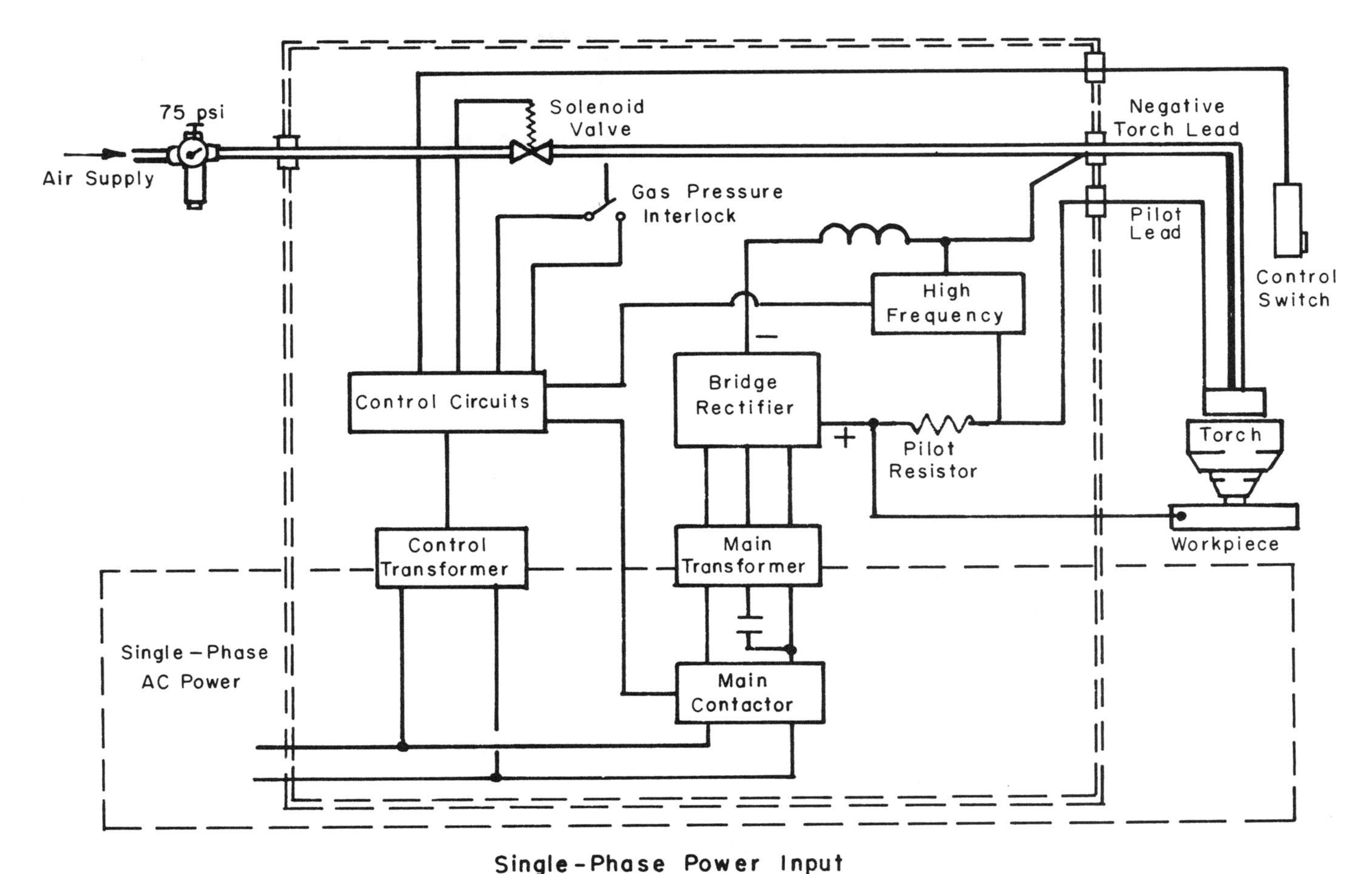

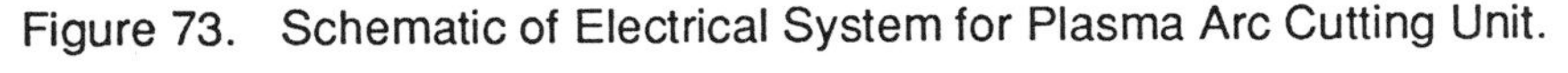
Figure 73. Schematic of Electrical System for Plasma Arc Cutting Unit.

Figure 74. The Plasma Arc Cutting Unit Cuts Stainless Steel with Ease.

WELDING PLASTICS

IDENTIFICATION OF PLASTICS FOR WELDING

The initial test before attempting to weld plastics material, is to determine whether the material is thermoplastic or thermosetting. Heat a stirring rod or similar implement to a temperature of approximately 500°F (260°C) and press against the material. **If the material softens, it is thermoplastic; if it remains hard, it is thermosetting.**

Burning test - The burning test is one of the most efficient and effective methods for identifying plastics. Equipment required to carry out the burning test consists of a match, or preferably a bunsen burner, and a pair of tongs to hold the sample to be tested. A piece of the material about 1/8" thick and 3" long is inserted into a flame and held there until it ignites (about 10 sec. maximum). The sample is then removed from the flame and observations are made of the flame characteristics and the effect upon the material. By comparing these observations with the characteristics listed in Table 10, identification can usually be made.

Caution should be exercised when smelling the fumes after the flames have been extinguished, since the fumes may be toxic. It is helpful to have a known sample for comparison when identifying odor.

It is sometimes necessary to perform additional testing, particularly to identify various types of a specific material. In the case of polyethylene, this may be done by a specific gravity test. For other materials it may be necessary to perform additional tests, such as tensile, impact or chemical resistance tests, in order to distinguish among the various grades.

Positive identification is rarely necessary for welding purposes. In most field welding operations, a match is all that is needed to determine if the material will burn.

It is sometimes possible to identify thermoplastic materials in the field by visual inspection and by feel: For example, the polyolefins (polyethylene and polypropylene) feel slippery and slightly greasy to the touch and usually have a very slippery surface. The common weldable plastics are listed in Table 11.

TABLE 10. THERMOPLASTIC BURNING TEST.

Material	Specific Gravity	Burning Characteristics			
		Rate*	Odor	Flame	Effect on Material
Cellulose Acetate butyrate (CAB)	1.15-1.22	0.5-1.5	Butyric Acid; (rancid butter)	Dark Yellow with slight,blue edges: some black (not sooty)	Melts: drips, drippings continue to burn
Polyamide (nylon)	1.09-1.14	Self-extinguishing	Burning wood or hair	Blue with yellow top	Melts, drips, and froths
Polyethylene					
1. Low density	0.910-0.925	1.0-1.1	Burning paraffin	Bottom blue, top yellow	Melts & drips
2. Medium density	0.926-0.940	1.0-1.1	Burning paraffin	Bottom blue, top yellow	Melts & drips
3. High density	0.941-0.965		Burning paraffin	Bottom blue, top yellow White smoke	Melts & drips
Polypropylene	0.90-0.91		Burning paraffin	Bottom blue, top yellow; white smoke	Melts & drips
Polyvinyl chloride (PVC)					
Type I	1.35-1.45	Self-extinguishing	Hydrochloric acid	Yellow, green on bottom edges; spurts green & yellow; white smoke	Softens; chars
Type II		Self-extinguishing	Hydrochloric acid		Softens; chars
Styrene copolymers (ABS)					
Type I	1.00-1.05	1.3	Artificial illuminating gas	Yellow; black smoke	Softens; chars
Type II	1.06-1.08	1.3	Artificial illuminating gas	Yellow; black smoke	Softens; chars

*ASTM Test No. D635-44 (rate in inches per minute)

TABLE 11. COMMON WELDABLE PLASTICS.

1. Acrylonitrile Butadiene Styrene - ABS
2. Cellulose Acetate Butyrate - CAB
3. Chlorinated Polyether
4. Polyacetal
5. Polyamide - Nylon
6. Polyethylene - PE
7. Polypropylene - PP
8. Ployvinyl Chloride, Rigid Unplasticized - PVC
   a. Normal Impact (Type I)
   b. High Impact (Type II)
9. Polyvinyl Dichloride - PVDC
10. Acrylics
11. Fluorocarbons
    a. TEE
    b. FEP
    c. CTFE
    d. Polyvinylidene Fluoride - PVF

Practice these identification methods thoroughly before attempting production welds. Compare your results with the illustrations and tests given. If necessary, your material supplier can furnish an analysis of your sample welds, or tests can be made by commercial laboratories.

WELDING EQUIPMENT AND MATERIALS

Shown in Figure 75 is a plastic welding unit. The welding unit consists of a heating element with various styles of tips available for heating the base material to be repaired and an air regulator and pressure gage to control the heated air. Heating elements range in size from 400 to 650 watts which will yield temperature ranges of 350 to 900 °F.

Welding rods are selected according to the type of material to be welded. Shown in Table 12 are five common plastic materials along with welding and forming temperatures, welding gas to be used and the recommended pressures.

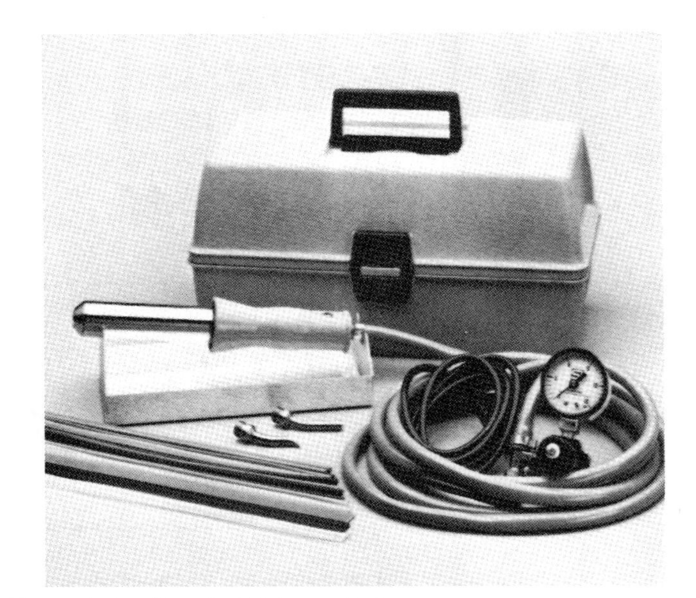

Figure 75. Plastic Welding Unit.

Welding rods are available in different sizes and shapes according to the welding tip selected for a specific job.

HOW TO ACHIEVE QUALITY PLASTIC WELDS

Types of weld joints are shown in Figure 76. Basically the same type of welds are performed in plastic welding as in metal welding. Select the correct weld joint based on the shape and position of the plastic piece to be repaired.

A good final weld, with either a round or high speed tip, is easy to achieve. Test yourself by using the round and speed tips. A good weld cannot be pulled apart, even when hot, and the final weld area can be stronger than the surrounding part. You will be an accomplished plastic welder if you observe and practice the following:

1. Small beads of "juicing" should form along each side of the weld where the rod meets the base material.
2. The welding rod should hold its basic round shape.
3. Neither the rod or base material should char or discolor.
4. Never stretch the rod over the weld. In other words, the length of the rod should match that of the weld.
5. Do not use oxygen or other flammable gasses. **Use air only.**
6. Practice proper welding methods and experiment with the tacking, round and high speed tips before you actually begin trying to repair a damaged part.

Compare your results with the illustrations shown in Figure 77.

1. No bond, weld can be pulled apart.
2. Proper weld, notice complete bonding.
3. Burned and charred material results in weak weld.

Follow carefully the operating instructions for your Electric Plastic Welder and any accessories. This will give you a successful welding technique and protect your investment in equipment.

TABLE 12. WELDING RODS FOR WELDING DIFFERENT PLASTICS.

	Polyvinylchloride	Polyethylene	Polypropylene	ABS	Polyurethane
Welding Rod Color Code	Light Grey	Brown	Black	Blue	Clear
Welding Temperature °F	525	550	575	500	500
Forming Temperature °F	300	300	350	300	300
Welding Gas	Air	Air	Air	Air	Air
Recommended Pressure,psi	3	3	3	3	4

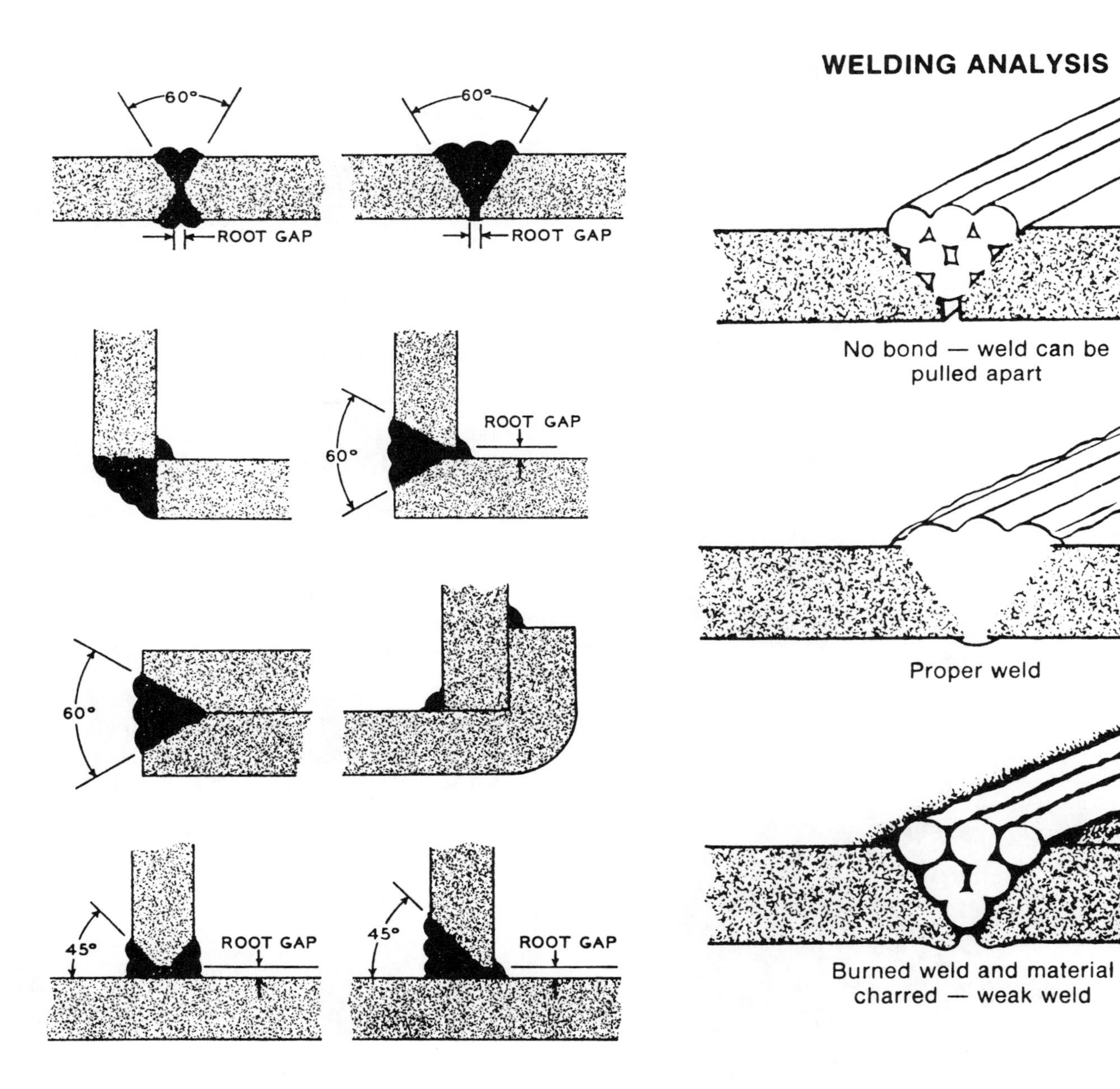

Figure 76. Common Plastic Weld Joints.

Figure 77. Welding Analysis.

CLASSROOM EXERCISE ARC WELDING

NAME________________________________
DATE_______________ GRADE_____________

1. Explain the purpose for standing on a wooden grate during the arc welding process.________________________

____

2. To avoid others in the area from getting a direct arc glare, the operator should say________________________to anyone in the area.

3. The correct shade number to have in the welding helmet for welding with 1/8" electrodes is a number__________________.

4. Identify the parts of the welding process.

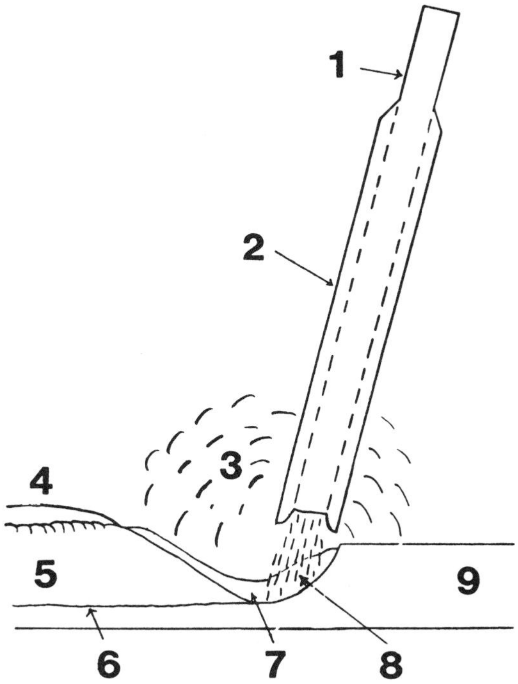

1 2 3 4 5 6 7 8 9

1. ____
2. ____
3. ____
4. ____
5. ____
6. ____
7. ____
8. ____
9. ____

5. One of the groups involved with classifying electrodes is the AWS which stands for:

   a. Always Weld Safely
   b. American Welding Standards
   c. Above Welded Surface
   d. American Welding Society

6. List seven factors that all successful arc welding depends upon.

   a. ________________________________
   b. ________________________________
   c. ________________________________
   d. ________________________________
   e. ________________________________
   f. ________________________________
   g. ________________________________

7. Identify the parts of the steel electrode classification system.

   E: ____________________

   60: ____________________

   1: ____________________

   3: ____________________

8. Electrodes can be identified by a color marking system. The three marking positions on the electrode are:

   a)____________, b) ____________and c) ____________.

9. Match the following electrodes identified by the number system with electrode characteristics.

________	a. E6010	1. Black end and brown spot marking
________	b. E6011	2. Flat and horizontal positions only
________	c. E6013	3. Orange, blue and white markings
________	d. E6014	4. No color markings
________	e. E7018	5. A popular fast-freeze electrode
________	f. E6024	6. Excellent low hydrogen electrode
________	g. E-Ni	7. Brown spot only color marking

10. Identify the two methods for striking the arc.

   A. ____________

   B. ____________

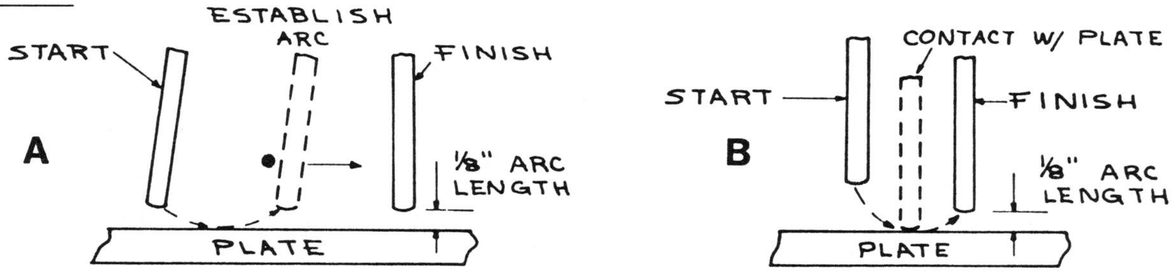

11. In flat position welding, the electrode should be at ________degrees from side to side and tilted at ________ degrees toward the direction of travel.

12. As a general rule, arc length should be ________ the diameter of the wire in the electrode.

   a. two times
   b. one-half
   c. equal to

13. Speed of travel in arc welding is determined by the

   ____

14. Explain what is meant by a "hot start" and list the procedures for carrying out a "hot start."

____

____

15. List four factors that need to be controlled or that influence undercutting in arc welding.

a. ______________________ c. ______________________

b. ______________________ d. ______________________

16. List four methods or practices to help control distortion in arc welding.

a. ______________________ c. ______________________

b. ______________________ d. ______________________

17. The AC arc welder is basically a single-phase electrical______________________.

18. List four methods used in arc welding machine design for controlling output amperage.

a. ______________________ c. ______________________

b. ______________________ d. ______________________

19. The AC-DC transformer welding machine changes single-phase AC current to DC current by using a

______________________.

20. The higher the open-circuit voltage the __________it is to strike the arc.

a. more difficult
b. same
c. easier

21. Arc voltage on the welding machine is defined as:______________________

____

22. One sure method to avoid arc blow is to select an ____________welding machine.

23. Power factor is the relationship between ____________ watts and ____________ watts.

24. A common duty cycle on limited input, AC transformer welding machines is ____________% which means the machine can weld at full amperage for ____________ minutes out of 10 minutes.

25. A 250 amp AC-DC welder with 30% duty cycle would have a 100% duty cycle at approximately

______________ amperage setting.

26. Temperature rise is the degrees C above room or ambient temperature that the ____________coil can withstand without causing damage to this coil.

27. Define the following industrial welding terms:

    a. GMAW - ______________________________

    b. GTAW - ______________________________

    c. SAW - ______________________________

    d. DCRP - ______________________________

    e. DCSP - ______________________________

    f. TIG - ______________________________

    g. MIG - ______________________________

28. The most common gases used in gas metal arc welding (GMAW) are ______________ and ______________ while ______________ and ______________ are the most common gases used for GTAW.

29. Identify the basic parts of the gas tungsten arc welder:

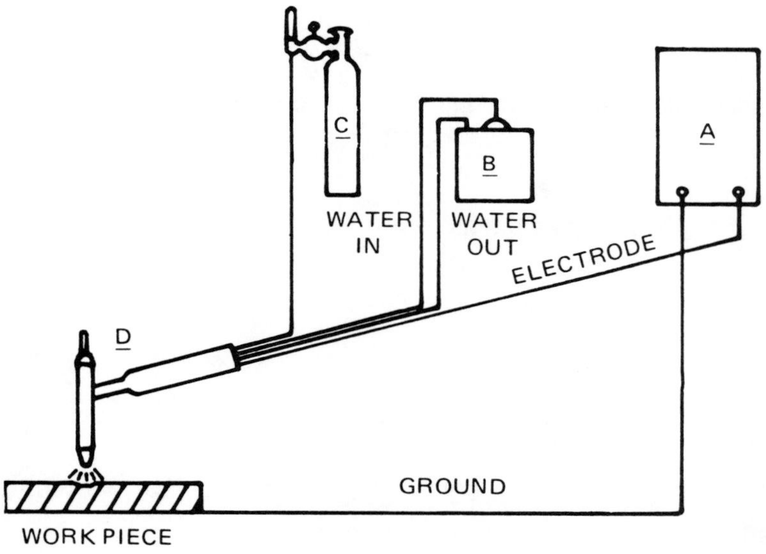

a. ______________________________

b. ______________________________

c. ______________________________

d. ______________________________

30. The GTAW uses a ______________________________ electrode.

    a. tungsten
    b. mild steel
    c. nickel
    d. aluminum

31. In submerged arc welding, the shield to protect the weld consists of:

    a. CO_2 gas
    b. granular flux
    c. argon gas
    d. cellulose potassium gas

32. List four conditions that the high frequency unit provides or controls during TIG welding.

    a. ______________________ c. ______________________

    b. ______________________ d. ______________________

33. Identify the five weld joints commonly used in arc welding.

    a. ______________________

    b. ______________________

    c. ______________________

    d. ______________________

    e. ______________________

A

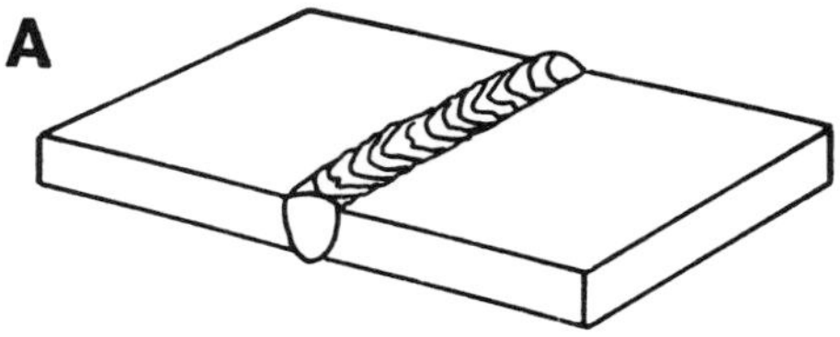

B

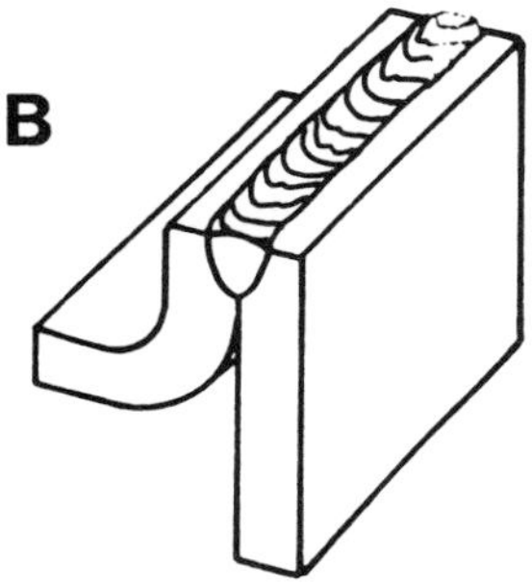

C

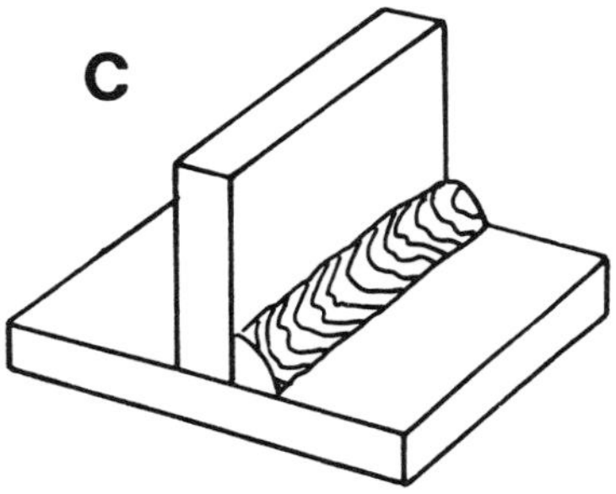

D

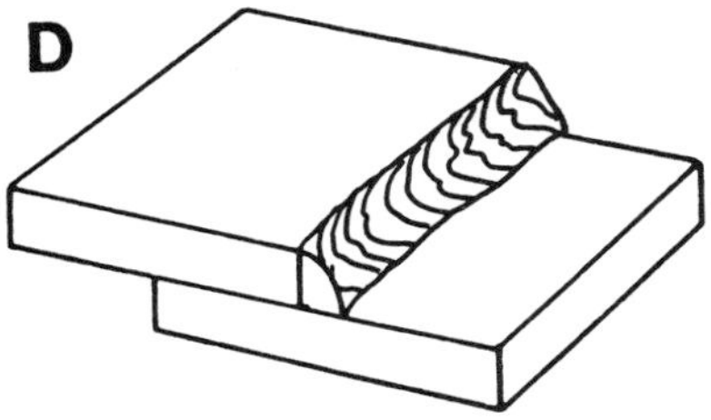

E

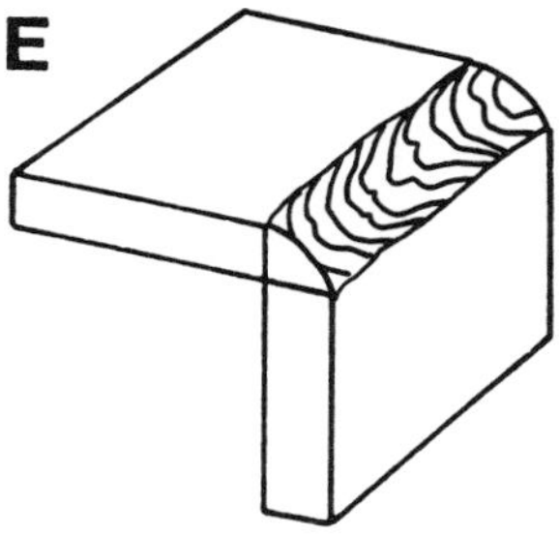

34. List the four common arc welding positions.

    a. ______________________ c. ______________________

    b. ______________________ d. ______________________

35. The main advantage of the plasma arc cutting (PAC) unit over oxyacetylene cutting is______________________

    ____.

36. The molten metal in the PAC operation is blown away by ______________________.

37. The most efficient and positive test for identifying plastics that are weldable is the________________________

________________________________ test.

38. When heated, if the plastic part softens it is ______________________, if it remains hard it is

____.

39. List five pieces of equipment or materials that are needed for welding plastics.

a. ______________________________ d. ______________________________

b. ______________________________ e. ______________________________

c. ______________________________

40. Weldable plastics melt at approximately ______________________degrees F.

Notes

CLASSROOM EXERCISE ARC WELDER NAMEPLATE

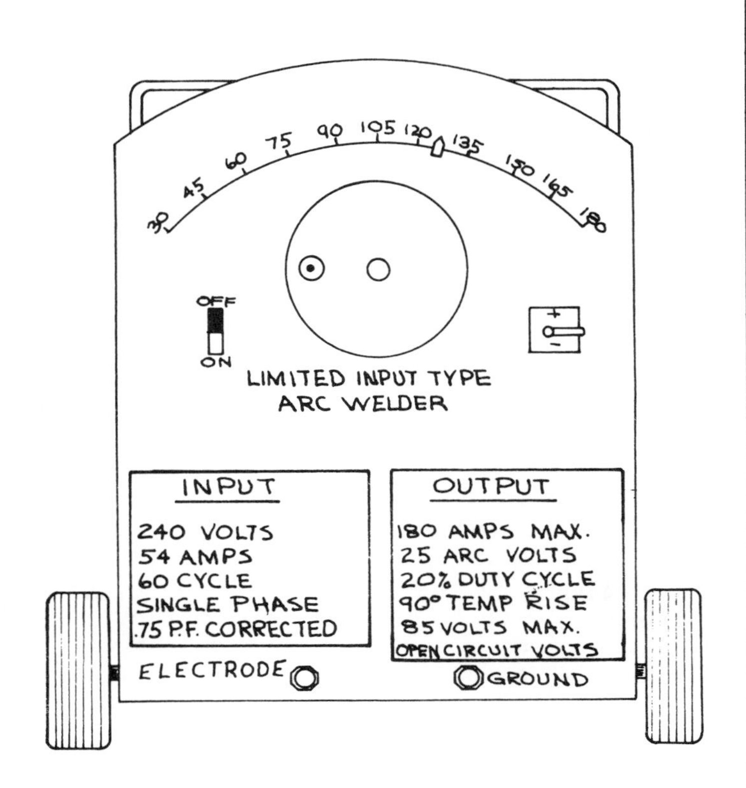

Operational Procedure:

A. Complete the welder nameplate information section for welder assigned.

B. Answer the following questions regarding the nameplate shown on this sheet.

1. Which has the higher voltage, the input or output circuit? ____________________
2. Explain arc volts. ____________________
3. Define open circuit voltage. ____________________
4. Is the output side the primary or secondary circuit in the transformer? ____________________
5. Define duty cycle. ____________________
6. Could the duty cycle ever be 100% for the machine pictured on this sheet? __________ Explain
7. Explain temperature rise. ____________________
8. What is the difference in the construction of a welding machine having 90°C rise and one having 110° C rise? ____________________
9. What is power factor P.F.? ____________________
10. What might the power factor be if it were not corrected? ____________________
11. Explain how power factor is corrected. ____________________
12. What size fuse would be needed to fuse the welder shown on this sheet?

    ______________ amps.

Material: Arc Welders with Nameplates

Welder Nameplate Information

Welding Machine No. __________

Input		Output	
______	Volts	______	Amps. Max
______	Amps	______	Arc Volts
______	Cycle	______	Volts, Max OCV
______	Phase	______	Temp. Rise
______	P.F.	______	Duty Cycle

Evaluation Score Sheet:

Item	Points Possible	Points Earned
1. Operational questions	24	_____
2. Information from welder	16	_____
3. Attitude and Work Habits	10	_____
Total Points	50	_____

Name: ____________________

Date: ____________________

Grade: ____________________

LABORATORY EXERCISES ARC WELDING

LAB EX-1 - RUNNING BEADS

Material - Piece of 1/4" mild steel plate or scrap, approximately 3" x 4". 1/8" or 5/32" electrodes.

Prodedure :

1. Practice striking arcs and running beads on a piece of scrap steel until a bead at least 3" long can be run.
2. Draw parallel lines 1/2" apart on metal with metal crayon or chalk and run beads.
3. Chip away slag and clean with a wire brush after running each bead.
4. Examine beads carefully and make any corrections necessary.
5. If arc is broken or new electrode is inserted in holder, strike arc on clean metal then move to the rear of crater before resuming forward direction.
6. The quality of the bead (weld) will depend on how well the following are controlled:

   a. **Amperage setting**
   b. **Length of arc**
   c. **Angle of electrode**
   d. **Speed of travel**

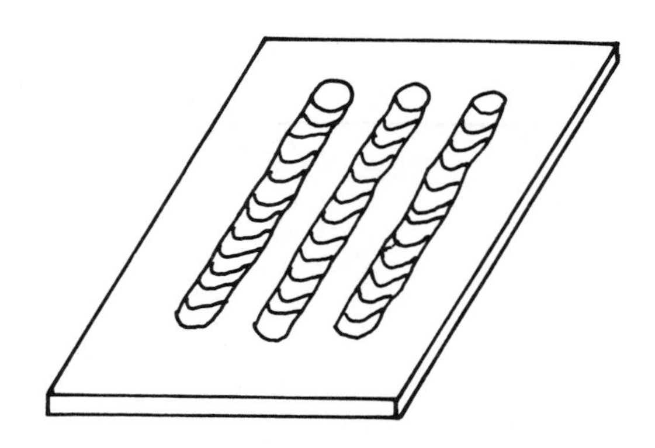

LAB EX-2 - PAD OR BOSS

Material - Same steel plate used in Job #1. 5/32" electrodes.

Procedure :

1. Deposit a series of adjacent parallel beads covering an area 2 x 2 inches.
2. Each bead must be run in contact, overlapping 1/4 to 1/2 way with the adjoining bead so that a continuous layer of metal is formed.
3. Chip and clean all slag thoroughly before depositing a second layer.
4. Deposit a second series of parallel beads over the first layer at right angles to the first.
5. Test quality by cutting through boss with hacksaw or grinding wheel.
6. Observe any air or slag pockets.

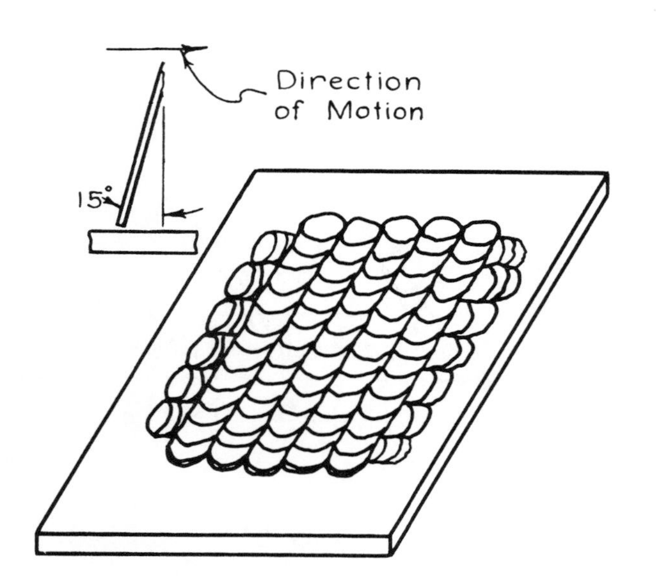

LAB EX-3 - BUTT WELD

Material - 2 - 1/4" x 1" mild steel plate.
1/8" and 5/32" electrodes.

Procedure :

1. Vee one edge of each piece to within 1/16 inch (single or double vee depending on thickness of metal).
2. Vee so that angle formed when two pieces are placed together is from 60 to 90 degrees.
3. Make a single pass in the bottom of vee.
4. Complete weld with two overlapping beads or one spread. Make completed weld slightly higher than the level of the plates.
5. Test weld using both the bend and cutting tests.

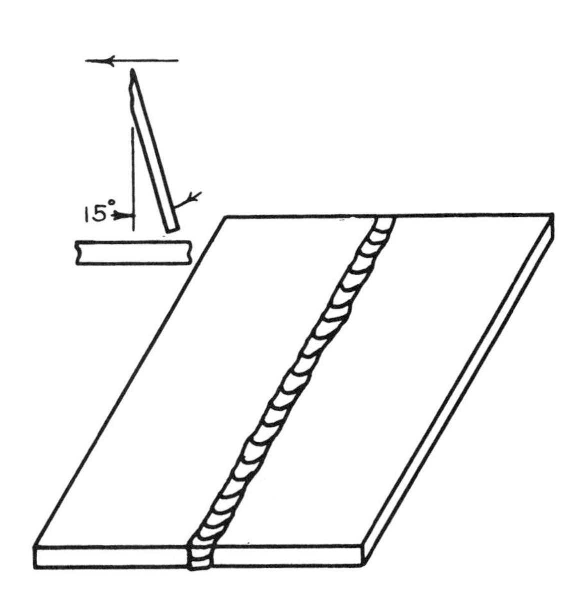

LAB EX-4 - LAP WELD

Material - Same pieces of metal used in jobs #1 and #3.
1/8" and 5/32" electrodes.

Procedure :

1. Place the two pieces of metal so they overlap 1/4" to 1/2". (Protect your finished work from spatter.)
2. Support the outer edge of the upper plate so that they will remain in this position.
3. Tack weld both ends of plates.
4. Fill in the vees formed by the overlapping pieces with single or spread beads.
5. Keep the electrode at an angle of from 30 to 45 ° with the lower plate.
6. Hold the arc slightly longer on the lower plate.
7. Test the quality of the weld.

REMEMBER - The quality of the weld will depend on how well the following have been controlled:

a. **Amperage setting**
b. **Length of arc**
c. **Angle of electrode**
d. **Speed of travel**

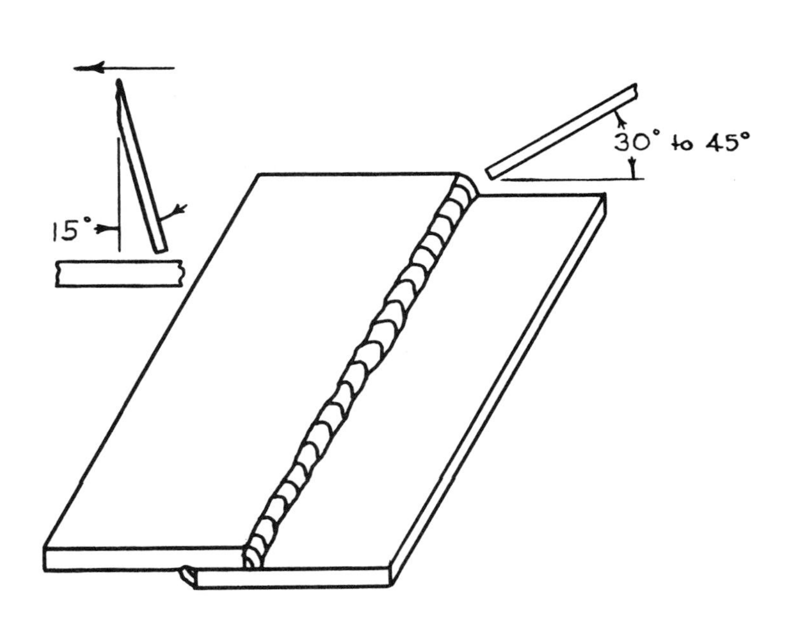

LAB EX-5 - FILLET WELD IN HORIZONTAL POSITION (TEE JOINT)

Material - 2 pieces mild steel 1/4" x 2" x 4", 5/32" E6011 or E6013 electrodes.

Procedure :

1. Set two pieces together at right angles forming a tee joint and tack weld both ends.

2. Place 5/32" electrode in electrode holder at a 45° angle.

3. Make a single pass fillet weld on each side:

   a. Maintain a short arc. A long arc will under cut the metal on vertical plate.
   b. Maintain a slightly wider arc gap between electrode and vertical plate than electrode and bottom plate.
   c. Hold electrode at 45° so that it divides angle between 2 plates.
   d. Tip electrode toward direction of travel 15 to 20 degrees.

4. Clean slag from welds.

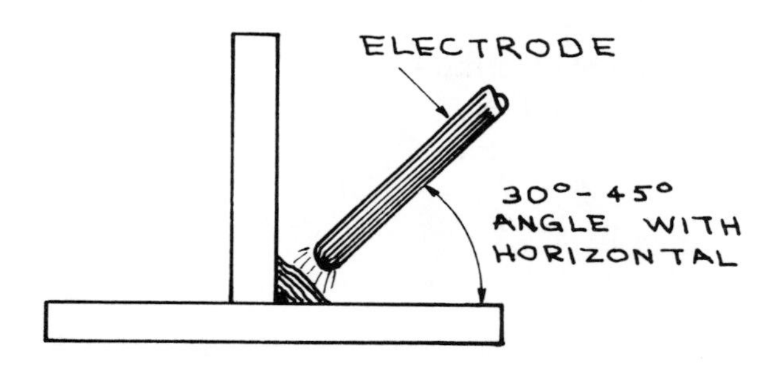

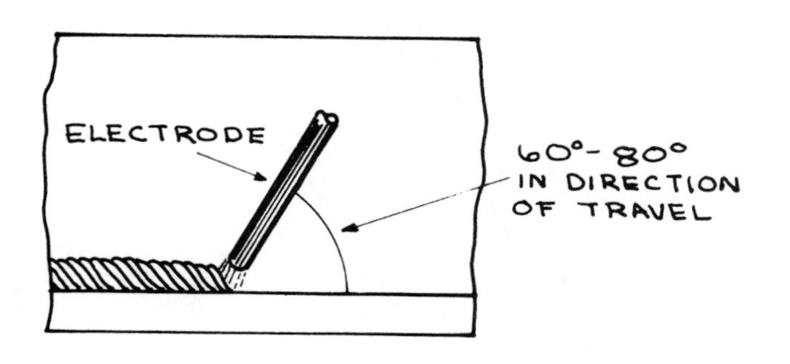

LAB EX-6 - ARC CUTTING AND PIERCING HOLES

Material - Piece of steel plate at least 1/4" in thickness.

Procedure :

1. Select 1/8" E6011 mild steel electrode and make an initial setting of about 180 amperes. If electrode is soaked in water for five minutes before use, it is not consumed as fast.

2. A piece of heavy metal may be used as a guide in cutting straight lines.

3. Short hacking motions are used in removing the molten metal from the cut. The metal is melted on the upward stroke and the waste forced out on the downward stroke. Metal may be cut in a vertical or flat position.

4. Holes pierced with the arc electrode are not as accurate as drilled holes but procedure is often preferred because of greater speed. Holes can be pierced in steels difficult to drill as spring leaves, harrow teeth and cast and malleable cast.

5. Maintain a long arc, rotate tip of electrode until base metal becomes molten, then force tip through metal. Hole may be extended to desired size.

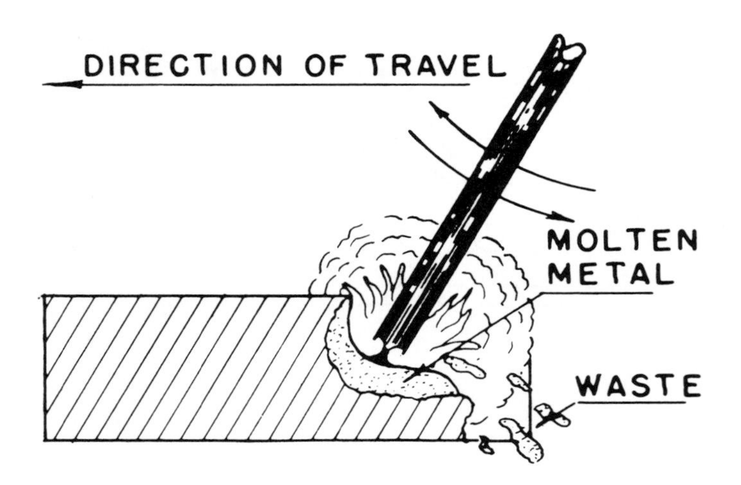

In arc cutting, the metal is melted on the upward stroke and the waste forced out on the downward stroke.

LAB EX-7 - ELECTRODE OPERATIONAL CHARACTERISTICS

The purpose of this lab is to learn electrode operational characteristics and uses by performance and observation. You will be given seven pieces of steel on which you are to weld a bead 2" + long. Start by tacking the seven pieces of metal together. Practice on scrap metal to learn how to use each electrode before running the bead. You are to use seven different 1/8" electrodes designated by AWS number. Complete the worksheet after completing the welds.

Arc Welds	AWS# Electrode	Mode AC DC- DC+	Approximate Amps	Observations		
				Flux Before Welding	Arc During Welding	Slag After Welding
	6010					
	6011					
	7024					
	6012					
	6013					
	7014					
	7018					

COMMENTS:

LAB EX-8 - ARC WELDED TROPHY

Welding Exercises

3 beads
2 butt welds
2 fillet welds
2 lap welds
1 plug weld

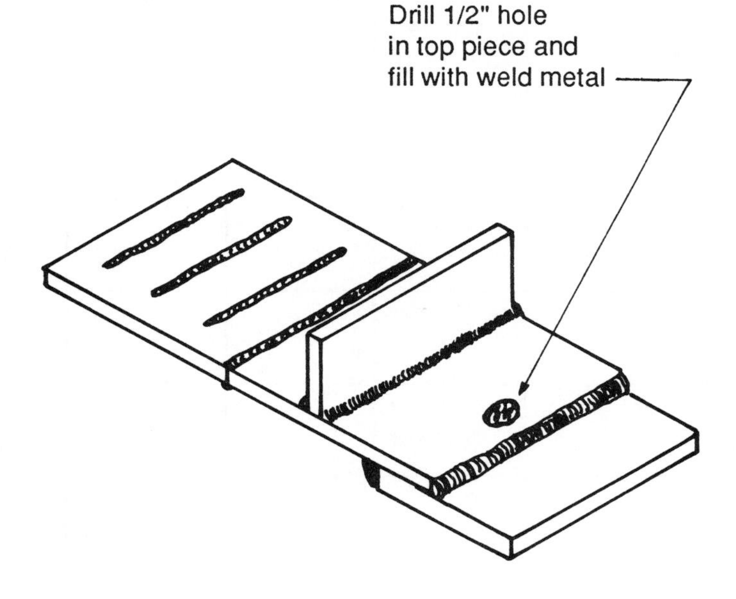

Materials :

3 - 4"x4"x1/4" Steel plates
1 - 2"x4"x1/4" Steel plate
E6011, E6013, or E7018 Electrodes 1/8" size

Evaluation Score Sheet :

	Item	Points Possible	Points Earned
1.	Three beads	25	______
2.	Two butt welds	5	______
3.	Two fillet welds	20	______
4.	Two lap welds	15	______
5.	Plug weld	10	______
6.	Freedom from distortion	20	______
7.	Work habits - attitude	5	______
	Total Points	100	______

Name______________________________

Date________________Grade________________

Procedure

A. Steps in Selection of Materials and Procedures in Welding:
1. Examine base metal Type of metal, carbon content, thickness, type of weld, type of joint, appearance desired, strength needed, position of the weld.
2. Select correct electrode Electrode number, electrode size.
3. Select correct amperage
4. Select correct joint preparation cleaning, veeing, clamping.
5. Select correct sofety equipment Safety glasses, helmet with No. 10 eye shield, leather gauntlet gloves, hard finish clothing.

B. Steps in Making the Weld:
1. Tack weld or clamp as needed
2. Strike the arc - Scratch or tap method
3. Pre-heat
4. Hold long arc until molten pool is desired size
5. Lay the bead - (a) Electrode angle — 15° from vertical in direction of travel; (b) Arc length - 1/16" to 1/8" makes frying sound; (c) Speed of travel - watch width (2 x electrode dia.), height and shape of molten pool. Fill the Crater: E6013 - shorten the arc, reverse direction, pull out from rear of pool; E6011, raise electrode slowly or whip out with long arc and back into pool with short arc.

Questions

1. How does the accumulated heat caused by welding on a small mass of metal affect bead quality?

____

____

2. How does spacing of parts and/or edge preparation affect penetration?

____

____

3. Can you weld equally fast with three electrodes?

____

____

4. Why does it require more time for fillet welds than for butt welds?

____

____

5. What should the operator do to control distortion while welding this assembly?

____

____

____

LAB EX-9 - DRILL PRESS V-BLOCK & DRILL GAGE

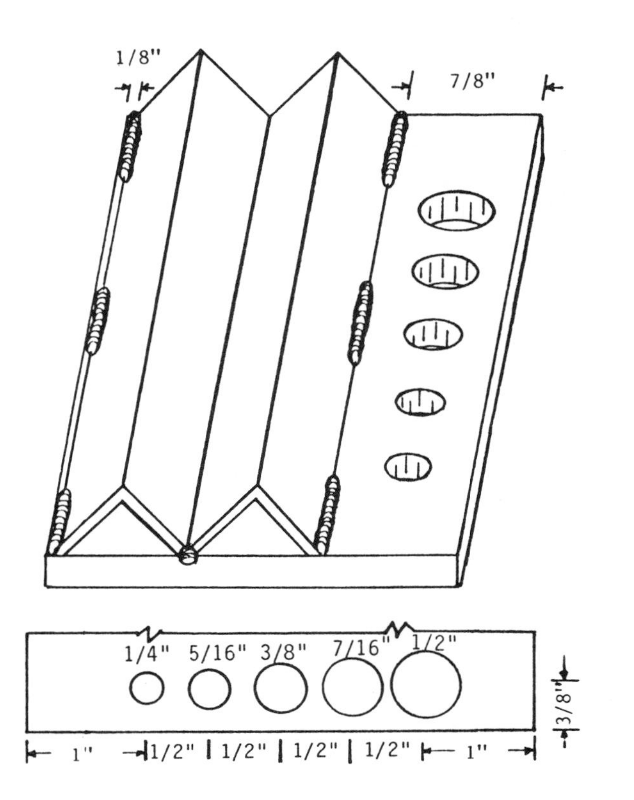

Materials :

1 - 1/4"x3"x4" steel plate
2 - 3/4"x3/4"x1/8" angle irons
1 - 1/8" E6011 or E6013 electrode
Drill bits 1/4", 5/16", 3/8", 7/16" and 1/2"
Layout tools, drill press, arc welder.
Safety equipment including eye and hand protection.

Operation Teaches : Ability to..

1. Select metal according to size.
2. Select arc welding electrodes.
3. Use layout tools.
4. Select drill bits by fractional sizes.
5. Complete quality arc welds.
6. Safely use drill press and related tools and equipment.

Construction Procedures :

1. Study the plan and select all materials needed for job.
2. Cut materials to size according to plan.
3. Layout base metal including positions for drill gage. Center punch drilling points.
4. Complete the arc welding phase of the job.
   a. Select electrode.
   b. Lay angle iron on base metal.
   c. Mark dimensions for angle iron on base.
   d. Tack angle irons in place, check layout.
   e. Complete arc welding using short string beads as shown on plan.
   f. Do not weld on inside of vee, use tack welds on each end of vee.
   g. Clean slag from welds and spatter from drill gage surface.
5. Check drill gage layout area - Do not drill into weld beads.
6. Select the 5 drill bits for the drill gage.
7. Secure V-block in drill press vise.
8. Drill holes in gage following all safety and operational procedures.
9. Remove burrs from bottom part of gage holes.
10. Check drill gage sizes with shank end of drill bits.
11. Clean all work areas and tools.
12. Submit completed activity to instructor for evaluation.

Evaluation Score Sheet:

	Item	Points Possible	Points Earned
1.	Layout of V-block angle irons	10	______
2.	Welds as specified on plan	10	______
3.	Quality of arc welds	15	______
4.	Layout of drill gage	15	______
5.	Correct drill sizes selected	10	______
6.	Drilling of gage holes	15	______
7.	Overall appearance	10	______
8.	Safety and work habits	15	______
	Total Points	100	______

Name ______________________________

Date: ____________ Grade: ______________

The repair of cast iron is a challenge to any weldor because the shapes of casting are seemingly endless. Cast iron is low in tensile strength therefore the expansion and contraction forces in arc welding must be carefully studied. Experienced weldors say that there are two ways to weld cast iron. One they call the "hot" method. The other the "cold" method. The hot method requires that the entire casting be heated dull red in a slow gas fired oven. The weld is then made and the part is cooled slowly. The cold method carefully avoids overheating the casting by welding short beads by back-stepping followed by peening. Backstepping makes a weldment from left to right however each short bead is from right to left. Peening is the mechanical stretching of the weld metal by striking the bead light blows with a pointed hammer as the bead cools. This stretching counteracts the shrinkage forces of cooling.

The preferred electrode for arc welding is the ENi (Nickel), color coded orange end, blue spot and white group. This electrode is easy to use. A soft machineable bead that seldom develops cracks at the bead boundaries can be made by careful application.

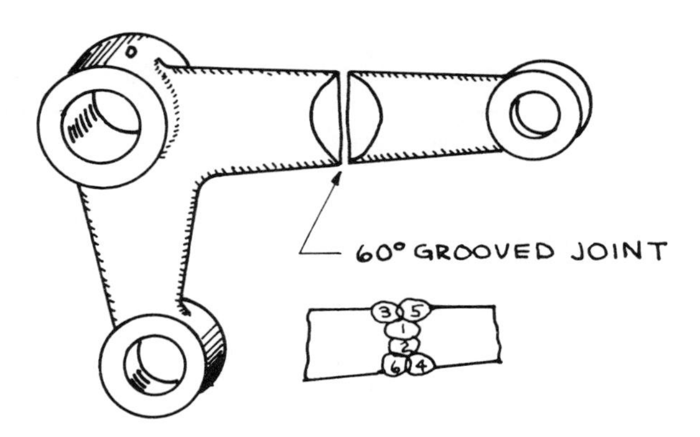

Figure 1. Edge Preparation and Weld Sequence.

Careful edge preparation as shown in Figure 1 is extremely important. A 60 - 90° angle leaving a 1/16" land for aligning the parts is necessary. Choose a 1/8" nickel electrode and set the welder for about 100 amps. Run a practice bead on scrap, holding a slightly long arc. The first passes should be stringer beads. The successive passes may weave slightly. Peening after each short bead is highly recommended.

Locked-in stress causes expansion and pressure-cracking when certain cast iron parts are heated during welding. Figure 2 shows how point heating at stressed areas will cause expansion thereby counteracting the expansion pressures at the weld site.

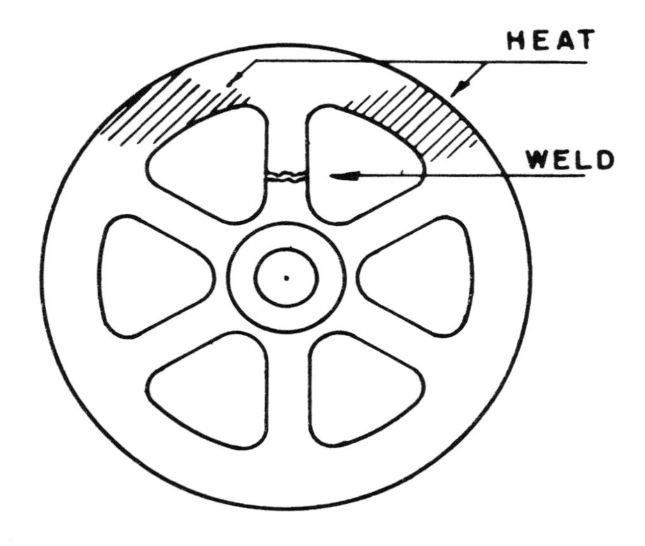

Figure 2. Point Heating Tends to Cancel Out Weld Pressures.

LAB EX-11 - HOW TO USE THE CARBON-ARC TORCH

The twin carbon-arc torch can extend the usefulness of an arc welder especially if any oxy-acetylene welder is unavailable. It can be used to heat metal for bending, loosen rusted-on nuts, apply hard surfacing powder or paste, braze weld light metal, and heat-treat metal.

The carbon arc torch, which sells for about $40.00 clamps 2 soft centered carbon copper coated electrodes in position. By hand adjustment the electrodes are brought together, then separated about 1/8" as shown in Figure 2. Amperage settings vary with the size of electrode as follows:

1/4" - 25-50
5/16" - 35-70
3/8" - 40-90
1/2" - 75-140

High amperages that tend to cause a noisy flame and overheat the electrodes more than 3/4" from the burned end also cause excess burn-off. If the twin carbon-arc torch is used on a DC welder, the positive carbon must be one size larger than the negative carbon. The greatest disadvantage of the carbon arc torch is that while the arc flame temperature is very hot the heat cannot be blown into a recessed area. It is a pressureless flame.

Braze welding can be done by using the carbon arc torch as shown in Figure 1. The usual care should be taken to clean the metal mechanically before applying flux and tinning with the brass rod. Care should be taken to avoid overheating the brass bècause the resulting zinc oxide white fumes can cause illness.

The application of hard surfacing powder or paste on a wearing surface can extend the wearing life of the part to a great extent. The process is shown in Figure 3. Care should be taken to clean the surface to be hard surfaced as for braze welding. Avoid overheating thin edges. Grounding the lead from one carbon to the work may help control heat in the application. Figure 3 shows further applications of hard surfacing on agricultural tools.

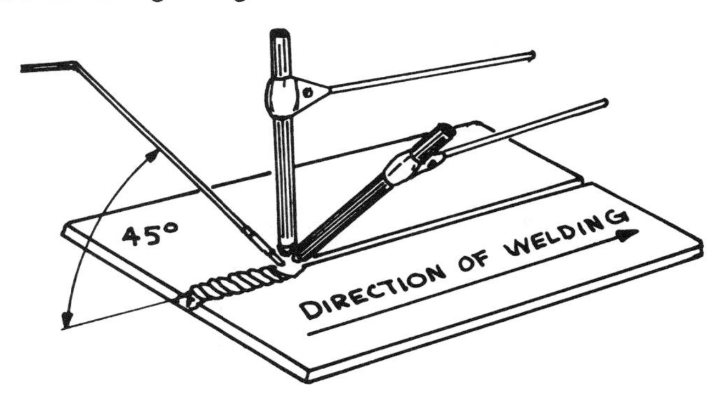

Figure 1. Braze Welding With the Carbon Arc Torch.

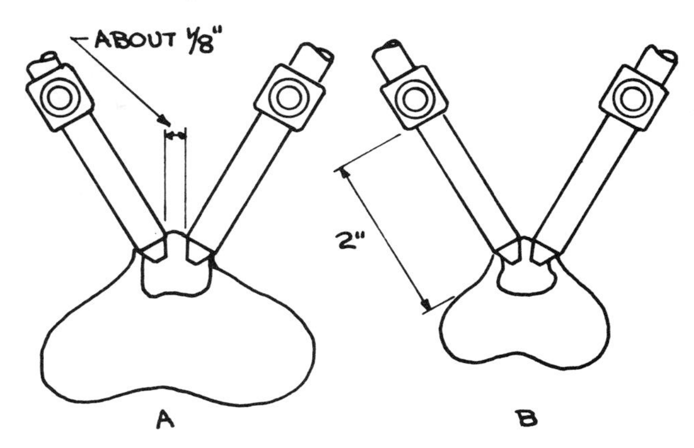

Figure 2. Positioning the Electrodes.

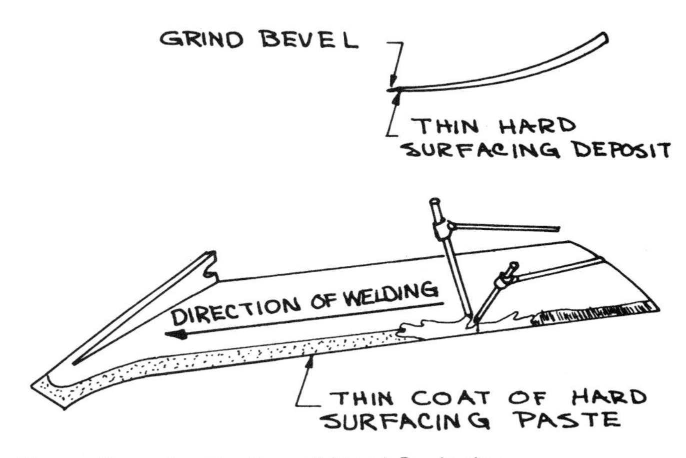

Figure 3. Application of Hard Surfacing.

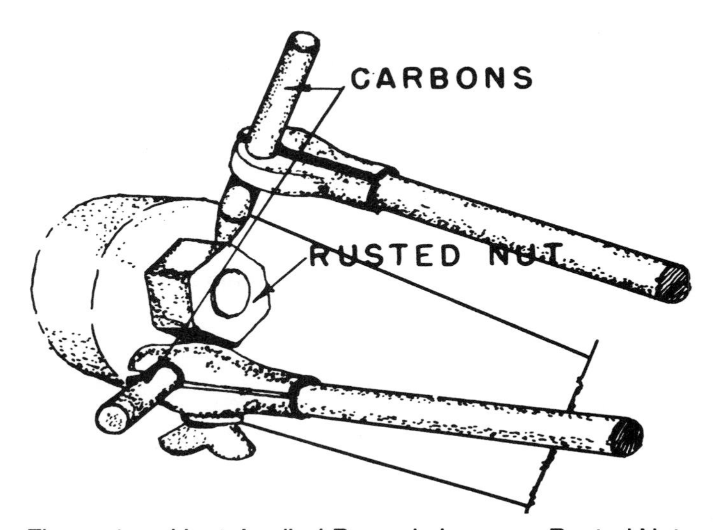

Figure 4. Heat Applied Properly Loosens Rusted Nut.

Probably one of the most successful uses for the twin carbon arc torch is shown in Figure 4. Short circuiting the carbons through a rusted, tight nut causes expansion of the nut allowing for easy removal.

HANDY HELPERS IN ARC WELDING

There are many repair jobs that require support of parts for alignment, support for weld metal for more accurate placement, and simple ways to make the job easier and more successful. Figure 1 shows the use of solid graphite (carbon) in various applications as back-up plates. Paste made from water mixed with graphite or powdered asbestos makes difficult welding jobs simple if properly applied.

A container of sand can be used to support parts for welding. Use of an arc striking plate grounded to the welding circuit helps considerably in avoiding disturbing aligned parts while tack-welding. Copper strips used as back-up plates provide support for weld metal but also act as heat sinks to remove excess heat that might damage the base metal.

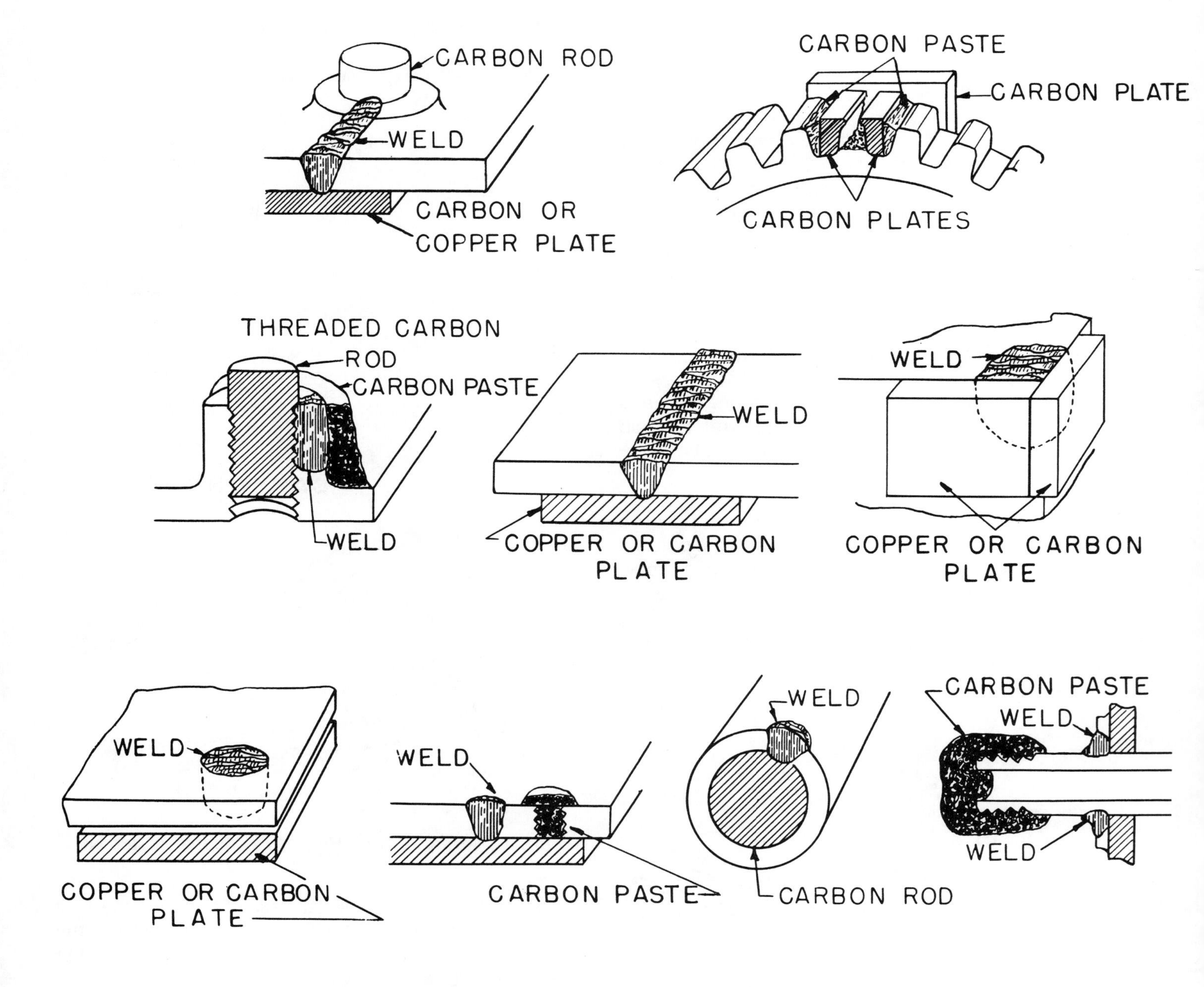

Figure 1. Paste Graphite, Solid Graphite and Copper Plate Help Make Welding Easier.

UNIT VII
CONSTRUCTION PROJECTS IN WELDING

CHIPPING HAMMER

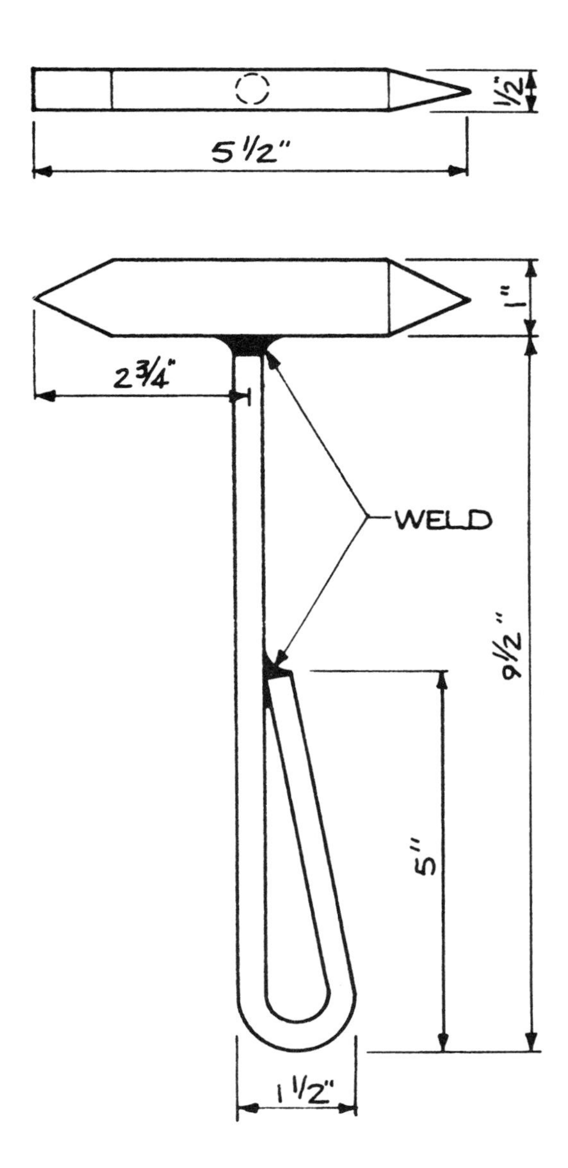

BILL OF MATERIAL

1 - 1/2" x 1" x 5 1/2" M.S.
1 - 3/8" x 15" H.R. round M.S.
(an old chisel or punch could be used for the head)

WELDED "C" CLAMP

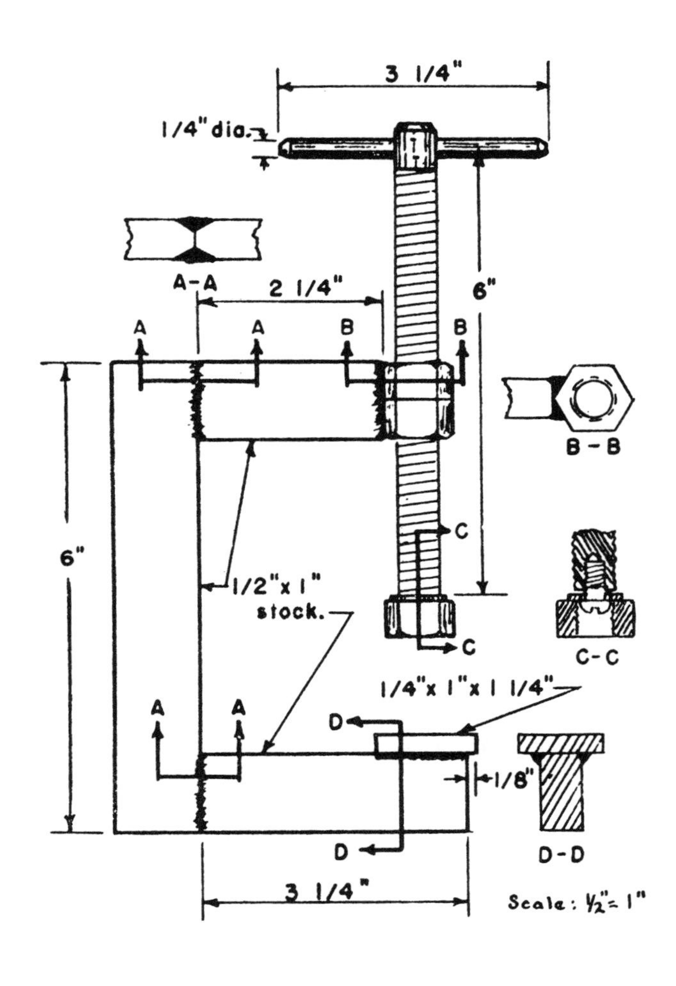

BILL OF MATERIAL

1 - 1/2" x 1" x 6" H.R. mild steel
1 - 1/2" x 1" x 2 1/4" H.R. mild steel
1 - 1/2" x 1" x 3 1/4" H.R. mild steel
1 - 1/4" x 1" x 1 1/4" H.R. mild steel
1 - 1/4" x 3 1/4" C.R. round
1 - 1/2" x 7" rod
3 - 1/2" nuts
1 - 3/8" washer
1 - #12 x 1/2" machine screw

TOOL HANGER

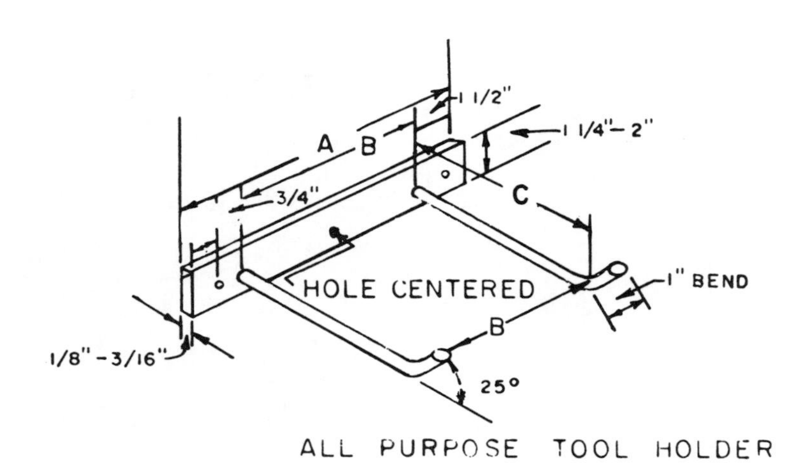

ALL PURPOSE TOOL HOLDER

DIMENSIONS

SIZE	A	B	C
SMALL	6"	3"	4"
MEDIUM	7"	4"	5"
LARGE	9"	6"	7"

ROD - 3/8" STEEL

BASE PLATE - MILD STEEL

CENTER SCREW HOLES TO WIDTH

SCREW HOLES: 3/16" - 1/4"

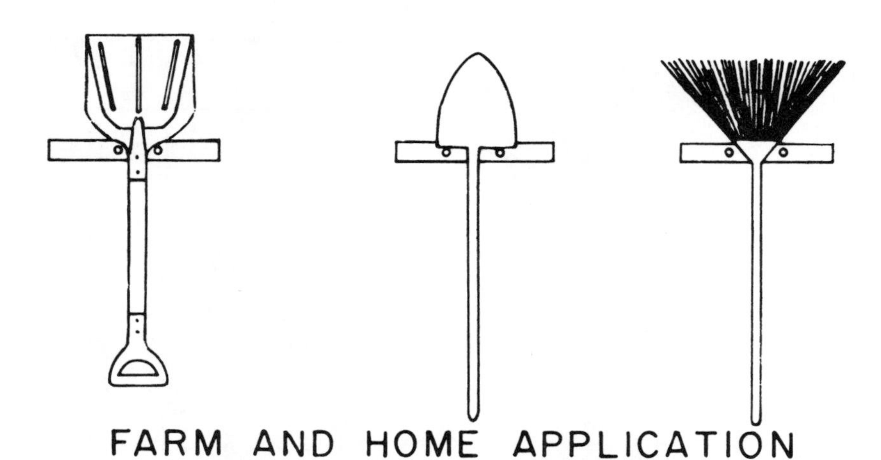

FARM AND HOME APPLICATION

SHOP BENCH STOOL

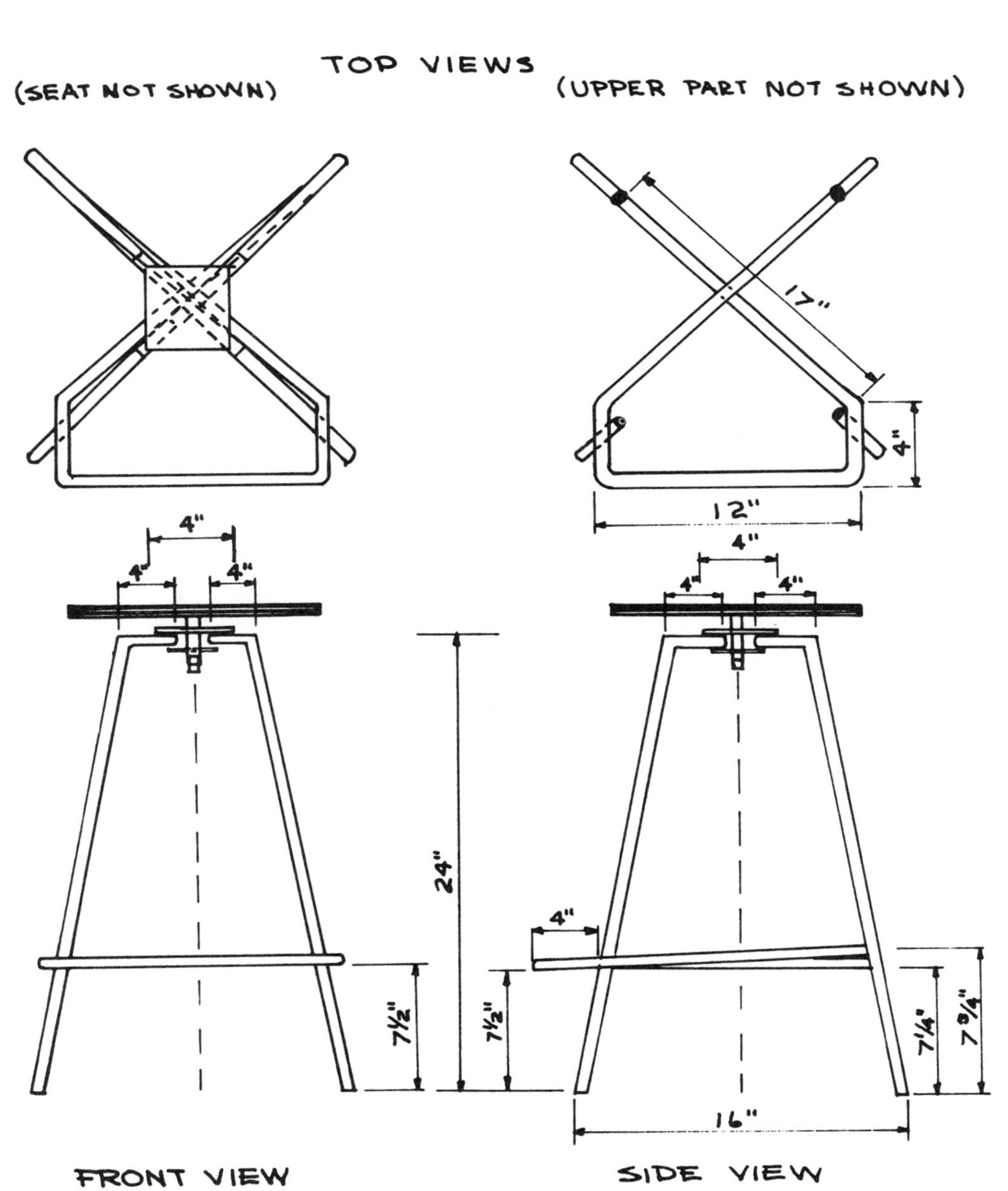

PORTABLE EXTENSION CORD

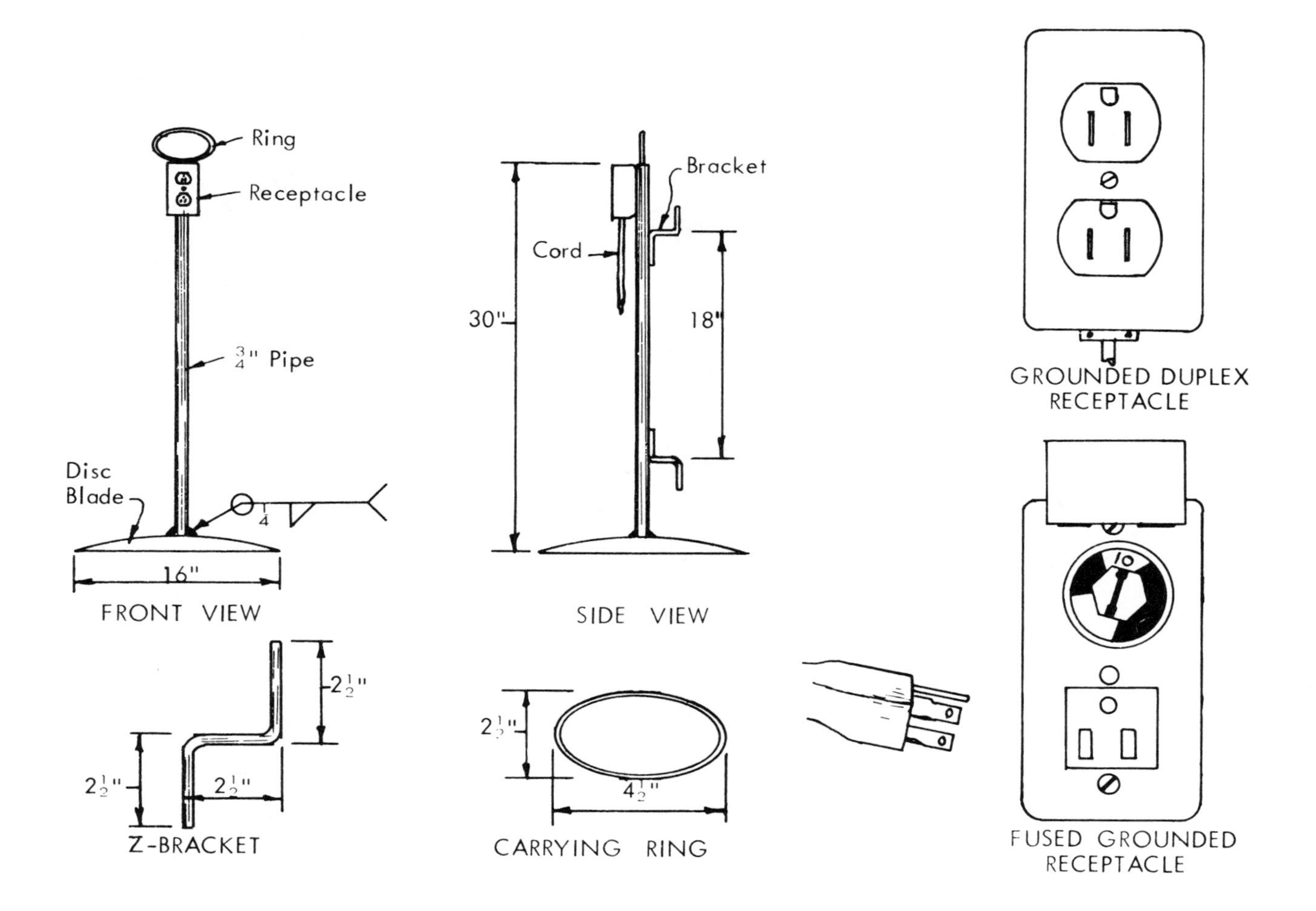

BILL OF MATERIAL

14" to 16" old disc blade
30"--3/4" pipe
4 1/2" wide by 2 1/2" oval carrying ring-5/16" rd.
2-z-shaped coiling brackets -- 2 1/2" x 2 1/2"--
5/16" rd.--Spaced 18" apart.

ELECTRICAL SUPPLIES

20' -25' No. 16, 3 wire rubber covered card. (10 amp--125v.)
OR
20' -25' No. 14, 3 wire rubber covered cord. (15 amp--125v.)
1 - Utility or handy box
1 - Box cable connector or clamp
1 - Grounded duplex outlet--receptacle with cover
OR
1 - Fused grounded receptacle
1 - 3 prong polorized plug

Weld disc blade to pipe with "Low Hydrogen" electrode-AWS-E7018. Use mild steel electrodes in other welding. Utility box may be spot welded to pipe or attached by boring and tapping pipe for 1/4" stove bolts.

OXY-ACETYLENE CART

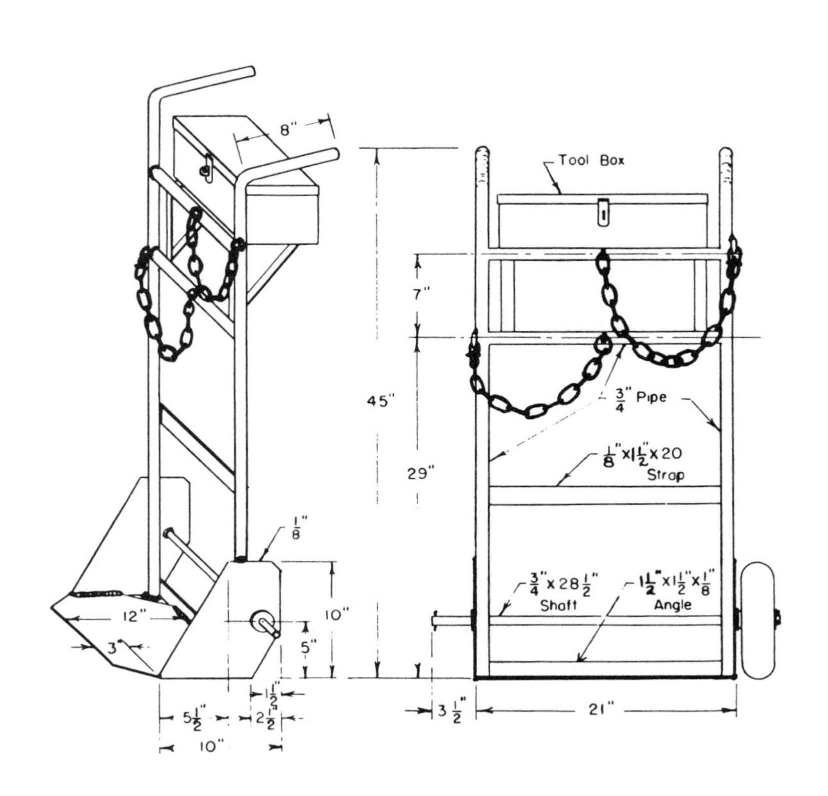

VISE STAND

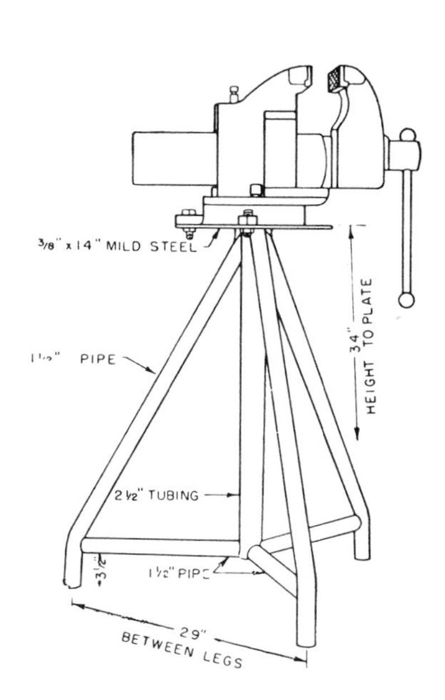

SHINGLE REMOVER

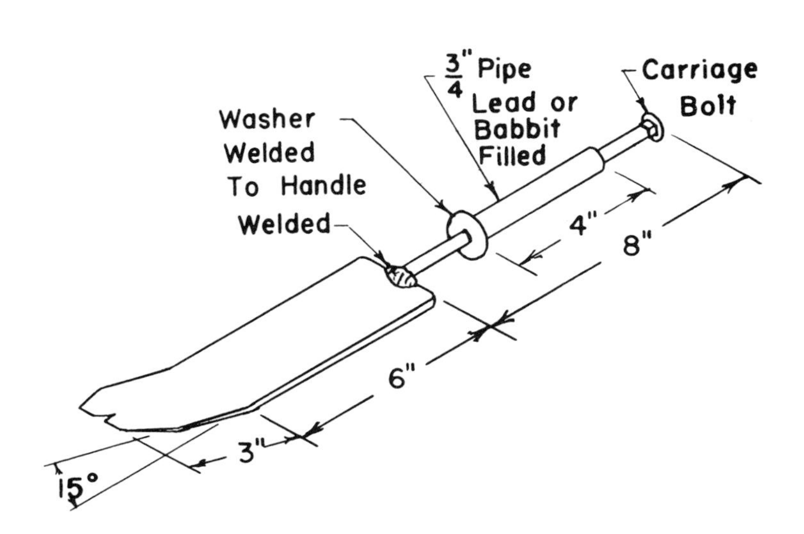

OXY-ACETYLENE WELDING CART

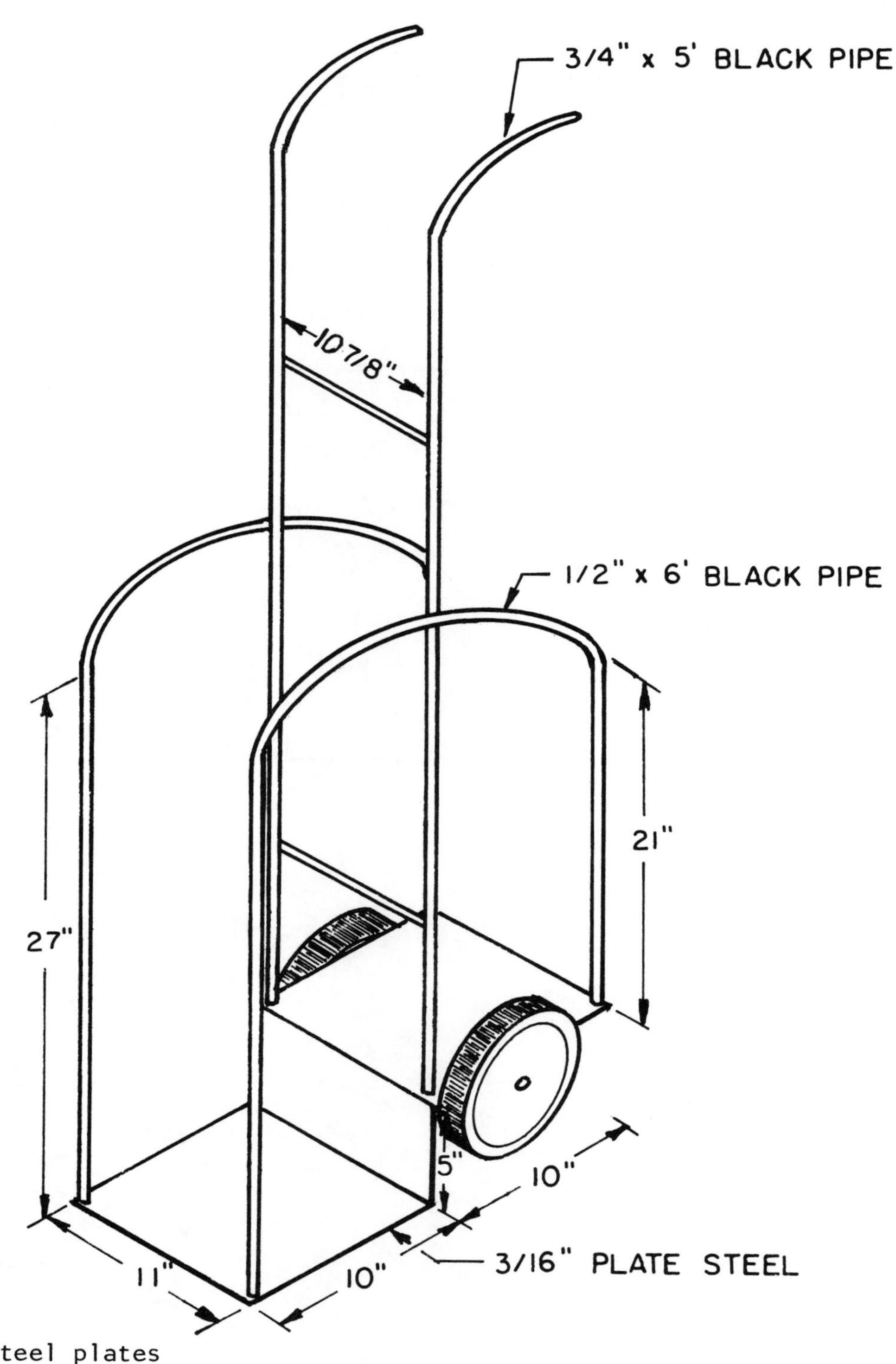

BILL OF MATERIAL

2 - 11" x 10" x 3/16" steel plates
1 - 11" x 5" x 3/16" steel plate
2 - 10" x 1" x 3/16" steel plates
2 - 3/4" x 5' black iron pipe
2 - 1/2" x 6' black iron pipe
4 - 3/8" x 1" carriage bolts
2 - 10" x 2 3/4" wheels, 1/2" or 5/8" arbor
1 - 18" x 1/2" or 5/8" axle
2 - Mounting brackets
4 - 2' chains

OXY-ACETYLENE CART

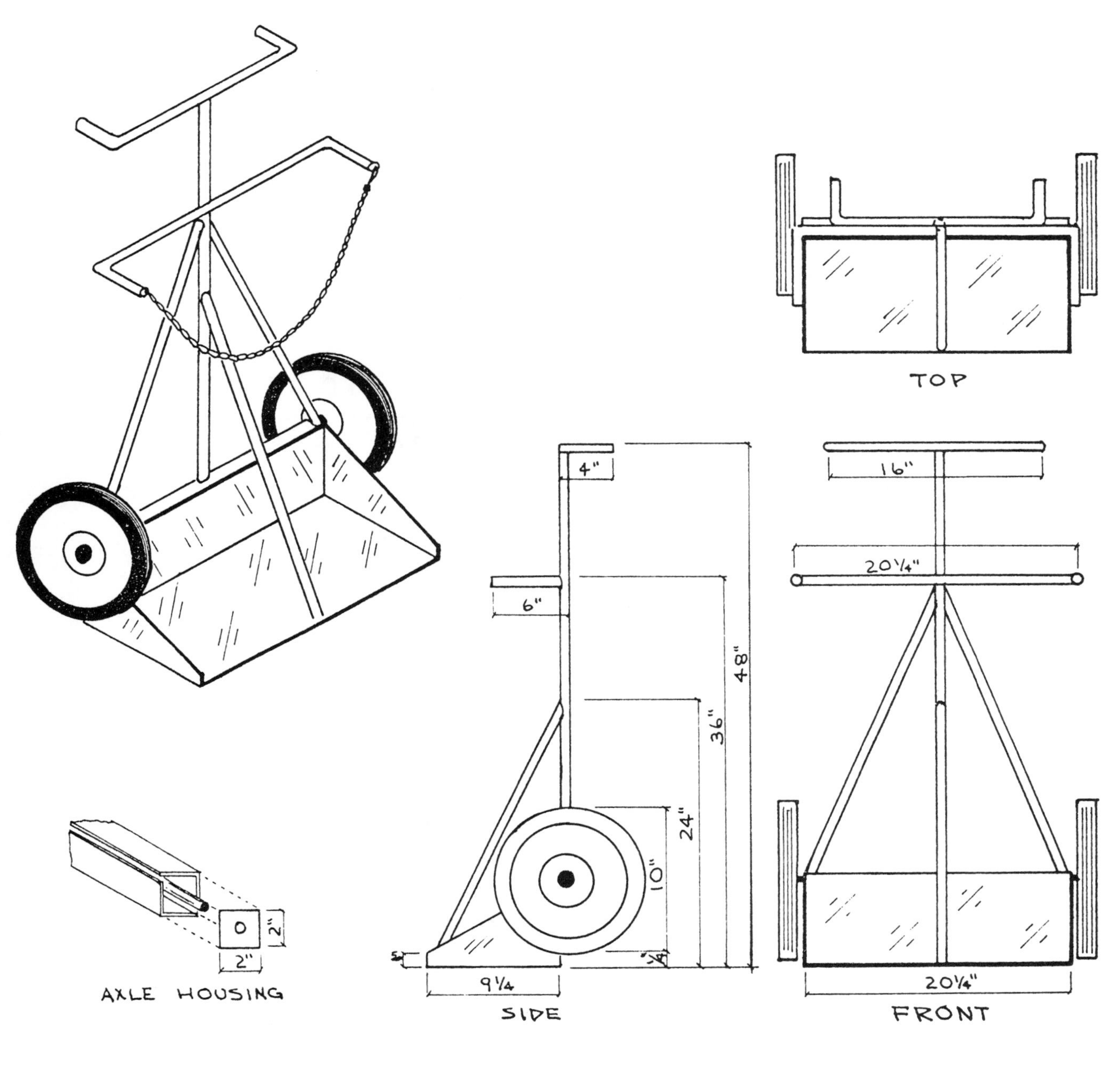

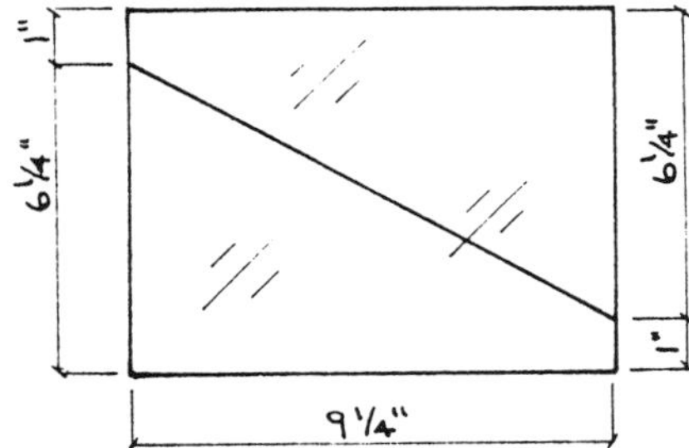

BILL OF MATERIAL

2 - 2" x 2" x 1/3" x 20 1/4" angle iron
1 - 9 1/4" x 20 1/4" x 1/4" steel plate
1 - 6 1/4" x 20 1/4" x 1/4" steel plate
1 - 7 1/4" x 9 1/4" x 1/4" steel plate
2 - 2" x 2" x 1/8" steel plates
1 - 1" x 40 3/4" black iron pipe
1 - 1" x 36" black iron pipe
2 - 1" x 21" black iron pipes
1 - 1" x 26" black iron pipe
1 - 1" x 36" black iron pipe
1 - 5/8" x 30" cold rolled steel
2 - 10" x 2 3/4" wheels, 5/8" arbor
1 - 30" chain for tank restraint

ADJUSTABLE HEIGHT WORK SUPPORT

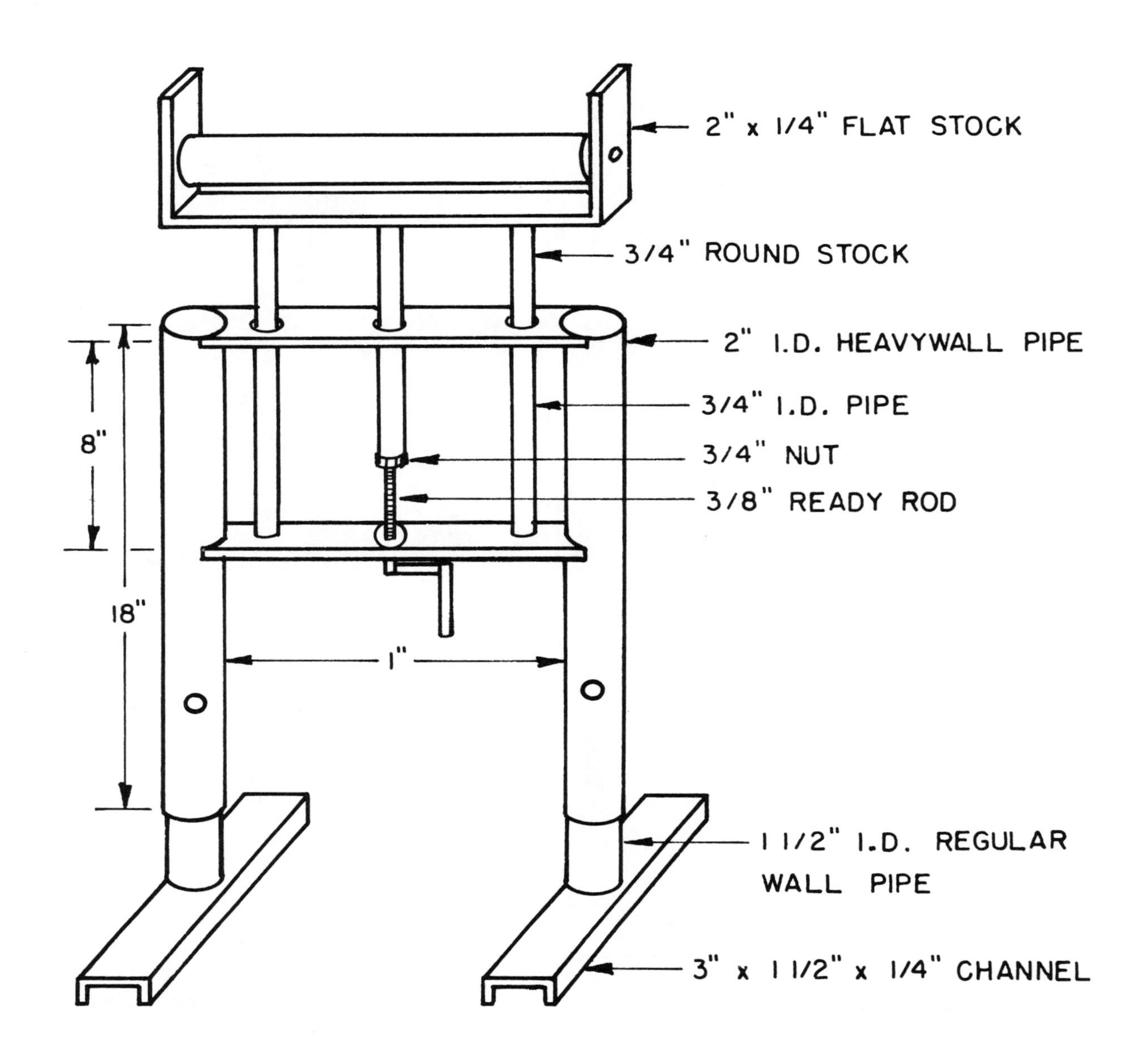

BILL OF MATERIAL

2 - 2" x 18" heavy wall pipe
2 - 1 1/2" x 18" regular wall pipe
1 - 1 1/2" x 12" heavy wall pipe
3 - 3/4" x 8" regular wall pipe
2 - 3/4" x 8" round stock
1 - 1/4" x 2" x 20" flat stock
2 - 1/4" x 2" x 14" flat stock
2 - 1 1/2" x 3" x 1/4" x 18" channel iron
1 - 3/4" x 18" ready rod
1 - 3/4" nut
1 - 3/4" washer
2 - 3/8" x 5" round stock

UTILITY CART

SIDE VIEW

FRONT VIEW

BILL OF MATERIAL

2 - 6" x 1.50 wheels for 1/2" axle
1 - 1/2" x 19-1/2" C.R. round for axle
1 - 3/16" x 9" x 14" plate for base
2 - 1/2" x 44" pipe for handles
1 - 3/16" x 3" x 3" plate, cut for front braces
2 - 3/16" x 4" x 9" plate, cut for rear braces
4 - 3/ 16" x 1" x 14" flat, handle supports
4 - 1/2" washers
2 - 1/8" cotter pins
1 - pint each of metal primer and enamel

ADJUSTABLE SAFETY JACK

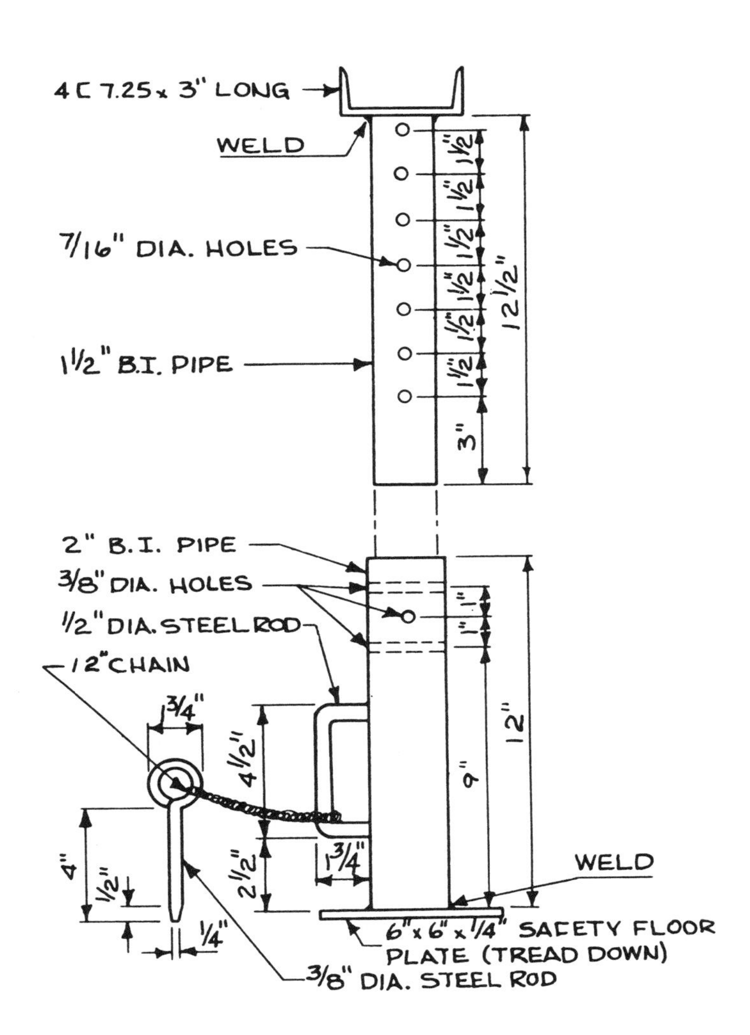

BILL OF MATERIAL

1 - 1/4" x 6" x 6" safety plate for base
1 - 2" x 12" standard black iron pipe, 2.375" O.D.
1 - 1 1/2" x 12 1/2" standard black pipe, 1.90" O.D.
1 - 3" structural channel 4" depth, 18" web thickness, 1.580" flange
1 - 1/2" x 8" mild steel rod for handle
1 - 3/8" x 8" mild steel rod for pin
1 - 12" chain
1 - pint desired color primer and paint

JACK STAND

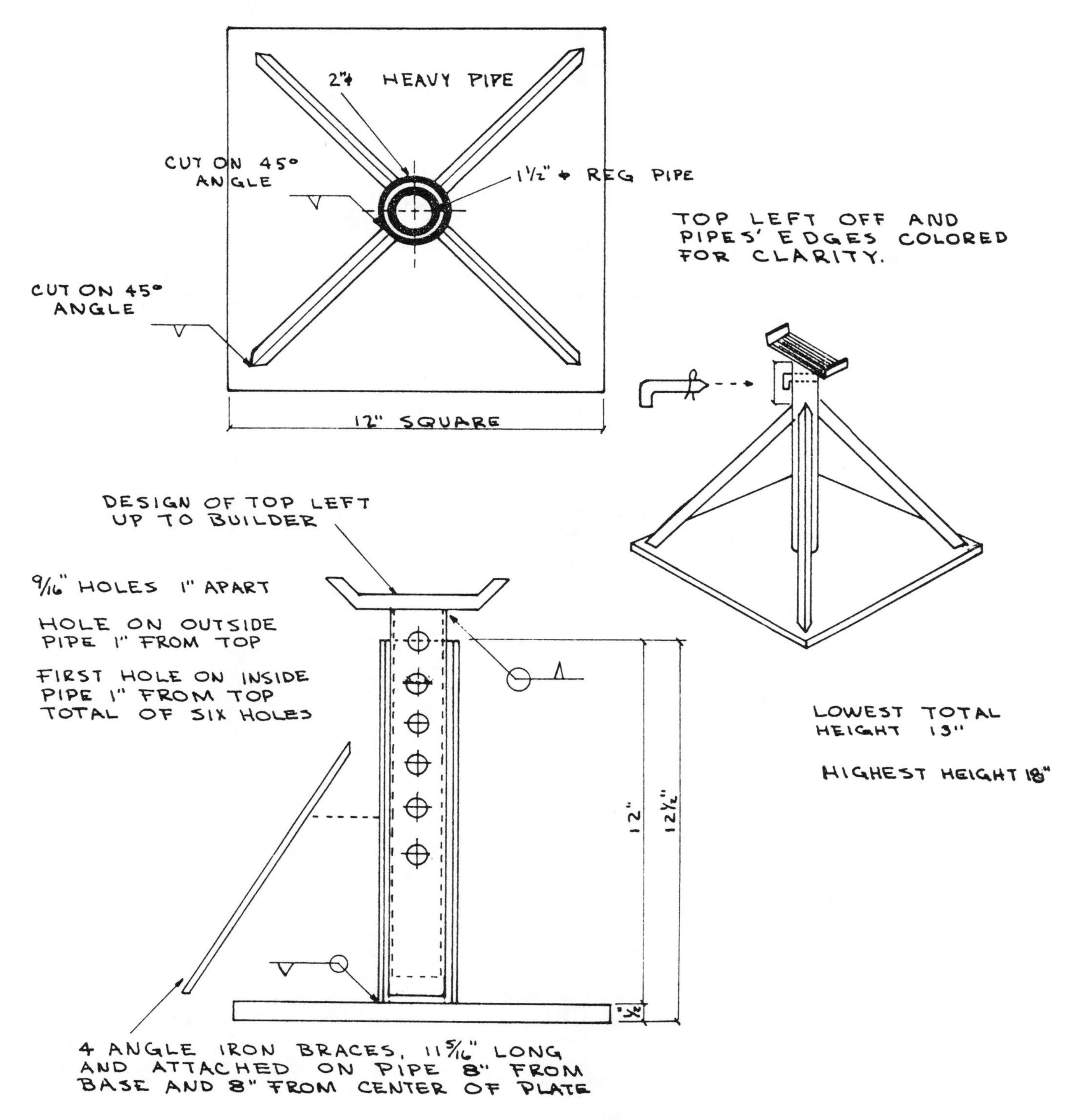

BILL OF MATERIAL

1 - 2" x 12 " black iron pipe
1 - 1 1/2" x 12" black iron pipe
4 - 1" x 1" x 3/16" x 11 5/16" angle iron
1 - 12" x 12" x 1/2" steel plate
1 - 3" x 6" x 1/2" steel plate
1 - 1/2" x 5" round stock

SAWHORSE

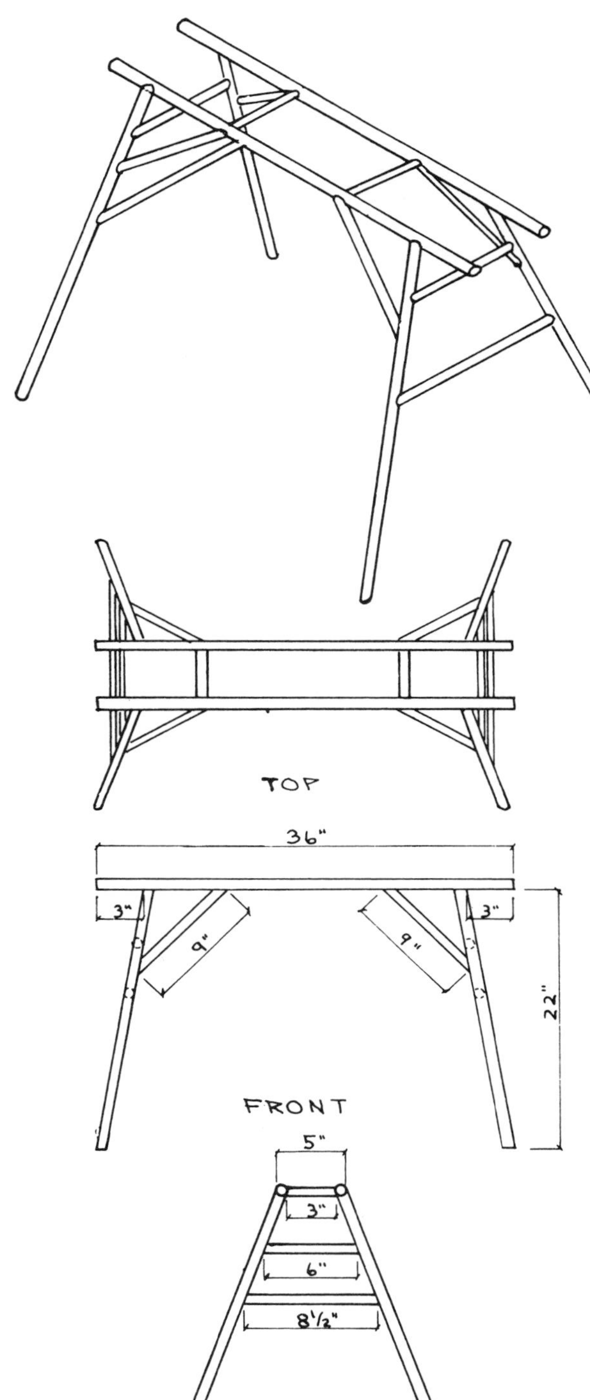

BILL OF MATERIAL

2 - 1'' x 36'' black iron pipes
4 - 1'' x 23 7/8'' black iron pipes
2 - 1'' x 3'' black iron pipes
2 - 1'' x 6'' black iron pipes
2 - 1'' x 8 1/2'' black iron pipes
4 - 1'' x 9'' black iron pipes

IMPLEMENT RAMP

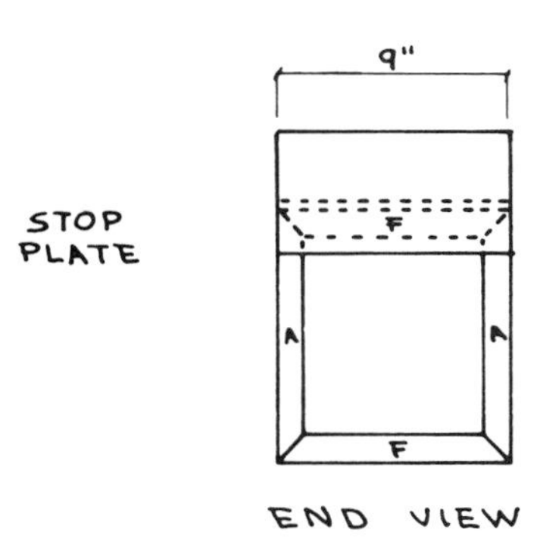

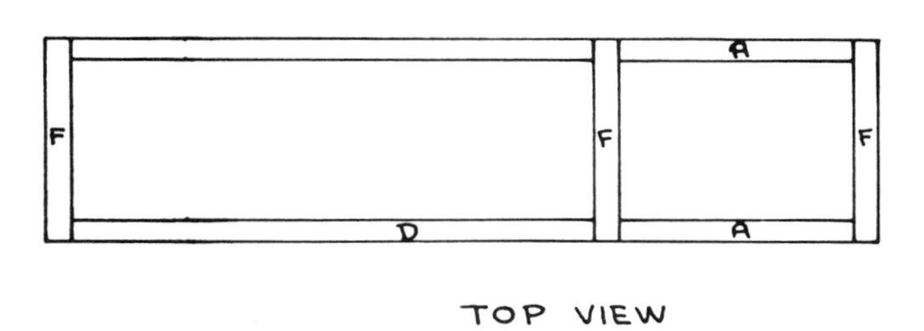

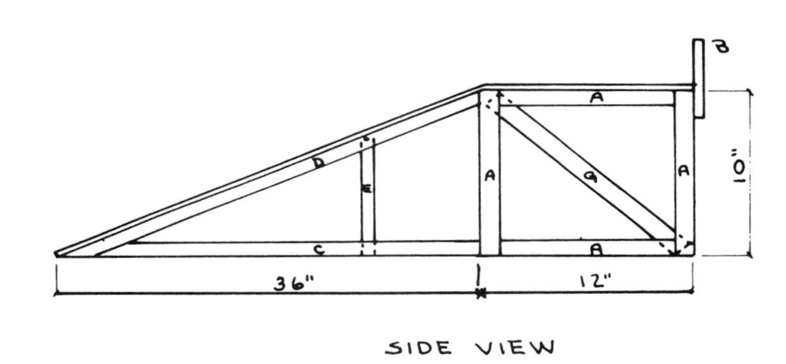

BILL OF MATERIAL

8 - 1'' x 1'' x 3/16'' x 10'' angle irons
4 - 1'' x 1'' x 3/16'' x 9'' angle irons
2 - 1'' x 1'' x 3/16'' x 14'' angle irons
2 - 1'' x 1'' x 3/16'' x 37 1/2'' angle irons
2 - 1'' x 1'' x 3/16'' x 33 3/4'' angle irons
2 - 1'' x 1'' x 3/16'' x 7'' angle irons
1 - 9'' x 12'' x 1/4'' steel plate
1 - 9'' x 37 1/2'' x 1/4'' steel plate
1 - 5'' x 9'' x 1/4'' steel plate

ENGINE HOIST

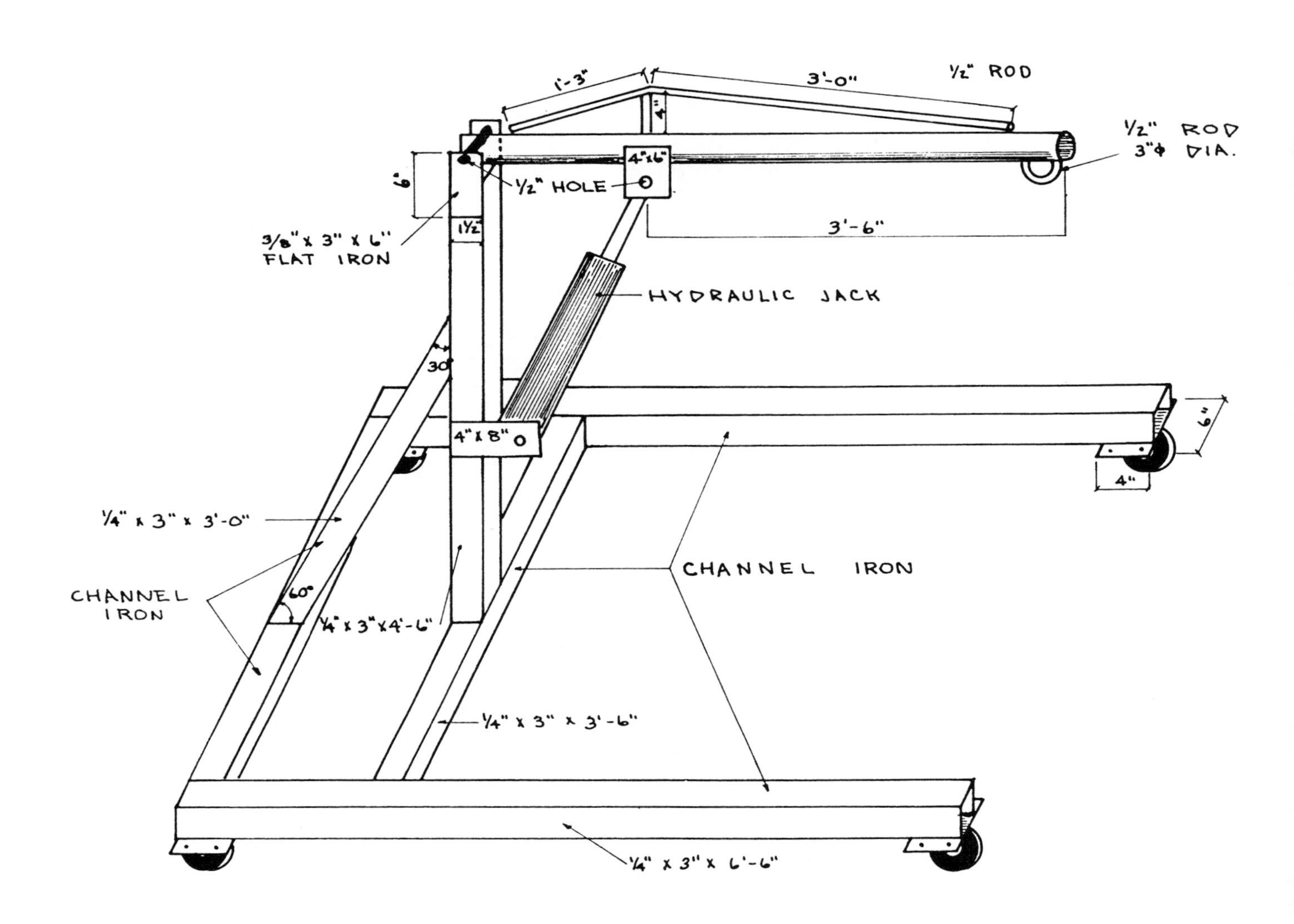

BILL OF MATERIAL

2 - 1 1/2" x 3" x 1/4" x 6 1/2' channel iron
2 - 1 1/2" x 3" x 1/4" x 3 1/2' channel iron
1 - 1 1/2" x 3" x 1/4" x 4 1/2' channel iron
1 - 1 1/2" x 3" x 1/4" x 3' channel iron
1 - 2 1/2" x 5' black iron pipe
4 - 4" x 6" x 1/4" steel plate
2 - 3" x 6" x 3/8" steel plate
2 - 4" x 6" x 1/2" steel plate
2 - 4" x 8" x 1/2" steel plate
1 - 1/2" x 4'-3" round stock
1 - 1/2" x 4" round stock
1 - 1/2" x 6 1/4" round stock
4 - 4" (2 stationary & 2 swivel) hvy. duty casters

COMBINATION WELDING TABLE

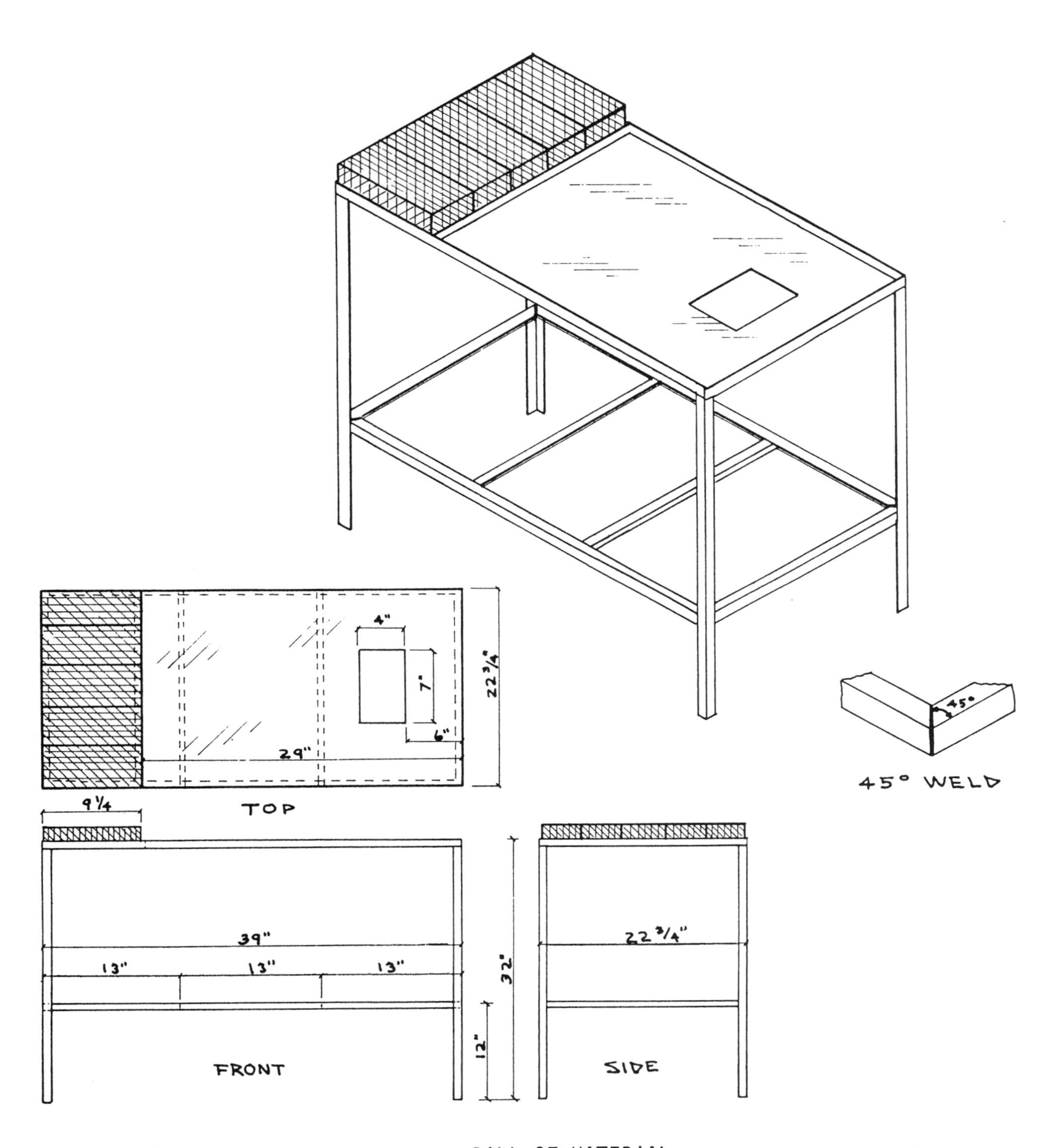

BILL OF MATERIAL

4 - 1 1/2" x 1 1/2" x 3/16" x 31" angle iron
2 - 1" x 1" x 3/16" x 39" angle iron
2 - 1" x 1" x 3/16" x 22 3/4" angle iron
2 - 1" x 1" x 3/16" x 38 5/8" angle iron
2 - 1" x 1" x 3/16" x 22 3/8" angle iron
4 - 1" x 1" x 3/16" x 19 3/4" angle iron
1 - 22 3/8" x 28 5/8" x 3/8" steel plate

GARDEN TRAILER

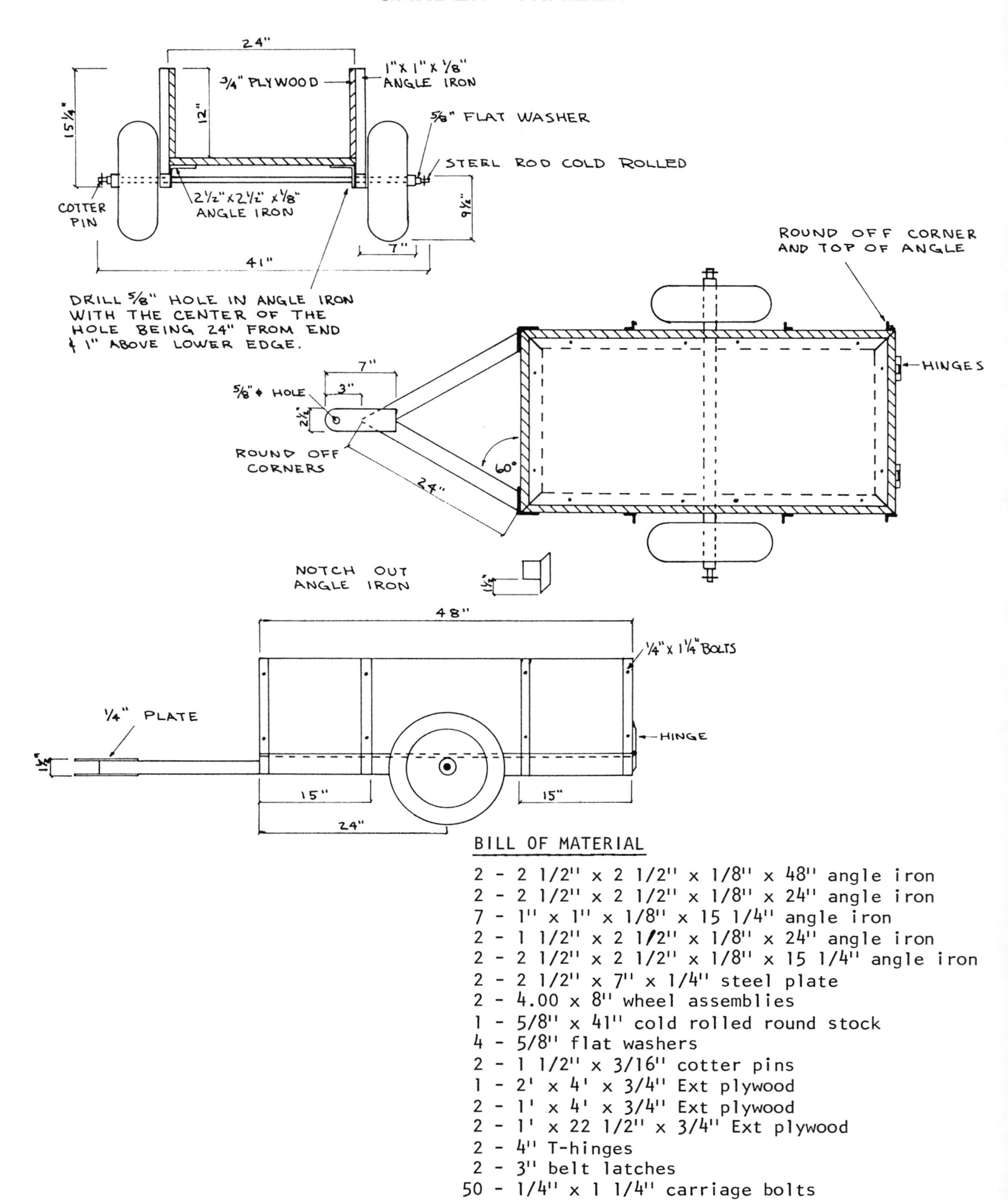

BILL OF MATERIAL

2 - 2 1/2" x 2 1/2" x 1/8" x 48" angle iron
2 - 2 1/2" x 2 1/2" x 1/8" x 24" angle iron
7 - 1" x 1" x 1/8" x 15 1/4" angle iron
2 - 1 1/2" x 2 1/2" x 1/8" x 24" angle iron
2 - 2 1/2" x 2 1/2" x 1/8" x 15 1/4" angle iron
2 - 2 1/2" x 7" x 1/4" steel plate
2 - 4.00 x 8" wheel assemblies
1 - 5/8" x 41" cold rolled round stock
4 - 5/8" flat washers
2 - 1 1/2" x 3/16" cotter pins
1 - 2' x 4' x 3/4" Ext plywood
2 - 1' x 4' x 3/4" Ext plywood
2 - 1' x 22 1/2" x 3/4" Ext plywood
2 - 4" T-hinges
2 - 3" belt latches
50 - 1/4" x 1 1/4" carriage bolts

3 - POINT HITCH CARRIER

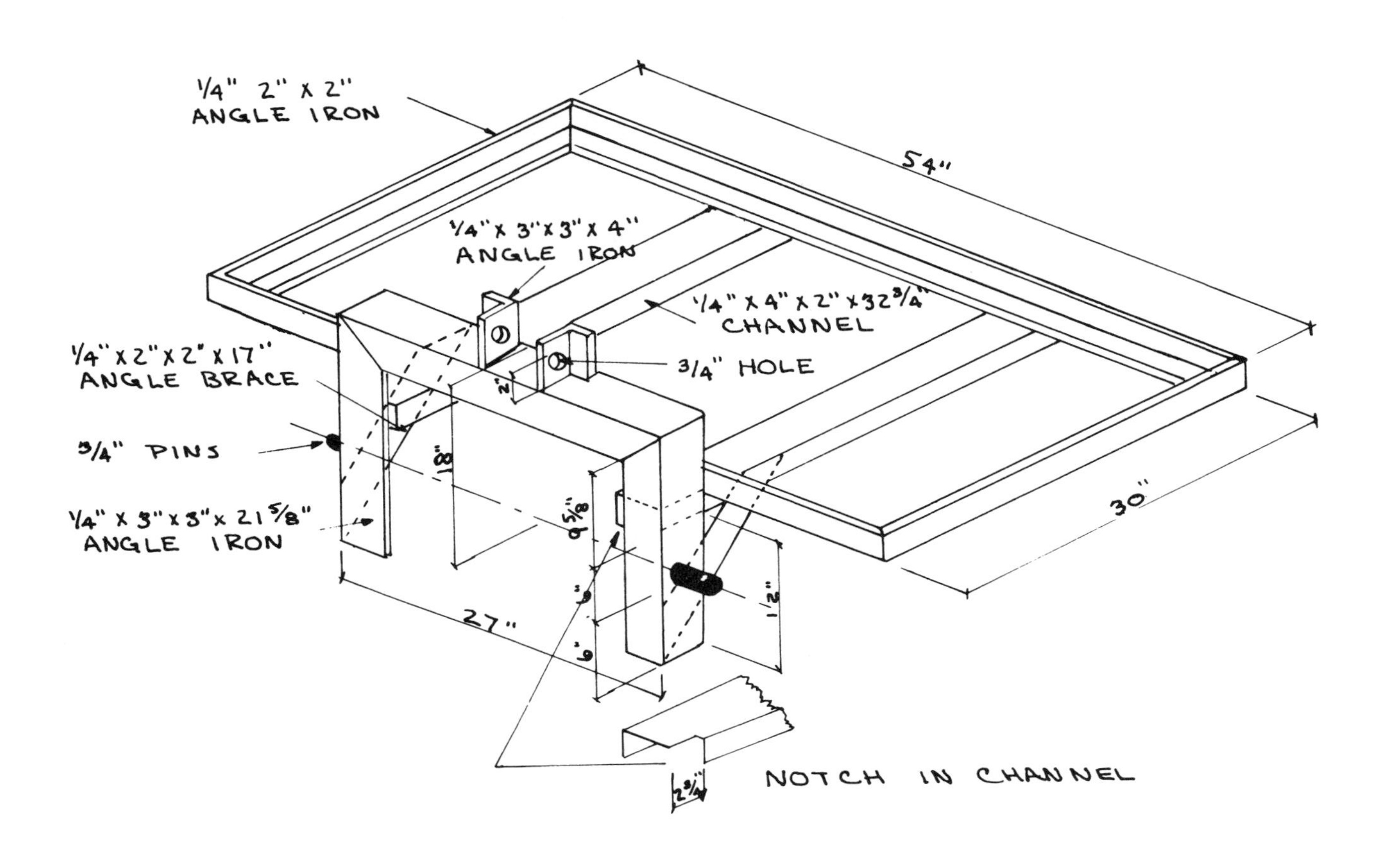

BILL OF MATERIAL

2 - 2'' x 2'' x 1/4'' x 54'' angle iron
2 - 2'' x 2'' x 1/4'' x 30'' angle iron
2 - 2'' x 4'' x 1/4'' x 32 3/4'' channel iron
1 - 3'' x 3'' x 1 1/4'' x 27'' angle iron
2 - 3'' x 3'' x 1/4'' x 21 5/8'' angle iron
2 - 2'' x 2'' x 1/4'' x 17'' angle iron
2 - 2'' x 2'' x 1/4'' x 4'' angle iron
2 - 3/4'' x 4'' cold rolled round stock

PICNIC TABLE

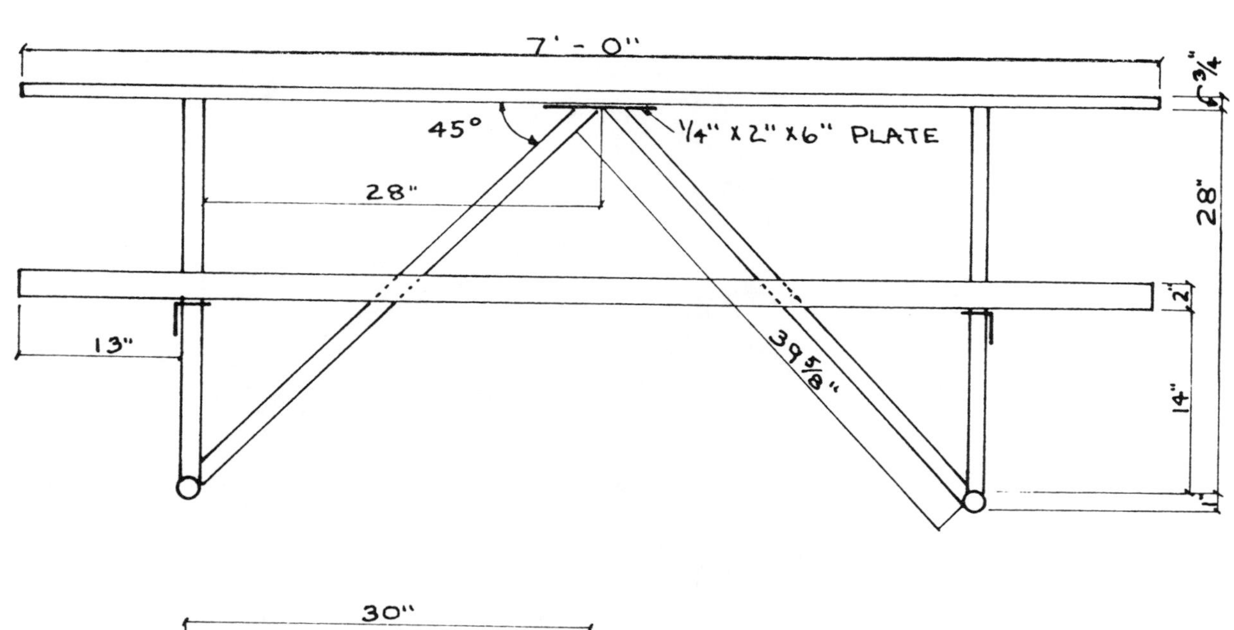

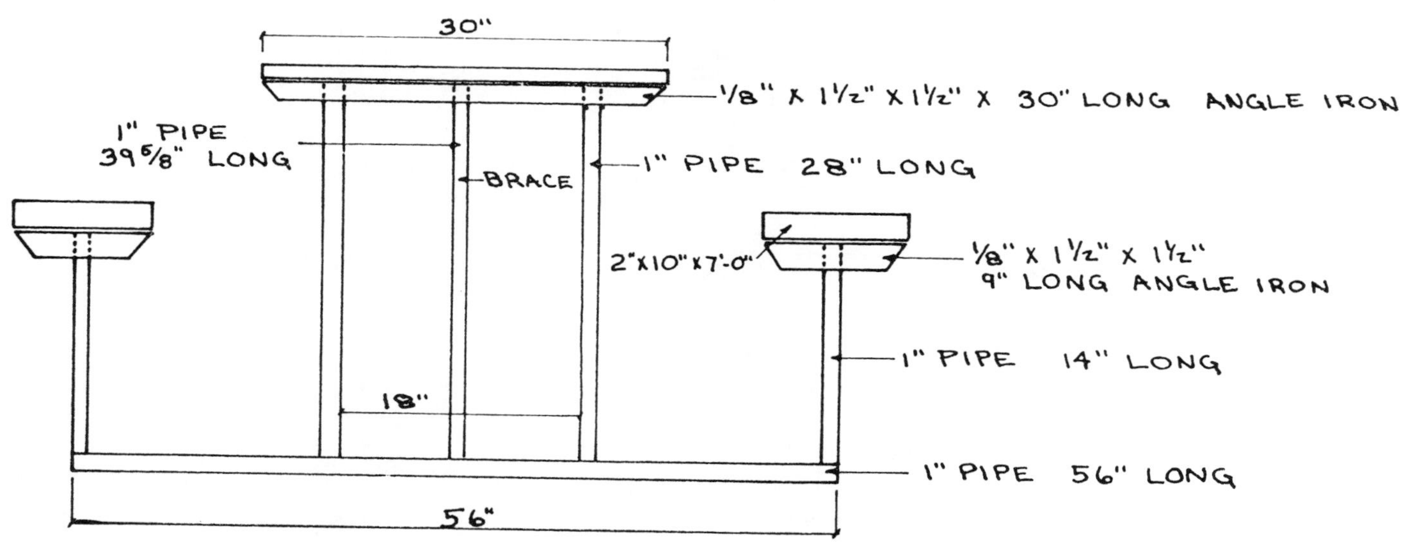

BILL OF MATERIAL

2 - 1" x 56" black iron pipe
4 - 1" x 14" black iron pipe
4 - 1" x 28" black iron pipe
2 - 1" x 39 5/8" black iron pipe
4 - 1 1/2" x 1 1/2" x 1/8" x 9" angle iron
2 - 1 1/2" x 1 1/2" x 1/8" x 30" angle iron
1 - 2" x 6" x 1/4" steel plate
1 - 30" x 7'-0" x 3/4" Ext plywood
2 - 2" x 10" x 7' constr. grade

SHOP WORK BENCH

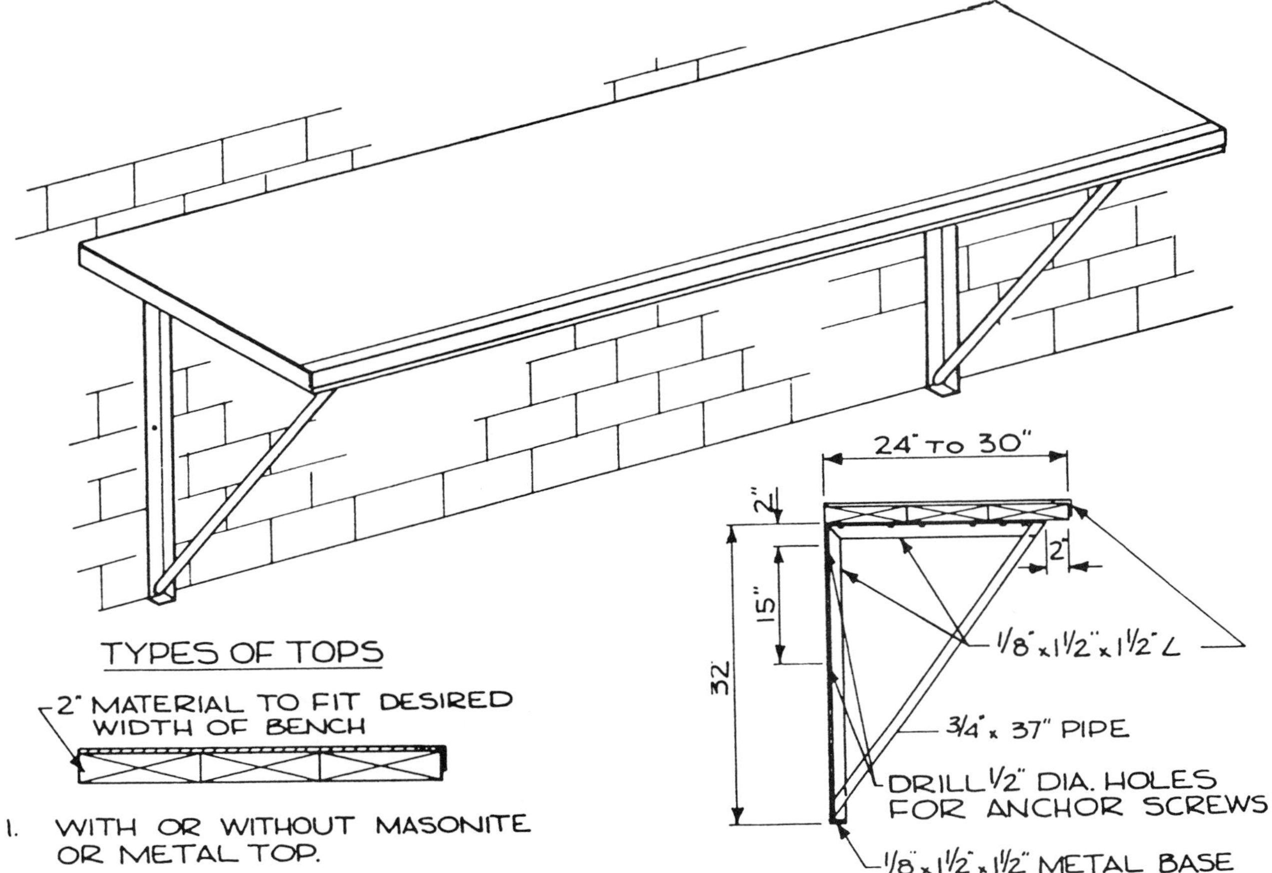

TYPES OF TOPS

2" MATERIAL TO FIT DESIRED WIDTH OF BENCH

1. WITH OR WITHOUT MASONITE OR METAL TOP.

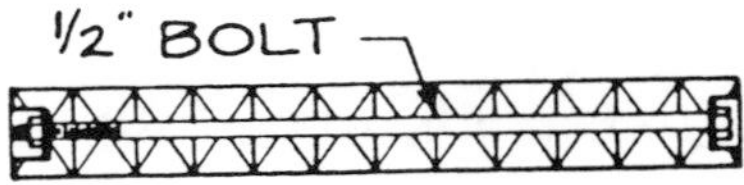

2. 2" x 6" MATERIAL RIPPED AND BOLTED TOGETHER.

3. SHIPLAP OR 2 x MATERIAL WITH OAK FLOORING TOP.

WATERPROOF OR WATER RESISTANT WOOD ADHESIVE TO BE USED FOR TOP ASSEMBLY.

NOTE:

SPACE WELDED SUPPORT BRACES 6'-0" o.c. ALONG BENCH WALL.

FASTEN BENCH BOARDS TO SUPPORTS WITH 1/4" x 1" LAGSCREWS.

FASTEN TOP & ∠ EDGE TO BENCH BOARDS WITH 1" #10 FLATHEAD WOOD SCREWS.

TYPES OF WALL FASTENINGS

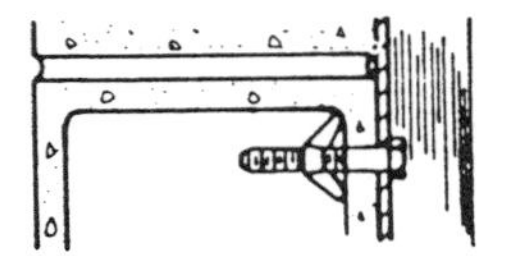

1. USE TOGGLE BOLT INTO HOLLOW TILE WALL.

2. USE LAGSCREW INTO EXPANSION ANCHOR.

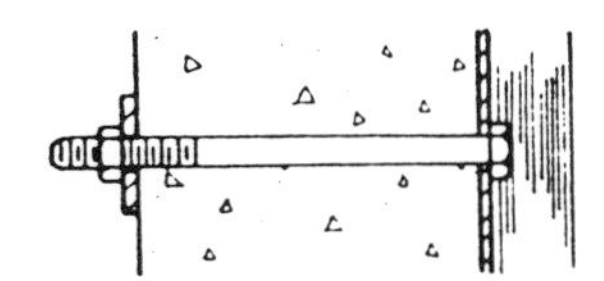

3. BOLT THROUGH WALL WITH METAL PLATE OUTSIDE.

ROLL-AWAY STORAGE UNIT

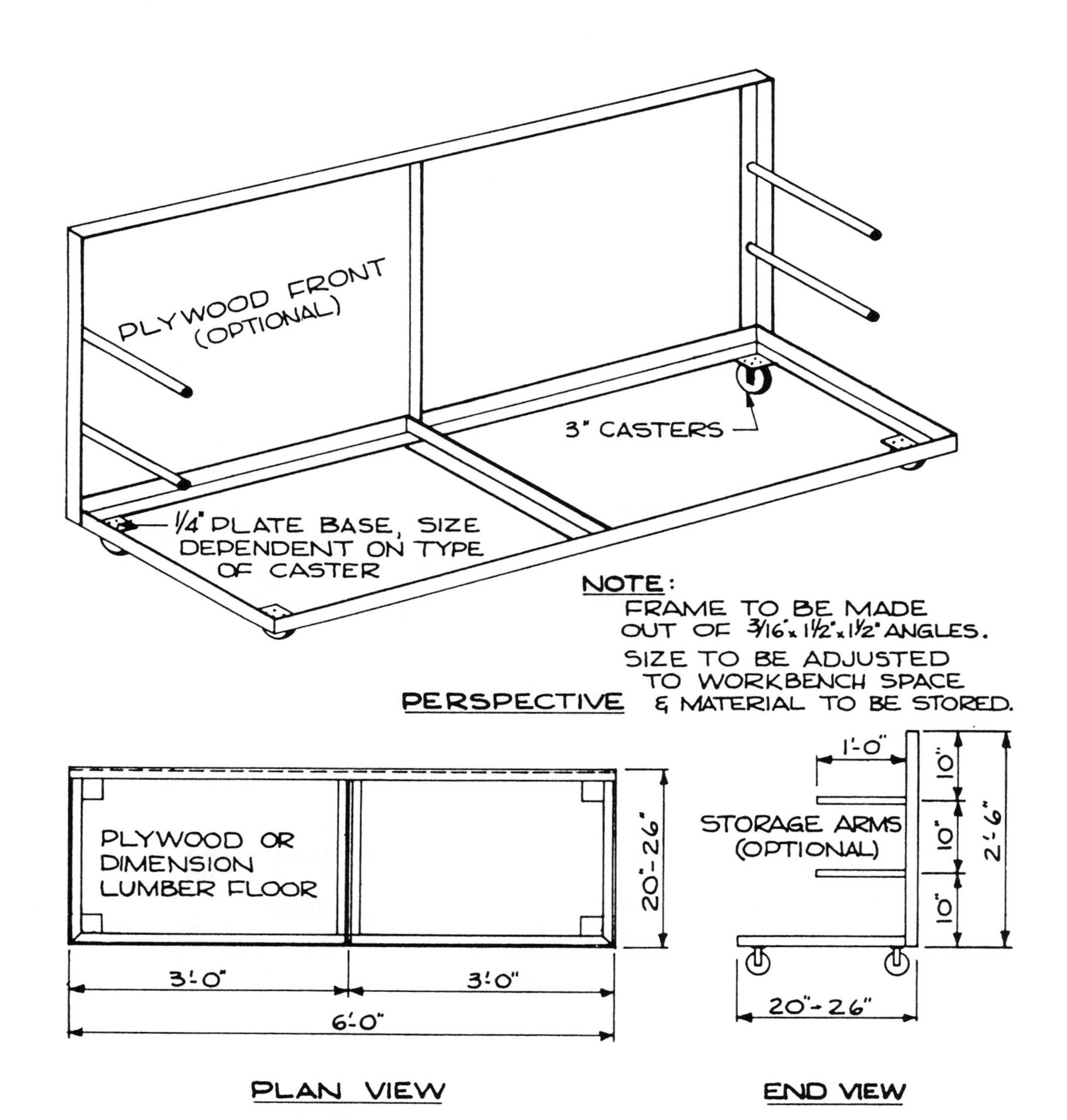

SHOP TOOL STORAGE CABINETS

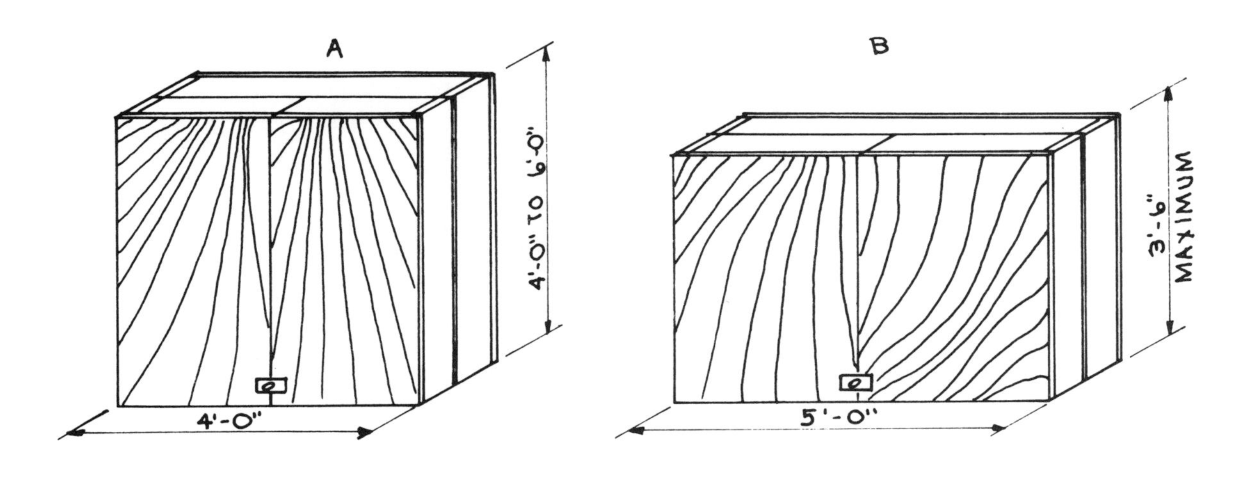

TWO TYPES OF STORAGE UNITS

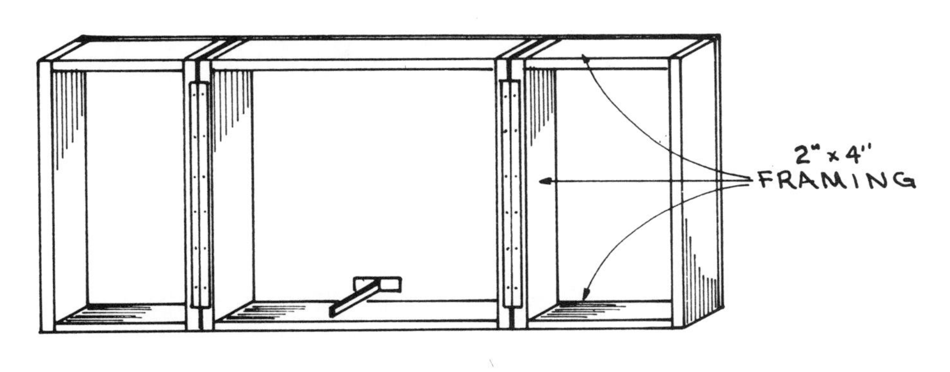

ISOMETRIC w/ DOORS OPEN

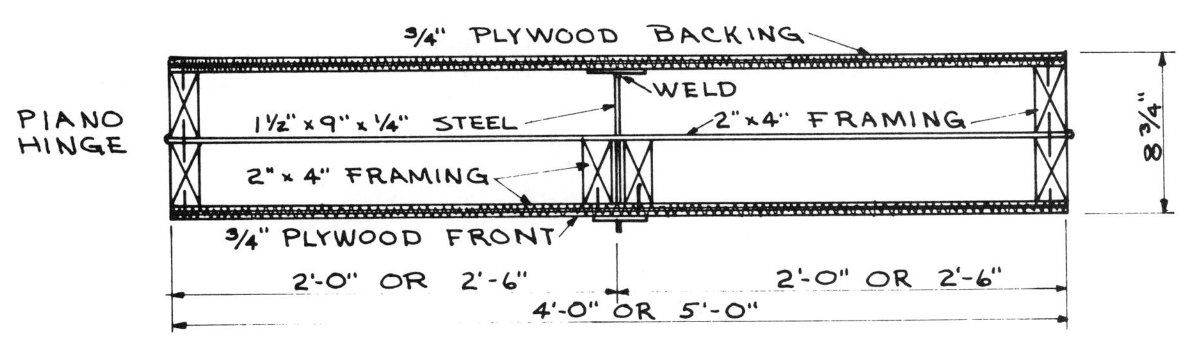

PLAN SECTION

Notes

APPENDIX

PROPERTIES OF METALS

	Melting Point Degrees Fahrenheit	Weight in Lbs. per Cubic Foot	Weight in Lbs. per Cubic Inch	Tensile Strength in Pounds per Square Inch
Aluminum	1140	166.5	.0963	15000 - 30000
Brass (average)	1500 - 1700	523.2	.3027	30000 - 45000
Copper	1930	552.	.3195	30000 - 40000
Iron, cast	1900 - 2200	450.	.2604	20000 - 35000
Iron, wrought	2700 - 2830	480.	.2779	35000 - 60000
Lead	618	709.7	.4106	1000 - 3000
Nickel	2800	548.7	.3175	
Silver (pure)	1800	655.1	.3791	40000
Steel	2370 - 2685	489.6	.2834	50000 - 120000
Tin	475	458.3	.2652	5000
Zinc	780	436.5	.2526	3500

FRACTION AND DECIMAL EQUIVALENTS

Fraction	Decimal	Fraction	Decimal
1/64...	.015625	33/64...	.515625
1/32..........	.03125	17/32...........	.53125
3/64...	.046875	35/64...	.546875
1/16..........	.0625	9/16...........	.5625
5/64...	.078125	37/64...	.578125
3/32..........	.09375	19/32...........	.59375
7/64...	.109375	39/64...	.609375
1/8................	.125	5/8.................	.625
9/64...	.140625	41/64...	.640625
3/32..........	.15625	21/32...........	.65625
11/64...	.203125	43/64...	.671875
3/16..........	.1875	11/16...........	.6875
13/64...	.203125	45/64...	.703125
7/32..........	.21875	23/32...........	.71875
15/64...	.234375	47/64...	.734375
1/4................	.25	3/4.................	.75
17/64...	.265625	49/64...	.765625
9/32..........	.28125	25/32...........	.78125
19/64...	.296875	51/64...	.796875
5/16..........	.3125	13/16...........	.8125
21/64...	.328125	53/64...	.828125
11/32..........	.34375	27/32...........	.84375
23/64...	.359375	55/64...	.859375
3/8................	.375	7/8.................	.875
25/64...	.390625	57/64...	.890625
13/32..........	.40625	29/32...........	.90625
27/64...	.421875	59/64...	.921875
7/16..........	.4375	15/16...........	.9375
29/64...	.453125	61/64...	.953125
15/32..........	.46875	31/32...........	.96875
31/64...	.484375	63/64...	.984375
1/2................	.5	1..................	1.000

BASIC THREAD DIMENSIONS AND TAP DRILL SIZES

AMERICAN NATIONAL COARSE THREAD (N.C.)

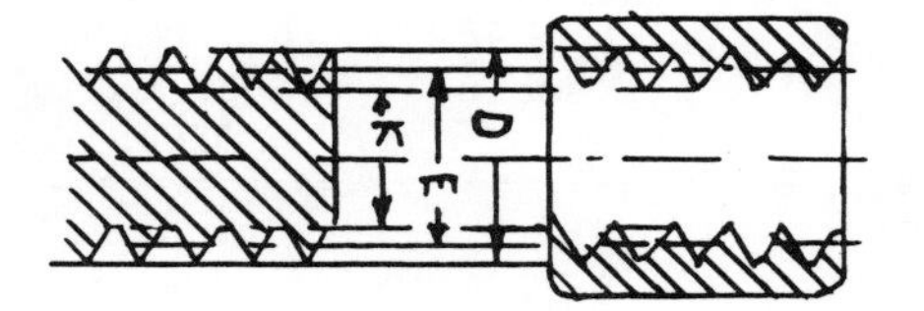

D. Major Dia.

E. Pitch Dia.

K. Minor Dia.

Size of Thread and Threads per inch	Major Diameter D Inches	Pitch Diameter E Inches	Minor Diameter K Inches	Commercial Tap Drill to Produce Approx. 75% Full Thread	Decimal Equivalent of Tap Drill Inches
1/4x20	.2500	.2175	.1850	No. 7	.2010
5/16x18	.3125	.2764	.2403	F	.2570
3/8x16	.3750	.3344	.2938	5/16	.3125
7/16x14	.4375	.3911	.3447	U	.3680
1/2x13	.5000	.4500	.4001	27/64	.4219
9/16x12	.5625	.5084	.4542	31/64	.4844
5/8x11	.6250	.5660	.5069	17/32	.5312
3/4x10	.7500	.6850	.6201	21/32	.6562
7/8x 9	.8750	.8028	.7307	49/64	.7656
1x 8	1.0000	.9188	.8376	7/8	.8750

AMERICAN NATIONAL FINE THREAD (N.F.)

Size of Thread and Threads per inch	Major Diameter D Inches	Pitch Diameter E Inches	Minor Diameter K Inches	Commercial Tap Drill to Produce Approx. 75% Full Thread	Decimal Equivalent of Tap Drill Inches
1/4x28	.2500	.2268	.2036	No. 3	.2130
5/16x24	.3125	.2854	.2584	I	.2720
3/8x24	.3720	.3479	.3209	Q	.3320
7/16x20	.4375	.4050	.3725	25/64	.3906
1/2x20	.5000	.4675	.4350	29/64	.4531
9/16x18	.5625	.5264	.4903	33/64	.5156
5/8x18	.6250	.5889	.5528	37/64	.5781
3/4x16	.7500	.7094	.6688	11/16	.6875
7/8x14	.8750	.8286	.7822	13/16	.8125
1x14	1.0000	.9536	.9072	15/16	.9375

STANDARD WEIGHTS AND DIMENSIONS OF WELDED AND SEAMLESS WROUGHT-IRON PIPE (ASA B36.10-1959)

Size: Nom. & (Outside Diam.), In.	Wall Thickness, In.	Weight per Foot, Plain Ends, Lb.	Identification	
			Schedule No.	Other*
1/8 (0.405)	0.069 0.096	0.24 0.31	40 80	STD XS
1/4 (0.540)	0.090 0.124	0.42 0.54	40 80	STD XS
3/8 (0.675)	0.093 0.129	0.57 0.74	40 80	STD XS
1/2 (0.840)	0.111 0.151 0.307	0.85 1.09 1.71	40 80 --	STD XS XXS
3/4 (1.050)	0.115 0.157 0.318	1.13 1.47 2.44	40 80 --	STD XS XXS
1 (1.315)	0.136 0.183 0.369	1.68 2.17 3.66	40 80 --	STD XS XXS
1 1/4 (1.660)	0.143 0.196 0.393	2.27 3.00 5.21	40 80 --	STD XS XXS
1 1/2 (1.900)	0.148 0.204 0.411	2.72 3.63 6.41	40 80 --	STD XS XXS
2 (2.375)	0.157 0.223 0.447	3.65 5.02 9.03	40 80 --	STD XS XXS
2 1/2 (2.875)	0.207 0.282 0.567	5.79 7.66 13.70	40 80 --	STD XS XXS
3 (3.500)	0.221 0.306 0.615	7.58 10.25 18.58	40 80 --	STD XS XXS
3 1/2 (4.000)	0.231 0.325 0.651	9.10 12.51 22.85	40 80 --	STD XS XXS
4 (4.500)	0.242 0.344 0.690	10.79 14.98 27.54	40 80 --	STD XS XXS
5 (5.563	0.263 0.383 0.768	14.61 20.78 38.55	40 80 --	STD XS XXS
6 (6.625)	0.286 0.441 0.884	18.97 28.57 53.16	40 80 --	STD XS XXS
8 (8.625)	0.283 0.329 0.511 0.895	24.70 28.55 43.39 72.42	30 40 80 --	--- --- XS XXS
10 (10.750)	0.313 0.372 0.510	34.24 40.48 54.74	30 40 60	--- --- XS
12 (12.750)	0.336 0.510	43.77 65.42	30 --	--- XS
14 (14.000)	0.383 0.510	54.57 72.09	30 --	STD XS
16 (16.000)	0.383 0.510	62.58 82.77	30 40	STD XS
18 (18.000)	0.383 0.510	70.59 93.45	-- --	STD XS
20 (20.000)	0.383 0.510	78.60 104.13	20 30	STD XS
24 (24.000)	0.383 0.510	94.62 125.49	20 --	STD XS

*The abbreviations in this column are: STD, standard; XS, extra strong; and XXS, double extra strong.

All sizes shown except 3 1/2" XXS are American Petroleum Institute (API) standard and bear designation 5L. Not shown are 10"--.284" wall thickness and 12"--.383" wall thickness API sizes.

WEIGHT OF STEEL BARS

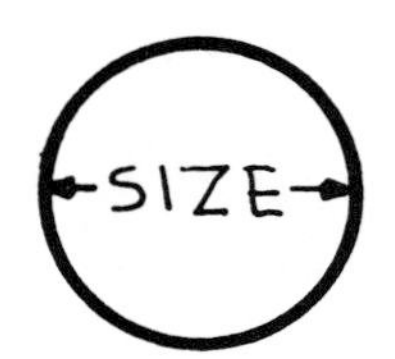

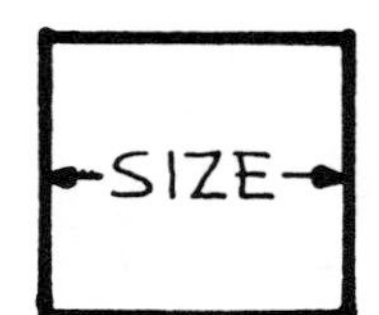

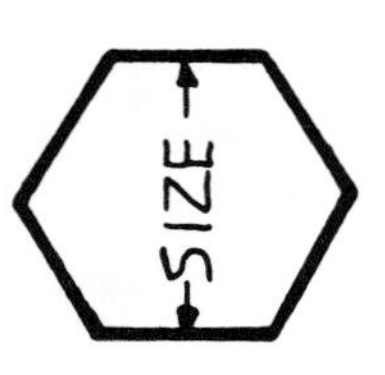

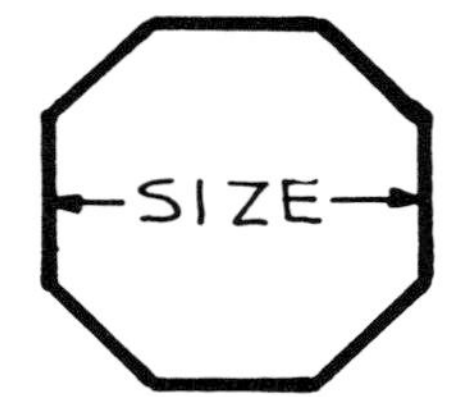

Size, Inches	Pounds per Inch	Pounds per Foot	Pounds per Inch	Pounds per Foot	Pounds per Inch	Pounds per Foot	Pounds per Inch	Pounds per Foot
1/8	.004	.042	.004	.053	.004	.046	.004	.044
3/16	.008	.094	.010	.120	.009	.104	.008	.099
1/4	.014	.167	.018	.213	.015	.184	.015	.176
5/16	.022	.261	.028	.332	.024	.288	.023	.275
3/8	.031	.376	.040	.478	.035	.414	.033	.396
7/16	.043	.511	.054	.651	.047	.564	.045	.539
1/2	.056	.668	.071	.850	.061	.737	.059	.704
9/16	.070	.845	.090	1.076	.078	.932	.074	.891
5/8	.087	1.040	.111	1.328	.096	1.150	.092	1.100
11/16	.105	1.260	.134	1.607	.116	1.393	.111	1.331
3/4	.125	1.500	.159	1.913	.138	1.658	.132	1.584
13/16	.147	1.760	.187	2.245	.162	1.944	.155	1.859
7/8	.170	2.040	.217	2.603	.188	2.256	.180	2.157
15/16	.196	2.350	.249	2.988	.216	2.588	.206	2.476
1	.223	2.670	.283	3.400	.245	2.944	.235	2.817
1- 1/16	.251	3.010	.320	3.833	.277	3.324	.265	3.180
1- 1/8	.282	3.380	.359	4.303	.311	3.727	.297	3.565
1- 3/16	.314	3.770	.400	4.795	.346	4.152	.331	3.972
1- 1/4	.348	4.170	.443	5.314	.383	4.601	.367	4.401
1- 5/16	.383	4.600	.488	5.847	.423	5.072	.404	4.852
1- 3/8	.421	5.050	.536	6.428	.464	5.567	.444	5.325
1- 7/16	.460	5.517	.586	7.026	.507	6.085	.485	5.820
1- 1/2	.501	6.010	.638	7.650	.552	6.625	.528	6.338
1- 9/16	.543	6.519	.692	8.301	.599	7.189	.573	6.877
1- 5/8	.588	7.050	.748	8.978	.648	7.775	.620	7.438
1-11/16	.634	7.604	.807	9.682	.699	8.385	.668	8.021
1- 3/4	.682	8.180	.868	10.414	.752	9.018	.719	8.626
1-13/16	.731	8.773	.931	11.170	.806	9.673	.771	9.253
1- 7/8	.783	9.390	1.000	12.000	.863	10.355	.825	9.902
1-15/16	.835	10.024	1.064	12.763	.921	11.053	.881	10.574
2	.892	10.700	1.133	13.600	.982	11.780	.939	11.267

WEIGHT OF SHEET STEEL

MANUFACTURERS' STANDARD GAGE FOR SHEET STEEL

Number of Gage	Approx. Thickness in Fractions of an Inch	Approx. Thickness in Decimal Part of an Inch	Weight per Sq. Ft. in Lbs. Avoirdupois
3	15/64	.2391	10.00
4	7/32	.2242	9.375
5	13/64	.2092	8.75
6	3/16	.1943	8.125
7	11/64	.1793	7.5
8	5/32	.1644	6.875
9	9/64	.1495	6.25
10	1/8	.1345	5.625
11	7/64	.1196	5.00
12	3/32	.1046	4.375
13	5/64	.0897	3.75
14	9/128	.0747	3.125
15	1/16	.0673	2.8125
16	9/160	.0598	2.5
17	1/20	.0538	2.25
18	3/64	.0478	2.00
19	3/80	.0418	1.75
20	11/320	.0359	1.50
21	1/32	.0329	1.375
22	9/320	.0299	1.25
23	1/40	.0269	1.125
24	7/320	.0239	1.00
25	3/160	.0209	.875
26	11/640	.0179	.75
27	1/64	.0164	.6875
28	9/640	.0149	.625
29	1/80	.0135	.5625
30	7/640	.0120	.500

Galvanized sheets weigh approximately 1 to 2-1/2 ounces more per square foot than uncoated sheets of the same gage.

WEIGHT PER FOOT OF FLAT BAR STEEL

Thickness in Inches	Width in inches 1	1-1/4	1-1/2	1-3/4	2	2-1/4	2-1/2	3	3-1/2	4
3/16	.638	.797	.957	1.11	1.28	1.44	1.59	1.91	2.23	2.55
1/8	.42	.53	.63	.74	.84	.95	1.05	1.26	1.47	1.68
1/4	.850	1.06	1.28	1.49	1.70	1.91	2.12	2.55	2.98	3.40
5/16	1.06	1.33	1.59	1.86	2.12	2.39	2.65	3.19	3.72	4.25
3/8	1.28	1.59	1.92	2.23	2.55	2.87	3.19	3.83	4.47	5.10
7/16	1.49	1.86	2.23	2.60	2.98	3.35	3.73	4.46	4.20	5.95
1/2	1.70	2.12	2.55	2.98	3.40	3.83	4.25	5.10	5.95	6.80
9/16	1.92	2.39	2.87	3.35	3.83	4.30	4.78	5.74	6.70	7.65
5/8	2.12	2.65	3.19	3.72	4.25	4.78	5.31	6.38	7.44	8.50
11/16	2.34	2.92	3.51	4.09	4.67	5.26	5.84	7.02	8.18	9.35
3/4	2.55	3.19	3.83	4.47	5.10	5.75	6.38	7.65	8.93	10.20
13/16	2.76	3.45	4.14	4.84	5.53	6.21	6.90	8.29	9.67	11.05
7/8	2.98	3.72	4.47	5.20	5.95	6.69	7.44	8.93	10.41	11.90
15/16	3.19	3.99	4.78	5.58	6.38	7.18	7.97	9.57	11.16	12.75
1	3.40	4.25	5.10	5.95	6.80	7.65	8.50	10.20	11.90	13.60

WEIGHT OF STEEL ANGLE IRON

Size in Inches	Wt. per Ft., Lbs.
1 x 1 x 1/8	0.80
1 x 1 x 3/16	1.16
1 x 1 x 1/4	1.49
1-1/4 x 1-1/4 x 1/8	1.01
1-1/4 x 1-1/4 x 3/16	1.48
1-1/4 x 1-1/4 x 1/4	1.92
1-1/2 x 1-1/2 x 1/8	1.23
1-1/2 x 1-1/2 x 3/16	1.80
1-1/2 x 1-1/2 x 1/4	2.34
1-3/4 x 1-3/4 x 1/8	1.44
1-3/4 x 1-3/4 x 3/16	2.12
1-3/4 x 1-3/4 x 1/4	2.77
2 x 2 x 1/8	1.65
2 x 2 x 3/16	2.44
2 x 2 x 1/4	3.19
2 x 2 x 5/16	3.92
2 x 2 x 3/8	4.70
2-1/2 x 2-1/2 x 3/16	3.07
2-1/2 x 2-1/2 x 1/4	4.10
2-1/2 x 2-1/2 x 5/16	5.00
2-1/2 x 2-1/2 x 3/8	5.90
3 x 3 x 1/4	4.9
3 x 3 x 5/16	6.1
3 x 3 x 3/8	7.2
3 x 3 x 7/16	8.3
3 x 3 x 1/2	9.4
3-1/2 x 3-1/2 x 1/4	5.8
3-1/2 x 3-1/2 x 5/16	7.2
3-1/2 x 3-1/2 x 3/8	8.5
3-1/2 x 3-1/2 x 7/16	9.8
3-1/2 x 3-1/2 x 1/2	11.1
4 x 4 x 1/4	6.6
4 x 4 x 5/16	8.2
4 x 4 x 3/8	9.8
4 x 4 x 7/16	11.3
4 x 4 x 1/2	12.8
4 x 4 x 5/8	15.7
4 x 4 x 3/4	18.5

WEIGHT OF STRUCTURAL I BEAM

Depth of Beam and Wt. per Ft.	Thickness of Web in Inches	Width of Flange in Inches
3" x 5.7 lbs.	.170	2.330
7.5 lbs.	.349	2.509
4" x 7.7 lbs.	.190	2.660
9.5 lbs.	.326	2.796
5" x 10.0 lbs.	.210	3.000
14.75 lbs.	.494	3.284
6" x 12.5 lbs.	.230	3.330
17.25 lbs.	.465	3.565
7" x 15.3 lbs.	.250	3.660
20.0 lbs.	.450	3.860
8" x 18.4 lbs.	.270	4.000
23.0 lbs.	.441	4.171
10" x 25.4 lbs.	.310	4.660
35.0 lbs.	.594	4.944
12" x 31.8 lbs.	.350	5.000
35.0 lbs.	.428	5.078
40.8 lbs.	.460	5.250
50.0 lbs.	.687	5.477

WEIGHT OF STRUCTURAL CHANNEL

Depth of Beam and Wt. per Ft.	Thickness of Web in Inches	Width of Flange in Inches
3" x 4.1 lbs.	.170	1.410
5.0 lbs.	.258	1.498
6.0 lbs.	.356	1.596
4" x 5.4 lbs.	.180	1.580
7.25 lbs.	.320	1.720
5" x 6.7 lbs.	.190	1.750
9.0 lbs.	.325	1.885
6" x 8.2 lbs.	.200	1.920
10.5 lbs.	.314	2.034
13.0 lbs.	.437	2.157
7" x 9.8 lbs.	.210	2.090
12.25 lbs.	.314	2.194
14.75 lbs.	.419	2.299
8" x 8.5 lbs.	.180	1.875
11.5 lbs.	.220	2.260
13.75 lbs.	.303	2.343
18.75 lbs.	.487	2.527
9" x 13.4 lbs.	.230	2.430
15.0 lbs.	.285	2.485
20.0 lbs.	.448	2.648
10" x 15.3 lbs.	.240	2.600
20.0 lbs.	.379	2.739
25.0 lbs.	.526	2.886
30.0 lbs.	.673	3.033
12" x 20.7 lbs.	.280	2.940
25.0 lbs.	.387	3.047
30.0 lbs.	.510	3.170
15" x 33.9 lbs.	.400	3.400
40.0 lbs.	.520	3.520
50.0 lbs.	.716	3.716

INDEX